ENVIRONMENTAL PERSPECTIVES

SECOND EDITION

ROGER P. EBERTZ
UNIVERSITY OF DUBUQUE

Kendall Hunt
publishing company

Cover image by Roger P. Ebertz.

Tree image used for interior design provided by Jessica Stepp.

Kendall Hunt
publishing company

www.kendallhunt.com
Send all inquiries to:
4050 Westmark Drive
Dubuque, IA 52004-1840

ISBN 978-1-4652-4018-7

Printed in the United States of America
10 9 8 7 6 5 4 3 2

I would like to dedicate this edition:

To the memory of Bill and Fern Ebertz, who introduced me to God's creation through many days of travel and nights of camping around the country,

And

To Mary, Dave, and Lois, who shared the back seat with me over the many miles, and who continue to share the memories.

Contents

SECTION 1

Introductory Essays

Chapter

1

Caring for Nature: What, Why and How?

What Is Nature?

Please come with me and think about nature and our relationship to it. As we make our way through this book we will encounter ideas from different times in history, different cultures, different religions, and different philosophies. Some of these ideas will sound familiar, others strange. But in the process of listening to voices speaking about the natural world, I hope you will begin thinking about the environment in new ways yourself. In the process of thinking about nature, I hope you will also come to believe that we as human beings should put more effort into taking proper care of our natural environment.

WHAT IS NATURE?

But what is nature, anyway? I have already used the words "nature" and "natural" multiple times. My guess is that you have a sense of what I mean by these words. But when we dig below the surface, we discover that "nature" is a slippery word. Lots of questions arise. The first, for example, is whether human beings are part of nature. Most people, I have found, will initially say "no." In fact, many define nature as anything that is not related to or influenced by human beings. But this is not always what we mean. When we talk of "natural" food, for example, we don't usually mean food that humans have done nothing to at all. For example, a box of granola might be labeled "all natural." This is clearly not meant to imply that it hasn't been processed by humans. It would not be granola at all if it had not been processed. It seems that the label is trying to say something else. Perhaps it means that all of the ingredients are "natural." But what does this mean? Any granola has one or more kinds of grain. This grain has been processed by humans. What kind of sweetener does it have? Sugar? Honey? Corn syrup? All of these are processed by humans. So, "natural" here must not mean "never touched by humans." Chances are, the label is saying that there are no "artificial" ingredients. But what makes an ingredient "artificial." One answer is that an artificial substance is "synthetic." But this doesn't get us too far. For clearly the granola was, in a sense, synthesized out of a number of other ingredients. Maybe "synthetic" is being used for a certain kind of synthesis, chemical synthesis, for example. But why is this different? Don't the ultimate building blocks of ingredients synthesized in this way come from the natural world? We could go on and on. It seems that answering the question "What is natural?" just raises lots more questions!!

How would you define nature?

Why is it so hard to define "nature" and "natural"? One reason is the fact that there is no one way of using the words "nature" and "natural." These words have many uses. Each use has its own definition. Allen Verhey[1], in fact, has identified sixteen different senses of the word "nature." He leaves off the list, "still counting." If the word has so many meanings, it is no wonder that we can't pin it down with one definition. Another reason the discussion gets so confusing is that "natural" is a popular advertising word these days. Like "green" and "organic," "natural" is perceived as a virtue, a quality that any good thing should have. Everyone in business wants their product to be seen as "natural." As it turns out, there is no standard! Unlike "organic," the business use of which is regulated by the government, "natural" can be defined in any way the marketer desires. So who knows what it means to say that a box of granola is "natural." It

[1] Allen Verhey, *Nature and Altering It.* (Grand Rapids, MI.: William B. Eerdmans Pub., 2010).

might mean nothing at all! Still another reason that it is so difficult to pin these very common words down is that their use and meaning varies from one culture to another. That, in fact is part of what this book is all about, exploring the different ways in which humans understand just what "nature" might mean.

To highlight one key issue in the discussion of "nature," let's go back to something that was suggested above. A common understanding is that nature is that domain that has not been affected by human beings in any way, or at least not in significant and direct ways. A house, for example, is not part of nature. The street that runs through the campus is not part of nature. The dam across the Mississippi River at Dubuque is not natural. Human beings constructed these things. But why does the fact that something was made by humans set it apart? Why, for example, is a human house not natural, while a fox's den is? Why do we say that a deer path through a forest is natural but that a human road is not? Why do we say that a dam built by humans in not part of nature, while a dam built by beavers is? One suggestion might be that the things that humans build change the environment, while the things that other animals build do not. But this doesn't seem right. After all, when a beaver builds a dam and creates a pond, the pond displaces and even kills some living things and creates a habitat for others. What else could it be that makes humans, and the things they make, so different? These questions bring us to an underlying assumption behind the "everything untouched by humans" definition of nature. This definition assumes that humans are in some sense radically different from the rest of the world. There is something about humans that sets them apart from the rest of reality. And because of this, the things that we create, the things we do, are different in a crucial way.

Some of the readings in this book affirm this view, that humans are set apart and different from the rest of the world. But we will also encounter perspectives in the pages ahead which argue that this assumption is wrong. In fact, in some readings it will be argued that there is really nothing that sets humans apart from nature. Humans are part of nature. Those who defend this view claim that the man vs. nature assumption is, in fact, a part of our problem. Because we don't realize that we are part of nature, we ignore it, misuse it, and cause all kinds of problems. If only we would realize that we are part of an intricately inter-connected system that must work together, we might begin to heal our world. Instead of trying to oversee nature, they say, we should live in harmony with nature, as part of nature ourselves.

What is nature? What is the human relationship to nature? How should we interact with the natural world? Is "nature" even the best word to use in speaking about the world around us? These are all questions that are raised by the ideas and views in this book.

NATURE, IMAGES, AND EMOTIONS

So far my discussion of "nature" has been in very abstract and conceptual terms. But our concepts and beliefs are not the only thing that affect our attitudes toward nature. Images and emotions affect those attitudes as well. When I say the word "nature" what images come into your mind? Do you think of a forest? a river? animals? The images that you associate with "nature" have an impact. If I show you a picture and ask you whether the scene in the picture is "natural," it doesn't take long for you to form an impression. This impression may or may not have anything to do with whether humans have been involved. Some things just look natural to us. They fit our images of what nature is. A similar thing can be said about the way we think about the earth. What image do you think of when I speak of the earth? For many of us living in the 21st century, the image that comes to mind is that of a photograph taken from space. From this perspective, earth appears to be a large stone, floating in a vast empty space. But this is not an image that would have come so readily to mind a hundred years ago, before the time of space travel. In some cultures, and in some times in history, speaking of earth would be more likely to evoke the image of a loving mother? In other times and places, the earth was conceived of as a kind of animal. Images like these, in which the earth is seen as a living thing, can have emotional overtones as well. "Mother Earth" suggests that the earth is that which nurtures, feeds and cares for us. Our relationship to the earth, on this way of viewing her, is a relationship in which love is appropriate. Images and emotions sometimes come together.

When I say the word "nature" what images come into your mind?

Consider another example: the notion of "wilderness." Does the word "wilderness" bring up emotional reactions for you? On the North American continent, it has been associated with different emotions. Early European Americans feared wilderness. They saw wilderness as a place of lurking danger. Two hundred years later, their descendants saw wilderness as a place of spiritual renewal and strength. In going to wild places, this later generation believed, one could escape the desperations of civilization and return to peace and tranquility. For one group the image of vast areas of mountains, woods and streams evoked a feeling of fear. For the other, it invited a feeling of joy, wonder, and release. Interestingly, those "Americans" who had lived on the continent for thousands of years, who we now call Native Americans, seem not to have thought of this type of image as wild at all. Concepts, images and emotions all influence how we see and treat the world around us.

Does the word "wilderness" bring up emotional reactions for you?

METAPHORS FOR EARTH

As I have said above, some cultures speak of earth as our mother. What do they mean by this? Sometimes, I believe, cultural groups use this expression in a fairly literal sense. They believe that the Earth has given birth to the living world, and that all Mother Earth's children are thus brothers and sisters in a diverse family. We are related. The expression "Mother Earth" is also used metaphorically. In fact, I think the "Mother Earth" or "Mother Nature" metaphor is a part of mainstream American culture. I grew up, for example, quite familiar with references to "Mother Nature." It was a way of suggesting that nature nurtured us and that we should value and care for nature. I did not believe that I, and other animals, had in any sense literally descended from the earth. Nevertheless, the metaphor of nature as a mother influenced my thinking. There are many metaphors that have influenced ways women and men think of nature. We will see that many ancient cultures understood the world to be an organism, either literally or metaphorically. Since at least the fourth century nature has frequently been seen as a book, from which humans could learn about God, parallel to the revealed scriptures. As science developed, understanding more and more the laws by which natural forces could be controlled, the image of a machine, and specifically of a clock, became a popular metaphor through which scientists and others saw the world. In the twentieth century, humans began voyaging into space, looking back at the tiny globe we call home. As they did, people began to talk of "spaceship earth" and "lifeboat earth," images that emphasize the limited capacity of the planet on which we live. All of these metaphors, mother, organism, clock, spaceship, lifeboat, have shaped humans ways of perceiving and behaving toward the earth. They continue to shape our beliefs, attitudes, and actions today.

WAYS OF SEEING NATURE

As we reflect on the many dimensions of our thoughts about nature, I would like to distinguish five ways of seeing nature.[2] Each "way of seeing nature" focuses its understanding of nature by means of a particular idea or image. It is not my intention to suggest that all worldviews can be squeezed into these categories or that these are the only options. I do not even want to suggest that they are mutually exclusive. But I do believe that these categories provide an initial framework that will help us ask questions about various views of nature. Perhaps when we finish our exploration you will want to modify and improve this list. If you do, I will be happy. That will show that you've been thinking about nature. But to start, I suggest that there are at least five general ways of conceptualizing nature:

- nature as a living organism,
- nature as physical resource,

---

[2]Neil Evernden, "Nature in Industrial Society." in *Cultural Politics in Contemporary America,* ed. I. Angus and S. Jhally, 151-64. In this essay, Evernden distinguishes 3 ways of seeing nature: nature as object, nature as self, and nature as miracle. I have been inspired by Evernden's helpful categories, but have expanded them to five for my purpose.

- nature as a community of persons,
- nature as sublime and mysterious, and
- nature as creation.

Nature as a Living Organism

In his creation myth, Plato describes a creator god who desires to create a world out of matter that is as good as possible. And since intelligence is supremely good, "the world became a living creature truly endowed with soul and intelligence by the providence of God." This view of the world as a living being, a person with intelligence and soul, is a prominent view in several schools of ancient Greek thought. The stoic philosophers after Aristotle, for example, see all of reality as a single living organism of which we are but parts. It is the "world soul" which infuses the organism with life and directs the universe according to reason. In these views, the earth is not just seen as a living organism, but as an intelligent being or person, with a rationality that exceeds even that of humans. In some cultures and worldviews, the earth is seen as a single God that encompasses all that exists, producing a pantheistic vision. References to "Mother Earth," which one finds in cultures in many parts of the world, also seem to suggest that the earth as a whole is a kind of person, even if not exactly a "god."

Nature as a Community of Persons

The view of nature as a living organism conceives of nature as a single living thing of which we are parts. The second category is similar, but slightly different. To see nature as a community of persons does not emphasize the unity of all of nature as much as the "relatedness" of all of nature, especially of all living things. As indicated previously, many native groups see earth as a mother. Even more pervasive in aboriginal cultures is the view that nature is a family. Some native North Americans see animals as "four-legged" brothers and sisters. In Native Hawaiian thought, the world is understood to be full of sentient beings who see and hear what one does. Other cultures have seen the forests and natural places as inhabited by spiritual beings. There are many variations. In some cultures, animals are gods or ancestors, in others, animals are simply brothers and sisters with a common mother, Earth. In others, trees and plants are the dwelling place of spirits which may or may not be identical to the plant itself. But in all of these worldviews, the world is seen as a community of living things which extends beyond the human community to include animals and plants. In a sense, all living things and humans are understood as living in complex network of life, each part having its own intrinsic importance and value.

The view of nature as a living organism conceives of nature as a single living thing of which we are parts. The second category is similar, but slightly different. To see nature as a community of persons does not emphasize the unity of all of nature as much as the "relatedness" of all of nature, especially of all living things.

Nature as a Material Resource

In contrast to seeing nature as a single living organism or as a community of persons, the perspective that is most typical of contemporary Western cultures is the perspective which sees the earth and the universe as a vast system of matter and physical energy. To think of nature in this way lends itself to seeing nature as a collection of resources to be used. In ethics, "persons" have a moral status of their own. They are not just valuable because someone else thinks they are valuable, they are valuable in themselves. If I am walking down the street, for example, and kick a rock, few would say that I have somehow violated the rock. It's possible that I have kicked some valuable stone and damaged it or moved it in a way that is not pleasing to the owner, but this only exemplifies the point that it is valuable to someone else,

To think of nature in this way lends itself to seeing nature as a collection of resources to be used.

not in itself. But if I were to kick a human baby that was lying along the way, moral outrage would result from those who witnessed my actions. And rightly so. A human baby has a moral status that a rock does not have. Since the early modern period in Western thought (around the 16th and 17th centuries), it has been very common to see the natural world as a collection of material things, with humans being a possible exception. To see nature in this way is to see its parts as like the rock that has no intrinsic value. The value of nature, in this view, is its value as a resource for human life. Later we will explore how this view has developed in Western thought. Although many people in developed nations of the world have other perspectives on nature, there is no doubt that the concept of nature as a resource to be used by humans has a large impact on how we relate to the environment.

Nature as Sublime and Mysterious

So far, I have suggested three general ways in which humans have "conceptualized" nature. Each of these views suggests that nature can be understood as like other things we understand. But sometimes nature simply seems to be beyond our understanding. There are aspects of nature that cannot be reduced to the concepts of science and other modes of understanding. Those who see in nature something that is beyond our understanding may be said to see nature as sublime and mysterious. Poets, in particular, have always been best at expressing the awe and wonder present when encountering wild nature. Standing out in a thunder storm, someone may be moved by the power and even the frightening beauty of it all. There is recognition that there is a mysterious reality, Nature itself, which is beyond the human ability to understand and control. Viewing nature as a collection of resources emphasizes usefulness of nature for humans and seems to suggest that these resources can be controlled by humans. In contrast, seeing nature as sublime and mysterious suggests that such human-centered thinking is naïve. Humans could never control nature, and to think that we could is small-minded and simply dangerous.

One aspect of the mystery and sublimity of nature is its beauty. Poets throughout history have been sensitive to this aspect of the mystery of nature. Men and women in various times have recognized natural beauty in the colors, shapes, movements and processes of nature. We walk in primitive forests and climb high peaks to experience the beauty of the smells, the sounds, the sights of nature. And for many, such experiences are spiritual, sources of mental and personal strength. Ironically, our science is at a loss to explain this aspect of human reality. Modern science reduces the world to physical processes in the world and in our brains, dismissing the subjective feel and smell and appearance of things as byproducts of physical processes. In its attempt to understand, science writes off the beauty of the world as subjective and unimportant, perhaps even an illusion to be eliminated by a fully scientific view of the world. But to one who has experienced the mystery of nature, there is something very unsatisfying about a purely scientific view. Nature is mysterious. The very experience of its mystery tells us something about ourselves that science seems to miss.

Nature is mysterious. The very experience of its mystery tells us something about ourselves that science seems to miss.

Nature as Creation

Finally, nature can be viewed as creation. Although nature as creation may be combined with nature as mysterious and sublime, the idea of nature as a creation highlights a different aspect of human experiences of nature. One may view nature as mysterious quite apart from any belief in God or gods. In its purist sense, a nature as mystery perspective would refuse to reduce or explain the mystery of nature even by appeal to divine reality. Although theists, believers in God, can appreciate the power and beauty in nature, they usually understand these as pointers to the creator God who made them. This is the nature as Creation view. Nature, the world of matter, plants and animals, human beings, the universe beyond the earth, are all the creation of God (or gods). Unlike the view mentioned earlier which sees nature itself as a god, nature as creation sees the world and the universe as separate from God, but created by God. I indicated before that the categories I am suggesting are not mutually exclusive. This is perhaps most obvious with regard to the present category. There is nothing inconsistent about combining the concept of nature as creation with several of the other views. Perhaps God created a world of matter and energy with no intrinsic value, created it to evolve into various life forms with the end goal of creating human beings and a world in which humans could live, full of resources that humans could use. There is nothing inconsistent about this. But it is also possible that God created a world of living things that all share God's value, in which humans are not the only beings that God cares for and about. Perhaps like an artist, God created a world of mysterious beauty.

CONCLUSION

In short, there are many ways to view the earth. We all see our world through the lenses of metaphors and images which influence the way we think about the world in which we live. As I asked at the beginning of this essay, how do you see nature? What images influence your view of the environment, its plants and animals? As you consider different views expressed in this text, ask yourself, "What images and metaphors does this author use to conceptualize the earth?" How do these images and metaphors influence how he or she values nature? Are these images helpful or unhelpful in developing a true and healthy environment. Chances are, you don't think about these questions very much, if ever. These days we think a lot about our environment and environmental problems we face, but we don't reflect on the way we think about the earth and nature. Yet the way we conceptualize the natural world greatly influences the way we live out our lives, which affects our natural environment. This book is an opportunity for you to think about how men and women, past and present, from many different cultures have thought about nature. It's an invitation to you to encourage you to reflect on how you think about the world in which we live and on your relationship to it.

Yet the way we conceptualize the natural world greatly influences the way we live out our lives, which affects our natural environment.

NAME ______________________

Worksheet 1.1 What Is Nature?

1. **In the Reading 1.1 several metaphors for the earth are mentioned: mother, book, machine, spaceship, lifeboat. What do you think each of these images conveys about the earth?**

   Mother Earth:

   The book of nature:

   Nature as a machine:

   Spaceship Earth:

   Lifeboat Earth:

2. **Which of the above metaphors do you like the most? Why?**

3. Later in the chapter, five general concepts for the earth are identified. What does each of these suggest about the earth?

   Nature as an organism:

   Nature as a community of persons:

   Nature as a collection of material resources:

   Nature as mystery:

   Nature as creation:

4. Again, which of these concepts do you like the best to describe nature? Briefly try to explain what it is about the concept you appreciate?

Environmental Stewardship: Why and How?

We have been exploring initial questions concerning the various meanings of "nature." But let's assume that we have come to an understanding of what nature is? Other questions arise: Why should we care about nature? Why should we humans care what happens to the environment? And if we do care, what does this mean? In this essay, we will open the way for exploration of these topics.

One way of thinking about caring for natural world involves the concept of "environmental stewardship." The terms "steward" and "stewardship" are not used as much as they used to be. What do they mean? "Steward" is derived from an Old English word for a person who looked after household affairs. The concept goes further back, however. "Steward" is used in translations of the Bible, for example, to refer to an individual who has been put in charge of a piece of property for a king or a rich person. Jesus, for example, in several of his parables speaks about a person who has been hired to take care of the affairs of another. In one story, a king hires a steward to take care of his vineyard. In another a steward has been given charge of a house. In still another, a steward has the responsibility of taking care of the master's financial affairs. These are good examples of what we mean by the concept of a "steward." A steward is simply a person who has the responsibility of taking care of something owned by someone else.

A steward is simply a person that who has the responsibility of taking care of something owned by someone else.

The concept of stewardship has many applications today. For those who have grown up attending church, the term "stewardship" may call to mind a time of the year when the priest or pastor preached sermons calling for members to give money to the church. At their best, the purpose of these sermons is to help people be good caretakers of the money God has given to them. Even in non-religious contexts, people sometimes speak about financial stewardship. The idea is that one should be a good caretaker of one's financial assets, or the assets of a company for which one works. The concept of stewardship is used in other ways also. Debates about mining and mineral rights may include references to the need to be stewards of our natural resources. Conservationists sometimes talk with farmers about the importance of land stewardship. A business may say it wants to be a good steward of its human resources. The idea of "stewardship," then, refers to the responsibility to "take care" of something. In its original use, it was about taking care of something that belonged to someone else.

ENVIRONMENTAL STEWARDSHIP

It has been common, over the last fifty years, to talk of environmental stewardship. This implies that we have a responsibility to take care of the earth. It also suggests that the earth we are charged with taking care of does not belong to us. Who does it belong to? People who believe in God, such as Jews, Christians, and Muslims, would answer that the earth belongs to God. In this view, it is God who has charged us with the responsibility of taking care of the world. But there are also ways one could conceive of our responsibility for nature in terms of stewardship without employing the concept of God. One might, for example, think of the world or earth as belonging to all human beings, both those living and future generations. Stewardship, on this view, would be taking care of something that belongs to others, as well as ourselves. One might even think of these "others" as including all living things, not just humans. In short, it might be possible to think of stewardship in ways that don't imply that there is a God who owns the world. Nevertheless, it should be noted that even among the religions of the world, there are belief systems that question the idea that we are "stewards" of the earth. Those who believe that we should see ourselves as part of nature, for example, sometimes argue that the concept of stewardship is too hierarchical. It suggests that we are above nature, rather than an equal part of it. As we explore the concept of nature, we will also be looking for the best way to express our task of caring for it. "Environmental stewardship" is one way of thinking about that task. There are others. For now, I will employ the concept, even though I hope you will continue to ask if it's the best way to articulate your own understanding of our relationship to the natural world around us.

ENVIRONMENTAL STEWARDSHIP: WHY?

Why should we be responsible environmental stewards? In other words, why should we care about and care for nature? As we consider different worldviews, we will see that different cultural and philosophical perspectives suggest different answers to this basic question. Let's begin, however, with one common idea of why we should care about nature. When I raise the question, "Why should we care about the environment?" in my classes, I get a variety of responses. But among them is always the suggestion that we should care for the environment because humans need it to survive and thrive. We need air and water just to live. Sometimes it is suggested that we should care for the sake of future generations of humans. In a similar vein, it is argued that our continued prosperity as a nation depends upon conserving our natural resources. From another direction, it is sometimes said that biodiversity is important because information is stored genetically in other species, information that my help us develop new medicines or find cures for our human diseases. All of these arguments for caring for nature echo one assumption: We should be stewards of the environment for the good of human beings, either now or in the future.

Why do you think humans care about the environment?

This way of thinking about the value of nature, and about the reasons why we should care, is called "anthropocentrism." "Anthropos" is a Greek word for "human." "Anthropocentric" means "human centered." An anthropocentric view of what is valuable puts humans at the center of value. An anthropocentric reason for being environmental stewards is a reason centered on human beings and human well-being. Anthropocentrism assumes that human beings are valuable in themselves in a way nothing else is. Is this true? As we study different perspectives, we will see that in Western culture, ethics has tended to be anthropocentric. We will also encounter voices that argue that we must expand our vision beyond human beings. We will consider both religious and philosophical perspectives which suggest there are non-anthropocentric reasons to care for the environment. These non-anthropocentric thinkers claim that as we weigh what is important in the world, it is not only humans that are valuable in themselves.

"Anthropocentric" means "human-centered." An anthropocentric view of what is valuable, puts humans at the center of value.

ENVIRONMENTAL STEWARDSHIP: HOW?

Let's assume that we have a responsibility to be stewards of our natural environment. What do we need to fulfill our task? Professor Calvin DeWitt, a professor of environmental studies at the University of Wisconsin, has suggested that to be a good steward one must be concerned with three things: science, ethics, and action.[1] DeWitt places these three crucial requirements for creation care on a diagram.

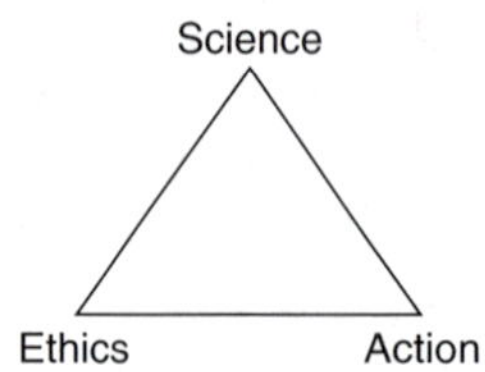

[1]Calvin DeWitt, "Science, Ethics and Praxis: Getting It All Together," in *Ecology and Religion: Scientists Speak* (Winona, MN: Anselm Academic, 2009). I have substituted the term "action" for DeWitt's "praxis."

Dewitt's triad is a helpful way to consider the dimensions of our work as environmental stewards. Imagine, for example, that I have inherited a piece of property on a lake in rural Iowa. Although it has not been used for crops for the past year, much of the land was at one time used as a cornfield. I have considered several options for my new tract of land. I could leave the property as it is and just let "nature take its course." I could subdivide it into smaller plots and build vacation cottages to lease or sell. I could build a cabin for my own use, and leave the rest of the property as it is. I could take steps to increase the biodiversity present on the land, reintroducing native plants that were more common before it was developed for agriculture. I could combine two or more of these options. What should I do? DeWitt's triad gives us important clues. Frist, good stewardship requires science. I may not be a scientist myself, but I can consult others who have the expertise to inform me of the science involved in my choices. Second, I should think about the ethical implications of my options. I should weigh what is most valuable. I should ask how the values of my worldview and my faith apply to the decision at hand. Bringing the science and ethics together, I should ask whether the plants, animals, and ecosystem involved are important to preserve. Yes, there is disagreement about these issues. But that is not a good excuse for not taking the questions seriously and coming to the best answers I can. Finally, I must make my decision and act upon it. I may decide that one of my choices is ethically and environmentally best, but if I do not act, my deliberation is pointless. Each of the corners of the triad is important in making a good decision.

The triad is also helpful on a public level. Whether it is a town, a city, a county, a state, or the nation, leaders make decisions that affect the environment. How should they make those decisions? They need to know the science, they need to think through the ethics, and they need to act on the basis of the best understanding to which they can come. The same is true for you and as we seek to influence our governmental leaders. Imagine that you and I have observed that a stream that runs through our village is polluted and dead. We want to work for change. What kind of work do we need to do? What kind of people do we need to consult and hire? First, we need scientists who can help us analyze the water, determine the sources of the pollution, and investigate various options for reviving the stream. We need ethicists to help us think about who will be affected by the changes we consider, how the well-being of both humans and animals will be impacted, and how we should we weigh the values involved. And we need activists who will work to put our research into action, lobbying county and state leaders, getting down and dirty to clean up the stream, and planting native plants that will more effectively absorb pollutants. In other words, environmental stewardship involves three things: science, ethics, and action.

Environmental stewardship involves three things: science, ethics, and action.

Finally, DeWitt's triad suggests various careers in environmental stewardship. A student who cares about nature and loves science can commit herself deeply to learning as much as she can about how the world of nature works, how humans interact with it, and what can be done to alleviate environmental problems and promote ecological health. She may well be committed to ethics and deeply committed to religious values that support her concern for the creation. But her calling is to be a scientist, doing the careful scientific work that will make environmental stewardship possible. Another student, who loves philosophy or religion can study hard and think deeply about the issues involved in environmental ethics. Since there are not a lot of jobs in environmental ethics, he may pursue a second major. His calling may be to incorporate environmental ethics into his position in business, education or ministry, helping others think about the value of nature. Finally, a third student may prepare to be an activist, studying government and how to work for change. Or like the second student, she may pursue a major with which she can be actively involved in influencing decisions for the environment. In short, on several different levels, DeWitt's triad illustrates a balance that is needed not just for individual stewardship decisions, but for our societal approach to environmental stewardship as well. What do we need to be responsible environmental stewards? Science, Ethics, action.

In this book it is primarily the ethical dimension of environmental thinking that is discussed. But we must remember that ethics alone is not enough. We also need the help of those who know the science to help us make decisions. And we need to put these two types of understanding together as a basis for responsible action.

NAME ____________________

Worksheet 1.2 Environmental Stewardship: Why and How?

1. What is the definition of an environmental steward?

2. This reading contains a triangle with three sides. Each corner represents a different element.

a. ______________

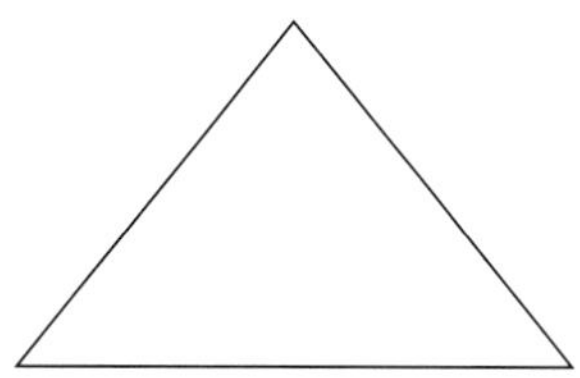

b. ______________ c. ______________

3. Why is scientific knowledge important if we are to care for the environment?

4. Why is ethical understanding important?

5. Why is action important?

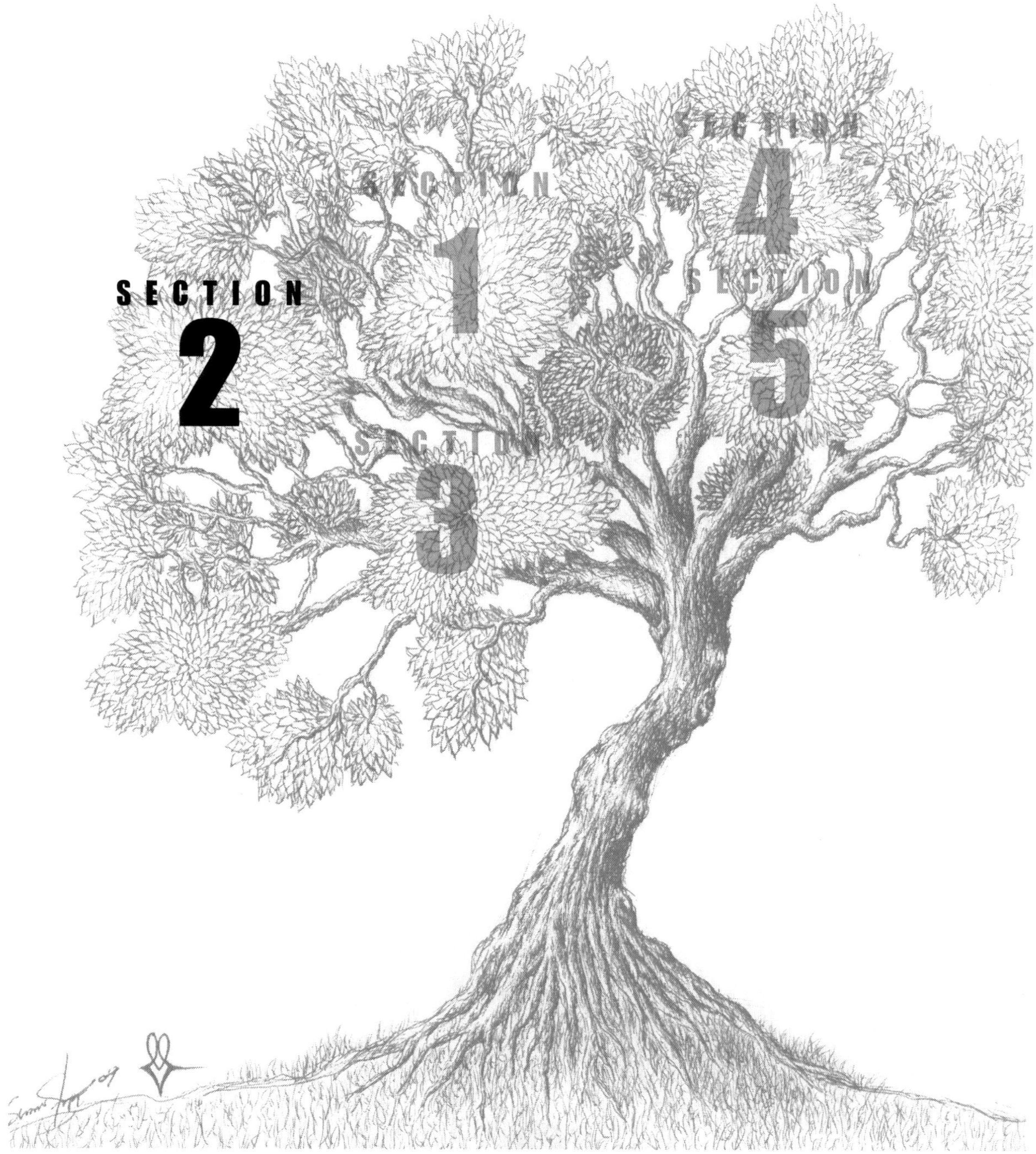

SECTION 2

Attitudes toward Nature in European-American Thought

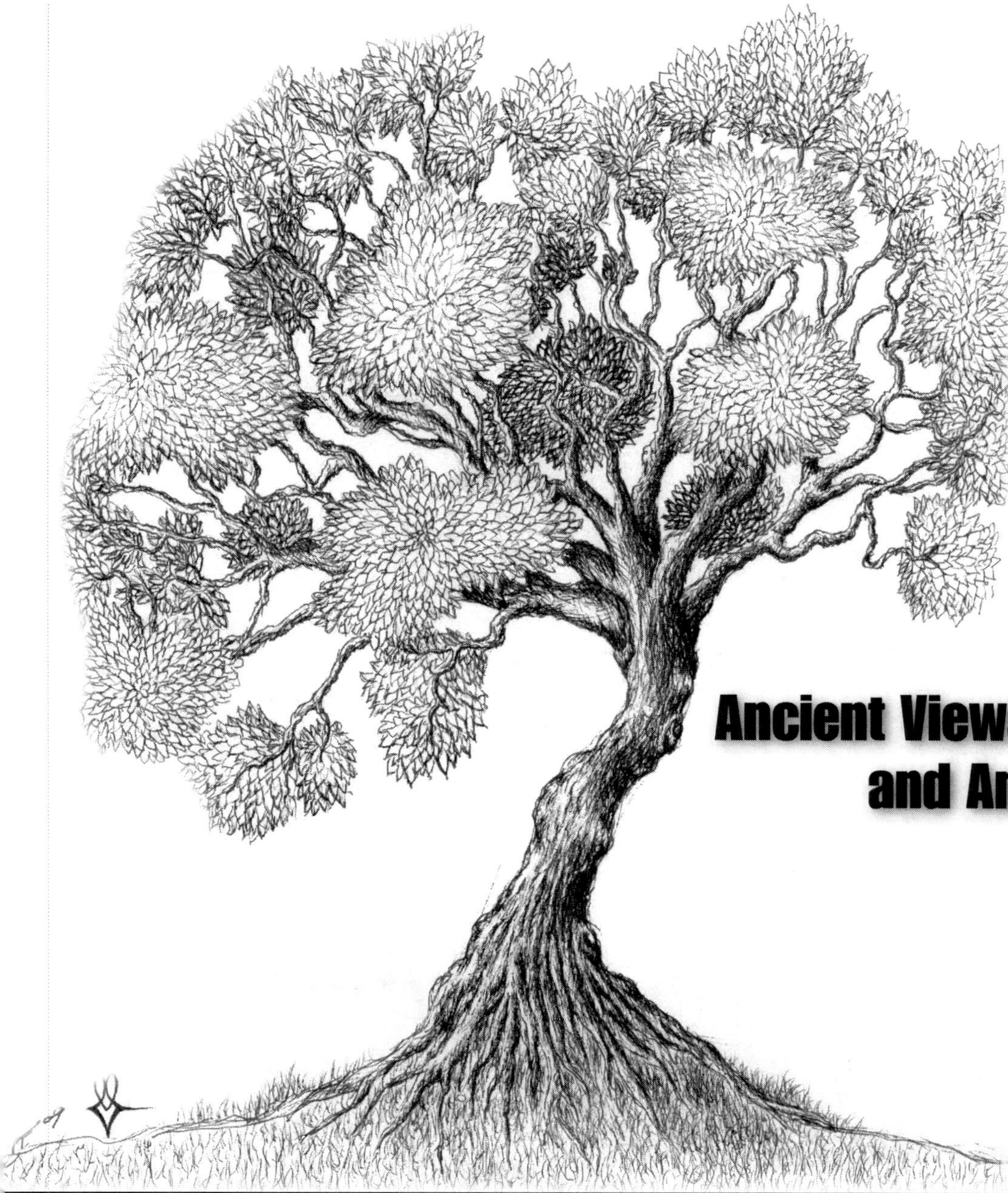

Chapter 2

Ancient Views of Creation and Animals

INTRODUCTION

In Chapter 1, we considered a number of metaphors that have been used for either the Earth or the cosmos as a whole. One point of that discussion is that the metaphor that one uses to picture nature or the Earth affects the way in which one values the natural world and the way in which one cares, or fails to care, for the natural environment in which we live. In addition to various metaphors, we have all grown up in cultures that have passed down to us deep currents of thought that shape the way we think about just about anything. "Thought-currents" are ways of thinking and acting that are passed down from generation to generation, through language, through informal and informal education, and even just through modeling as we mature. Certainly they evolve and change over time, but they also carry ideas down through the generations for thousands of years. To understand how you and I view the natural world today, we need to look back at the sources of these idea-currents in the past.

Each of us has been influenced by a variety of these thought-currents. But anyone growing up in Europe, North America, and even South America, has been deeply influenced by the European collection of intellectual traditions that is often described as "Western thought." In Chapters 2 to 5, we will consider perspectives on the non-human world as they developed and were passed down through Western European culture. We will follow these ideas from Europe to the United States and consider their development there. Although cultural streams never come from nowhere, always evolving from what has gone before, it is customary to trace mainstream Western European cultures back to Greece. I believe it is also important to consider the Hebrew and Christian ideas that have influenced the tradition as well. Other streams of thought have flowed into the currents of Western European thought and culture, but these two streams of thought, the Greek and the Hebrew-Christian, have been the largest. As the Roman empire spread, Greco-Roman ideas and Hebrew-Christian ideas mixed in various ways, picking up elements of local cultures along the way as well. When we think of Europe, it is obvious that there is much diversity. The culture of Denmark is radically different than culture of Spain. But there is also commonality. From the North to the South, from East to the West, the ideas of the Abrahamic these two major streams of thought and culture mixed in various ways, sometimes incorporating the thought of native peoples in various places.

I have said that Western culture began around the Mediterranean and in what we now call "the Middle East." This, of course, is an oversimplification. Cultures never arise out of nothing. Cultures don't really "begin" as much as evolve. They always have roots in earlier forms of life and thought. Similarly, to speak as if there is one "Western culture" is itself inaccurate. What I am referring to is more like a family of cultures, diverse within itself. The Spanish and Portuguese cultures that traveled to South America and southern North America, for example, are significantly different from the English, French, German, and Scandinavian cultures that significantly influenced the United States and Canada. Yet, there are commonalities as well. Christianity, for example, had a real influence on both northern and southern Europe and was, in turn, highly influential in the colonies of all of the European nations around the world. Historically, Latin American Christianity may have manifested itself in styles and forms that are quite different from those common to the history of Christianity in the United States and Canada, but nevertheless, Christian ideas have had a large impact on both the north and the south. If one thinks of a Western culture as a large tree, there are many roots that feed and nourish the life above ground. Some of the roots are large, others small, some go deep, and others are shallow.

If one thinks of a Western culture as a large tree, there are many roots that feed and nourish the life above ground. Some of the roots are large, others small, some go deep, and others are shallow.

We will not be able to consider all of the various roots that have contributed to the Western thought. At the risk of great over-simplification, in this chapter we will consider two large roots of Western thought, each of which are rooted even deeper in cultures of the ancient world: the Hebrew and the Greek. To explore Hebrew concepts of the environment, we will read several passages from the Hebrew Bible. Perhaps the most well known piece of ancient Hebrew literature is the passage that contains the accounts of creation found in the first three chapters of Genesis. Shared by both Christians and Jews, and influencing Islamic thought as well, Genesis 1 to 3 includes what most scholars believe are two accounts of the creation of the cosmos. We will begin our investigation of representative Hebrew writings with these creation accounts. They affirm that the created world is "very good," and that humankind is created in the image of God. Human beings are told to "rule over" (traditionally translated "have dominion over") other parts of the created world. Recent environmental philosophers and theologians have pointed to these ideas, perhaps more than any other in the Bible, as both the cause of our environmental crisis and the basis for a positive response. The Genesis creation stories, however, should be seen in the wider context of the Bible. So we will also consider several other biblical passages, in which God tells his people that "the earth is the Lord's," and in which all of creation is called to praise the creator.

A second major root system is grounded in Ancient Greece. Although there are common themes in Greek philosophy, there is really no one ancient Greek perspective. As these ancient thinkers began seeking a systematic perspective on the world, they develop quite a diversity of views. Our exploration will include readings from two major thinkers. First, we will examine two excerpts from the ancient philosopher Plato, one of the most famous philosophers in the Western tradition. In the selection from Plato's dialogue, *Timeaus*, we find the character Timeaus spelling out his own creation story. This story is similar to, but also significantly different from, the accounts of creation we find in the Bible. We will also take a look at a brief passage from another of Plato's dialogues, *Phaedo*, in which the character Socrates discusses the relationship between the body and the soul. Finally, we will turn to Plato's student Aristotle, also another well-known Greek philosopher. Like Plato, Aristotle believed that human beings are set apart from other living things by their ability to reason. We will consider a very brief reading in which Aristotle argues that because humans have the ability to reason, while other animals do not, all other living things are "for humans."

The readings from Hebrew and Greek sources provide only snapshots of two very rich cultural and intellectual traditions. They do not represent all of the intellectual root-systems through which Western thought draws its nourishment. But we will encounter some important ancient ideas that have significantly influenced Western attitudes toward the non-human world around us. As you read, ask yourself, "What ideas fit with the way I have learned to understand nature? What is different?"

worksheet 2.1

The Hebrew Bible on Creation

As we have observed in the introduction to this chapter, Western ways of thinking about nature are influenced by two great strands of ancient thought: the Hebrew-Christian strand and the Greek-Roman strand. The Hebrew Bible, the collection of ancient Hebrew literature that Christians refer to as "the Old Testament," is an important source for understanding ancient Hebrew thought. Since this collection of writings is considered sacred Scripture by both Judaism and Christianity, it has continued to exert an influence on cultures that have been touched by them. This includes cultures in the Middle East, Europe, African, North America and South America.

The Hebrew Bible includes many types of literature, including religious accounts of the origins of the world, narratives about the interaction between humans and God in history, explanations of the religious and cultural heritage of the Hebrew people, instructions for rituals they were to perform, laws and rules for virtually every aspect of life, prophetic speeches, love poetry, proverbs and expressions of both faith and doubt. In accepting these writings as Scripture, both Christianity and Judaism affirm that these works were inspired by God. Neither tradition, however, denies that they were written by human beings, influenced by the times and cultures in which they lived. With the possible exception of a few passages, the Hebrew Bible itself does not claim to be dictated directly by God. The mainstream of both the Jewish and the Christian traditions from there beginnings, have believed that the Scriptures were written by human authors to the people of their own times, in language and images these people could understand. As they wrote, these authors were inspired by God, enabled to speak God's Word to God's people in their own times. Jews and Christians believe that these writings, when rightly interpreted, also speak God's Word to humankind today. Whether you accept the Hebrew Bible as inspired or not, these passages provide windows on how ancient Hebrews viewed the natural world, and how their beliefs have influenced the views of others ever since.

The first passage for the Hebrew Bible we consider contains the stories by which the Hebrews explained the creation of the world. It is much debated, and little understood, in contemporary America. As you read this text, try to put aside the question of how this relates to scientific accounts of the origins of the cosmos. Instead, look for things you have not noticed before. And in particular, ask what this passage says about the Hebrew understanding of the world, human beings and the rest of creation.

GENESIS 1-3

1 In the beginning when God created the heavens and the earth, 2 the earth was
a formless void and darkness covered the face of the deep, while a wind from
God swept over the face of the waters. 3 Then God said, 'Let there be light'; and
there was light. 4 And God saw that the light was good; and God separated the
light from the darkness. 5 God called the light Day, and the darkness he called
Night. And there was evening and there was morning, the first day.

God said, 'Let there be light'; and there was light.

6 And God said, 'Let there be a dome in the midst of the waters, and let it separate the waters from the
waters.' 7 So God made the dome and separated the waters that were under the dome from the waters that
were above the dome. And it was so. 8 God called the dome Sky. And there was evening and there was
morning, the second day.

9 And God said, 'Let the waters under the sky be gathered together into one place, and let the dry land
appear.' And it was so. 10 God called the dry land Earth, and the waters that were gathered together he
called Seas. And God saw that it was good. 11 Then God said, 'Let the earth put forth vegetation: plants
yielding seed, and fruit trees of every kind on earth that bear fruit with the seed in it.' And it was so. 12
The earth brought forth vegetation: plants yielding seed of every kind, and trees of every kind bearing fruit
with the seed in it. And God saw that it was good. 13 And there was evening and there was morning, the
third day.

14 And God said, 'Let there be lights in the dome of the sky to separate the day from the night; and let them
be for signs and for seasons and for days and years, 15 and let them be lights in the dome of the sky to give
light upon the earth.' And it was so. 16 God made the two great lights—the greater light to rule the day and
the lesser light to rule the night—and the stars. 17 God set them in the dome of the sky to give light upon
the earth, 18 to rule over the day and over the night, and to separate the light from the darkness. And God
saw that it was good. 19 And there was evening and there was morning, the fourth day.

20 And God said, 'Let the waters bring forth swarms of living creatures, and let birds fly above the earth
across the dome of the sky.' 21 So God created the great sea monsters and every living creature that moves,
of every kind, with which the waters swarm, and every winged bird of every kind. And God saw that it was
good. 22 God blessed them, saying, 'Be fruitful and multiply and fill the waters in the seas, and let birds
multiply on the earth.' 23 And there was evening and there was morning, the fifth day.

24 And God said, 'Let the earth bring forth living creatures of every kind: cattle and creeping things and
wild animals of the earth of every kind.' And it was so. 25 God made the wild animals of the earth of every
kind, and the cattle of every kind, and everything that creeps upon the ground of every kind. And God saw
that it was good.

26 Then God said, 'Let us make humankind in our image, according to our likeness; and let them have
dominion over the fish of the sea, and over the birds of the air, and over the cattle, and over all the wild
animals of the earth, and over every creeping thing that creeps upon the earth.'

27 So God created humankind in his image,

in the image of God he created them;

male and female he created them.

28 God blessed them, and God said to them, 'Be fruitful and multiply, and fill the earth and subdue it; and
have dominion over the fish of the sea and over the birds of the air and over every living thing that moves
upon the earth.' 29 God said, 'See, I have given you every plant yielding seed that
is upon the face of all the earth, and every tree with seed in its fruit; you shall have
them for food. 30 And to every beast of the earth, and to every bird of the air, and
to everything that creeps on the earth, everything that has the breath of life, I have
given every green plant for food.' And it was so. And there was evening and there
was morning, the sixth day. 31 God saw everything that he had made, and indeed,
it was very good.

God saw everything that he had made, and indeed, it was very good.

2 Thus the heavens and the earth were finished, and all their multitude. 2 And on the seventh day God
finished the work that he had done, and he rested on the seventh day from all the work that he had done.
3 So God blessed the seventh day and hallowed it, because on it God rested from all the work that he had
done in creation.

4 These are the generations of the heavens and the earth when they were created.

In the day that the Lord God made the earth and the heavens, 5 when no plant of the field was yet in
the earth and no herb of the field had yet sprung up—for the Lord God had not caused it to rain upon
the earth, and there was no one to till the ground; 6 but a stream would rise from the earth, and water the
whole face of the ground— 7 then the Lord God formed man from the dust of the ground, and breathed
into his nostrils the breath of life; and the man became a living being. 8 And the Lord God planted a garden

in Eden, in the east; and there he put the man whom he had formed. 9 Out of the ground the Lord God made to grow every tree that is pleasant to the sight and good for food, the tree of life also in the midst of the garden, and the tree of the knowledge of good and evil.

10 A river flows out of Eden to water the garden, and from there it divides and becomes four branches.

11 The name of the first is Pishon; it is the one that flows around the whole land of Havilah, where there is gold; 12 and the gold of that land is good; bdellium and onyx stone are there.
13 The name of the second river is Gihon; it is the one that flows around the whole land of Cush. 14 The name of the third river is Tigris, which flows east of Assyria. And the fourth river is the Euphrates.

The Lord God took the man and put him in the garden of Eden to till it and keep it.

15 The Lord God took the man and put him in the garden of Eden to till it and keep it. 16 And the Lord God commanded the man, 'You may freely eat of every tree of the garden;

17 but of the tree of the knowledge of good and evil you shall not eat, for in the day that you eat of it you shall die.'

18 Then the Lord God said, 'It is not good that the man should be alone; I will make him a helper as his partner.' 19 So out of the ground the Lord God formed every animal of the field and every bird of the air, and brought them to the man to see what he would call them; and whatever the man called each living creature, that was its name. 20 The man gave names to all cattle, and to the birds of the air, and to every animal of the field; but for the man there was not found a helper as his partner. 21 So the Lord God caused a deep sleep to fall upon the man, and he slept; then he took one of his ribs and closed up its place with flesh. 22 And the rib that the Lord God had taken from the man he made into a woman and brought her to the man. 23 Then the man said,

'This at last is bone of my bones

and flesh of my flesh;

this one shall be called Woman,

for out of Man this one was taken.'

24 Therefore a man leaves his father and his mother and clings to his wife, and they become one flesh.

25 And the man and his wife were both naked, and were not ashamed.

3 Now the serpent was more crafty than any other wild animal that the Lord God had made. He said to the woman, 'Did God say, "You shall not eat from any tree in the garden"?' 2 The woman said to the serpent, 'We may eat of the fruit of the trees in the garden; 3 but God said, "You shall not eat of the fruit of the tree that is in the middle of the garden, nor shall you touch it, or you shall die." ' 4 But the serpent said to the woman, 'You will not die; 5 for God knows that when you eat of it your eyes will be opened, and you will be like God, knowing good and evil.' 6 So when the woman saw that the tree was good for food, and that it was a delight to the eyes, and that the tree was to be desired to make one wise, she took of its fruit and ate; and she also gave some to her husband, who was with her, and he ate. 7 Then the eyes of both were opened, and they knew that they were naked; and they sewed fig leaves together and made loincloths for themselves.

8 They heard the sound of the Lord God walking in the garden at the time of the evening breeze, and the man and his wife hid themselves from the presence of the Lord God among the trees of the garden. 9 But the Lord God called to the man, and said to him, 'Where are you?' 10 He said, 'I heard the sound of you in the garden, and I was afraid, because I was naked; and I hid myself.' 11 He said, 'Who told you that you were naked? Have you eaten from the tree of which I commanded you not to eat?' 12 The man said, 'The woman whom you gave to be with me, she gave me fruit from the tree, and I ate.' 13 Then the Lord God

said to the woman, 'What is this that you have done?' The woman said, 'The serpent tricked me, and I ate.'
14 The Lord God said to the serpent,

'Because you have done this,
cursed are you among all animals
and among all wild creatures;
upon your belly you shall go,
and dust you shall eat
all the days of your life.

15 I will put enmity between you and the woman,
and between your offspring and hers;
he will strike your head,
and you will strike his heel.'

16 To the woman he said,
'I will greatly increase your pangs in childbearing;
in pain you shall bring forth children,
yet your desire shall be for your husband,
and he shall rule over you.'

17 And to the man he said,
'Because you have listened to the voice of your wife,
and have eaten of the tree
about which I commanded you,
"You shall not eat of it",
cursed is the ground because of you;
in toil you shall eat of it all the days of your life;

18 thorns and thistles it shall bring forth for you;
and you shall eat the plants of the field.

19 By the sweat of your face
you shall eat bread
until you return to the ground,
for out of it you were taken;
you are dust,
and to dust you shall return.'

Cursed is the ground because of you.

20 The man named his wife Eve, because she was the mother of all who live.
21 And the Lord God made garments of skins for the man and for his wife, and
clothed them.

22 Then the Lord God said, 'See, the man has become like one of us, knowing good and evil; and now,
he might reach out his hand and take also from the tree of life, and eat, and live for ever'— 23 therefore
the Lord God sent him forth from the garden of Eden, to till the ground from which he was taken. 24 He
drove out the man; and at the east of the garden of Eden he placed the cherubim, and a sword flaming and
turning to guard the way to the tree of life.

PSALM 50

The following two passage are from the book of Psalms in the Bible. The Psalms are ancient songs, many of which were used as prayers, either for individual meditation or in gathered worship. Psalm 50 is written as a message from God. God is upset with the people. They are very faithful in performing the ritual sacrifices, offering bulls and goats on the alter, but they have not listened to the way God wanted them to live. To use the term as it is used these days, they are very "religious," but they live unjust, deceitful, slanderous, backbiting, adulterous lives. And so God says, I don't want your sacrifices. "After all," he says, "I own the cattle and goats. I don't need you to give them to me. What I want is for you to live righteously with thanksgiving."

A Psalm of Asaph.
1 The mighty one, God the Lord,
speaks and summons the earth
from the rising of the sun to its setting.

2 Out of Zion, the perfection of beauty,
God shines forth.

3 Our God comes and does not keep silence,
before him is a devouring fire,
and a mighty tempest all around him.

4 He calls to the heavens above
and to the earth, that he may judge his people:

5 'Gather to me my faithful ones,
who made a covenant with me by sacrifice!'

6 The heavens declare his righteousness,
for God himself is judge.

Selah

7 'Hear, O my people, and I will speak,
O Israel, I will testify against you.
I am God, your God.

8 Not for your sacrifices do I rebuke you;
your burnt-offerings are continually before me.

9 I will not accept a bull from your house,
or goats from your folds.

10 For every wild animal of the forest is mine,
the cattle on a thousand hills.

11 I know all the birds of the air,
and all that moves in the field is mine.

> **For every wild animal of the forest is mine,the cattle on a thousand hills.**
>
> **I know all the birds of the air, and all that moves in the field is mine.**

12 'If I were hungry, I would not tell you,
for the world and all that is in it is mine.

13 Do I eat the flesh of bulls,
or drink the blood of goats?

14 Offer to God a sacrifice of thanksgiving,
and pay your vows to the Most High.

15 Call on me in the day of trouble;
I will deliver you, and you shall glorify me.'

16 But to the wicked God says:
'What right have you to recite my statutes,
or take my covenant on your lips?

17 For you hate discipline,
and you cast my words behind you.

18 You make friends with a thief when you see one,
and you keep company with adulterers.

19 'You give your mouth free rein for evil,
and your tongue frames deceit.

20 You sit and speak against your kin;
you slander your own mother's child.

21 These things you have done and I have been silent;
you thought that I was one just like yourself.
But now I rebuke you, and lay the charge before you.

22 'Mark this, then, you who forget God,
or I will tear you apart, and there will be no one to deliver.

23 Those who bring thanksgiving as their sacrifice honour me;
to those who go the right way
I will show the salvation of God.'

PSALM 148

The ancient Hebrew people used the following song in their worship. As you read it, ask yourself what it says and implies about the natural world, about animals and other natural phenomena. What are they told to do? How do you think they do this?

1 Praise the Lord!
Praise the Lord from the heavens;
praise him in the heights!

2 Praise him, all his angels;
praise him, all his host!

3 Praise him, sun and moon;
praise him, all you shining stars!

4 Praise him, you highest heavens,
and you waters above the heavens!

5 Let them praise the name of the Lord,
for he commanded and they were created.

6 He established them for ever and ever;
he fixed their bounds, which cannot be passed.

7 Praise the Lord from the earth,
you sea monsters and all deeps,

8 fire and hail, snow and frost,
stormy wind fulfilling his command!

9 Mountains and all hills,
fruit trees and all cedars!

10 Wild animals and all cattle,
creeping things and flying birds!

11 Kings of the earth and all peoples,
princes and all rulers of the earth!

12 Young men and women alike,
old and young together!

13 Let them praise the name of the Lord,
for his name alone is exalted;
his glory is above earth and heaven.

14 He has raised up a horn for his people,
praise for all his faithful,
for the people of Israel who are close to him.
Praise the Lord!

Two Selections from the Christian New Testament

The focus of the Bible, both the Hebrew Bible (Old Testament) and the Christian New Testament is on human beings and their relationship with God and one another. We have seen, however, that in the Hebrew accounts of the creation of the world, God gives human beings responsibility to care for the his the created world. We have also seen that as a result of human sin, the created world in "cursed." In other words, it is not just human beings who are affected by sin, but the rest of creation as well. Creation has been broken; it is bent out of shape. The New Testament does not often speak directly about the non-human living world. Nevertheless, Jesus does draw from the world of plants and animals to illustrate his teaching. In the first of the following selections from the New Testament, we see one example of ths. The second selection is from the letter of Paul to the Romans, in which he makes an interesting claim concerning the future of the creation.

MATTHEW 6:25-34

25 'Therefore I tell you, do not worry about your life, what you will eat or what you will drink, or about your body, what you will wear. Is not life more than food, and the body more than clothing? 26 Look at the birds of the air; they neither sow nor reap nor gather into barns, and yet your heavenly Father feeds them. Are you not of more value than they? 27 And can any of you by worrying add a single hour to your span of life? 28 And why do you worry about clothing? Consider the lilies of the field, how they grow; they neither toil nor spin, 29 yet I tell you, even Solomon in all his glory was not clothed like one of these. 30 But if God so clothes the grass of the field, which is alive today and tomorrow is thrown into the oven, will he not much more clothe you—you of little faith? 31 Therefore do not worry, saying, "What will we eat?" or "What will we drink?" or "What will we wear?" 32 For it is the Gentiles who strive for all these things; and indeed your heavenly Father knows that you need all these things. 33 But strive first for the kingdom of God and his righteousness, and all these things will be given to you as well.

34 'So do not worry about tomorrow, for tomorrow will bring worries of its own. Today's trouble is enough for today.

ROMANS 8:18-25

18 I consider that the sufferings of this present time are not worth comparing with the glory about to be revealed to us. 19 For the creation waits with eager longing for the revealing of the children of God; 20 for the creation was subjected to futility, not of its own will but by the will of the one who subjected it, in hope 21 that the creation itself will be set free from its bondage to decay and will obtain the freedom of the glory of the children of God. 22 We know that the whole creation has been groaning in labor pains until now; 23 and not only the creation, but we ourselves, who have the first fruits of the Spirit, groan inwardly while we wait for adoption, the redemption of our bodies. 24 For in hope we were saved. Now hope that is seen is not hope. For who hopes for what is seen? 25 But if we hope for what we do not see, we wait for it with patience.

The creation itself will be set free from its bondage to decay and will obtain the freedom of the glory of the children of God.

How the World Began

—Plato

In the reading from Genesis 1 to 3 from the Bible, we saw that creation stories provide important clues to how a culture understands the natural world. For example, in the case of the creation accounts in the Bible, the created world is repeatedly said to be "good," and after the creation of human beings, the Bible affirms that the creation is "very good." This tells us that in ancient Hebrew culture, the created world, including what we might call "nature," was understood to be good. For the Hebrews, the created world declared the wonderful power and beauty of God. In the next two readings we turn to Greek culture. The following selections are from two of the works of Plato, an ancient Greek philosopher who lived from 427 to 347 BCE. Many of Plato's works take the form of dialogues. A Platonic dialogue is an account of a philosophical discussion between two or more individuals. Often, Plato used his teacher, Socrates, as one of the characters, which is true for both of the excerpts included here. The first passage is from a dialogue entitled Timaeus. It is named after the dialogue's main character. Just prior to the point at which our passage begins, Socrates has asked Timaeus to give his view of the nature and creation of the world. The following excerpt is a paraphrase of part of Timaeus' response. Although the story is told through the character Timaeus, this creation account is often seen as representing the view of Plato. Unlike the Biblical creation accounts, the story expressed by Timaeus is the view of a single individual, not a religious tradition. Earlier Greek poetry had suggested several accounts of the creation of the world. In Plato's time, philosophers were questioning these traditional accounts and suggesting their own speculative stories of the world's beginning. Plato's account should be seen as one of these.

Timaeus: If we're going to discuss the nature of the universe and how it started, we first need to make a distinction between two kinds of things. On the one hand, there are things which are eternal, which, in other words, have always existed and have no beginning. On the other hand, there are things which have a beginning, and are always changing, always in the process of becoming one thing or another. Since these things are always changing, they never really exist in a finished state. They are always coming to be and never really are.

It is only things in the first category, the things which are eternal and never change, that our rational minds can have true understanding of. On the other hand, we can only have opinions about constantly changing things, the things which we observe with our senses (eyes, ears, etc.). After all, how could we really know anything about something that will change into something else by tomorrow. Real knowledge can only be about things that never change.

It is only things in the first category, the things which are eternal and never change, that our rational minds can have true understanding of.

Everything that comes into being, or in other words, which has a beginning, must come to be because of some cause, for without a cause nothing would come to be. Imagine an artist at work. What does the artist use as a guide and model for what he or she creates. The artist may look to eternal reality as a model , or the artist may look at changing things as a model. When the artist looks at eternal reality, and shapes his or her work to be similar to this unchanging pattern, the things created are beautiful and perfect. But if a creator looks at changing things, then the things created will not be beautiful and they will be imperfect.

This excerpt from Plato's Timeaus has been paraphrased by Roger P. Ebertz.

In our discussion of the nature of the world, the first question to ask, then, is whether the world itself is eternal or created. Is the world something that came to be? Or is something that has always been. It seems obvious that the world is a created thing, a thing which has come to be, and not an eternal thing, since the world consists of things that a person can see and touch; it has a body that can be experienced with the senses. The kind of things which can be seen and touched with our physical senses are created things, things which are always coming to be, always changing. We can have opinions about these types of things, but since they are always changing, we cannot pin them down enough to have true understanding or knowledge about them. (Thus, since the world is not eternal, we can have opinions about it, but not knowledge.)

So, what about the creation of the world? We have just concluded that the world is not eternal, but came to be. In other words, it was created. But what can we say about its cause and its creator? This is a difficult question. Whatever created the world is beyond human understanding. It is beyond the ability of human language to fully describe this creator or the creator's work. Nevertheless, there is still a question we can ask: What pattern did the creator of the world have in view when he created the world, 1) the unchanging pattern of eternal reality, or 2) the constantly shifting pattern of changing things? Since the world is the most beautiful thing in the realm of changing things, the creator must have been a good creator and the model the creator used must have been the eternal pattern. Thus the world has been formed in the likeness of unchanging reality, the kind of thing that our rational minds can come to understand, not in the likeness of changing and changeable things. This is what we can say about the world: The world is a *likeness of the eternal,* created by a most excellent creator. Since it is a likeness, however, and not the eternal itself, our understanding of the world will never rise to the level of knowledge, but will always involve opinion. Similarly, our story of how the world came to be will not be literally true in every respect. So if you want me to give an explanation of the world and how it began, the best I can do is an approximation.

The world is *a likeness of the eternal,* created by a most excellent creator. Since it is a likeness, however, and not the eternal itself, our understanding of the world will never rise to the level of knowledge, but will always involve opinion.

Socrates: That's a great introduction, Timaeus. I understand that your account will not be literal or exact. Still, I want to hear what you think, even if it can only be an approximation of the truth. Please, by all means, continue with your account of the beginning of the world.

Timaeus: Okay, here's my approximation of how the constantly changing world, the "world of becoming," began, and why. The creator of the world was good, and because the creator was good, he wanted everything else to be as much like him as possible. In other words, he wanted everything to be good, and nothing evil, as much as that was possible. As God began the work of creation, he found the whole visible realm to be nothing but a disorderly, constantly moving, mess made up of the material elements in disarray. Order is clearly better than disorder, so God transformed the disorderly, ever changing, mess into an ordered world. But the God who created the world was not just good; he was supremely good, so he wanted the result of his work to be as good as possible. Now it is clear that intelligent beings, in other words, rational beings, are superior to unintelligent beings. So if the world was going to be the best possible type of being, the world itself would have to be an intelligent being. But the world could not be intelligent if it did not have a soul. So the creator gave the world a rational soul. Since the rational is superior to the irrational, the world God created is better and more beautiful than any other possible creation. Thus, still speaking in the language of approximation, we can say that God created the world as a beautiful, living creature, having a soul and intelligence.

We can say that God created the world as a beautiful, living creature, having a soul and intelligence.

So far, I have said that God created the world as a living thing, or in other words as a kind of rational animal. But what other living things was it like? What was the specific "model" by which it was created? One idea might be that it was like one of the other living things. This would mean that the world as a whole was similar to a part of something. But a part is less good and beautiful than the whole of which it is a part.

Thus, if the world was as good and beautiful as possible, it must be something whole and not just a part. The world, then, must be modeled after that which is eternal and includes within itself everything in the eternal realm, that is, everything that is intelligible, can be understood with the mind. The eternal pattern to which God looked as he created the world was the Living Thing that incorporates within itself everything that is eternal and understandable by the mind, and was thus the best of all rationally understandable things. In the same way, the visible world incorporates within itself all kinds of visible beings. The world then, is a living being that includes within itself all other visible, living things.

So God, wanting to make a world patterned after the most perfect of all rationally understandabl, eternal beings, made the world as one visible animal that included within itself all other animals and their kinds. A question arises here: Did God create one world or an infinite number of worlds? I argue that God only created one world. For the model God used in creation, the all-encompassing eternal and rationally understandable living being, could not have had a twin. If there were two such beings, neither one would be all encompassing. In that case there would have to be another, greater, being that encompassed both of them. There can be only one all-encompassing thing. Thus the model God used was unique and singular. If there was only one most excellent model, then there could be only one world created in the image of that model. It follows that God created only one world, not many worlds and certainly not an infinite number of such worlds. There is, and always will be, only one world in the realm of visible, changing things.

So now, let's discuss what the visible world is made from. We have decided that the world must have been created. Anything that is created is made from physical matter. It is visible and tangible. All visible things contain fire. All tangible things have solidity, and no solid thing is without earth. So in the beginning of creation, God made the body of the universe out of the elements of fire and earth. But if two things are put together, there must be something to bind them together. The strongest bond is formed when the right proportions of elements come together in such a way that the things bound together and the thing which binds them all fuse together. So God placed water and air in between earth and fire, in just the right proportion, to bind them all together. Thus, the earth was created from these four elements, earth, water, air, and fire. The result was a harmonized world, made from the best possible proportion of elements. So the world was reconciled to itself and was characterized by a spirit of friendship. The bond was so complete that it could not be dissolved by anyone other than the creator.

The visible world that God created used up all of each of the four elements. In other words, the creator compounded the world out of all the fire and all the water and all the air and all the earth, leaving no part of any of them. Furthermore, God used all physical power or energy, so that there was left no power or energy outside of this creation. His intention, first of all, was that this living animal, the world, should be as far as possible a perfect whole made up of perfect parts. Secondly, he intended it to be one being, leaving no matter or energy out, so that no other world could be formed from left over stuff. Third, God did not want the created world to be susceptible to old age and disease. Disease and aging are caused when heat and cold and other powerful forces attack a body from the outside when the body is unprepared. These forces cause decomposition, bring disease and aging, and make a body waste away. So God included all of the elements and powers within the creation, so there would be nothing to attack if from the outside.

In addition, God gave the world a suitable and natural shape. The most suitable shape for an animal which includes all other shapes is a perfect globe, which has all of its edges equidistant from the center. The sphere or globe is the shape most uniform and consistent with itself. God considered a shape that was uniform and consistent with itself the most excellent and beautiful shape. So God created the world as a globe.

Finally, God finished the world off by making the surface of the globe smooth. God had many reasons for doing this. First, this all-encompassing, living being did not need eyes or ears because there was nothing outside of itself to be seen or heard. There was no need for it to breathe, eat, or remove it's digestive wastes. In short, this animal had no need of the organs needed by other creatures. Nothing went into this creature and nothing came out, because this living animal, the world, included everything that existed in the physical world within itself. The Creator designed the world so that the world's waste was its own food, and all that the world did or experienced, was done and experienced by itself. The world was self-sufficient. It had no need of anything else. For the Creator believed that a being which was self-sufficient was far more excellent than one which needed input from the outside. God did not give the world hands because it did not need to take anything for itself from outside or to protect itself from attack. And since the world moved in a circular pattern appropriate to a sphere, the most perfect of motions, and set within specific limits, the world was created without legs or feet. Thus, since it had no need whatsoever of appendages of any kind, the surface of the world was smooth.

This, then, was the plan the eternal God implemented in creating the world. The world that was created was itself a kind of god. The creator gave it a body, smooth and even, having a surface in every direction equidistant from the center; it had a complete and perfect body, formed out of perfect physical parts. Into this body, he put a soul, which he diffused throughout the body, so that the soul both penetrated the body and became the body's exterior environment. He made the universe a circle, moving in a circle. As we have said, God created only one world or universe, but this world was so excellent that it had the ability to converse with itself, and thus had no need of companions or friends. In short, God carried out this plan and created the world. The world that God created was nothing less than a blessed god.

The world that God created was nothing less than a blessed god.

The Soul's Imprisonment and Escape

—Plato

Like the previous reading, this excerpt is taken from one of Plato's dialogues. Here, Plato focuses on what happens at death, and on the relationship between the body and the soul. It may seem odd that a reading on the body and the soul is included in this book on views of nature. Yet, in this passage we see a Platonic theme that is in tension with the creation story we have just read. While the creation story seems to suggest that the physical world is good, this discussion of the relationship of the soul and the body suggests that the physical body, a part of the physical world, has a strong negative influence on the soul. The goal of a true philosopher is to escape the physical world altogether. While the soul is at home in the eternal, unchanging realm, the body is part of the temporal, changing realm. Plato suggests that while the eternal, rational world is good, the physical world is a dangerous realm of temptations and traps to be escaped.

This dualistic understanding of reality, dividing reality into good vs. evil, non-physical vs. physical, influenced subsequent philosophers and theologians, including thinkers from the Jewish, Christian and Muslim traditions. The physical world came to be seen as a negative influence on the spiritual lives of humans. True spirituality was seen to involve turning away from the natural world and focusing on the spiritual, eternal world. Taken to its extreme, this suggests that there is no reason to care about physical world, through which we are passing. We should focus our care on the soul, instead. Thus Plato's view of the body and soul seems to have implications for whether humans should be concerned for the natural world. The influence of Plato's dualism has echoed down through European and North American intellectual history, having a negative impact on our concern and care for the earth. On the other hand, it could be argued that much of the damage humankind has caused to our physical environment is the result of the gratification of the desires of the body. We are consumers in the extreme and our consumerism is driven by our desires for physical satisfaction and pleasure. If this is true, perhaps Plato's philosophical way of life, turning away from the body and focusing on the soul, would be environmentally helpful.

Socrates: Now, my friends, I'm going to try to prove that a real philosopher should be cheerful when in the face of death and that after death the philosopher can hope to obtain the greatest good in the other world. I will explain to you, Simmias and Cebes, how this can be so. For it seems that people often misunderstand the one who truly practices philosophy. They do not see that the philosopher is always pursuing death, and if he or she is always pursuing death, then he or she has had the desire to die all through life. If this is so, then why would the philosopher be afraid or resistant when the thing that has consistently been desired and pursued is near.

Simmias (laughingly): You make me laugh, Socrates, even though I am not in a laughing mood. I can't help but think how many people, when they hear what you just said, will agree that you have accurately described philosophers. They will say that a philosopher's life is really death, and that philosophers deserve the death that they desire!!

This excerpt from Plato's *Phaedro* has been paraphrased by Roger P. Ebertz.

Socrates: Oh yes, Simmias, in a sense they are right. But they do not understand the nature of the death which the true philosopher deserves. And they do not understand in what sense the philosopher deserves or desires death. But who cares what they think, let's discuss the matter among ourselves. Do we believe that there is such a thing as death?

Simmias: Certainly we do.

Socrates: So what is death? Isn't it when the soul and the body separate? One is dead when this separation is complete. In death the soul exists independently in itself, it is released from the body, and the body is released from the soul. Isn't this what death is?

So what is death? Isn't it when the soul and the body separate?

Simmias: Yes, that's exactly what death is.

Socrates: To throw light on our discussion of death, we must ask another question: Should a philosopher care about the pleasures – if they are to be called pleasures – of eating and drinking?

Simmias: Certainly not.

Socrates: And what about sexual pleasure?

Simmias: By no means.

Socrates: And will the true philosopher care about other ways of making the body happy, such as buying expensive clothes, or shoes, or other bodily adornments? Instead of caring about these things, does not the true philosopher despise anything more than what nature requires? What do you say?

Simmias: I should say that the true philosopher would despise them.

Socrates: Don't you agree that the true philosopher is entirely concerned with the good of the soul and not with physical pleasures which are associated with the body? The philosopher's goal is to get away from the body, as far as possible, and to turn to the soul.

The philosopher's goal is to get away from the body, as far as possible, and to turn to the soul.

Simmias: Quite true.

Socrates: So philosophers, above all others, may be observed in every sort of way to separate the soul from interaction with the body and resist desires for physical pleasures in favor of goods of the soul.

Simmias: Very true.

Socrates: But the rest of the world, Simmias, believes that the people who turn away from physical pleasures are living lives that are not worth having. In the eyes of many, people who live without physical pleasures are as good as dead! That's what most people think, right?

Simmias: You're right there, Socrates.

Socrates: Let's approach from another direction and talk about how we learn, how we acquire knowledge. Does the body help or hinder when it comes to acquiring knowledge? In other words, do sight and hearing lead us to truth? Are they not, as the poets are always telling us, inaccurate witnesses. And if these two senses are inaccurate and indistinct, the other senses are even worse.

Simmias: Certainly.

Socrates: Then how does the soul or mind really come to learn truth, if the physical senses are often inaccurate in what they tell us. If we are misled by the senses, it must be in careful thinking that we come to know what is true. So true reality is revealed to the mind or soul through thought, if at all.

Simmias: Yes.

Socrates: Now tell me, isn't thought best when it is able to concentrate on itself, and is not distracted by sounds and sights, and pain and pleasure. In other word, the mind works best when it puts aside the body, and has as little to do with it, when it is not subject to physical senses and desires, but is focusing on thinking alone.

Simmias: Certainly.

Socrates: In other words, The philosopher who is seeking truth "dishonors" the body. The soul or mind of the philosopher tries to get away from the body and its desires and to be alone by itself.

The philosopher who is seeking truth "dishonors" the body. The soul or mind of the philosopher tries to get away from the body and its desires and to be alone by itself.

Simmias: That is true.

Socrates: Okay, let's consider the matter of the body and the soul from still another angle. Are not the most important things that can be known general truths about what is really good and really just and really beautiful. In other words, are not absolute justice, and absolute goodness, and absolute beauty the things the soul wants to understand above all else?

Simmias: Yes, that is so.

Socrates: But did you ever behold the reality of goodness in general, or justice in general, or beauty itself with your eyes? In other words, did you ever see absolute justice, or absolute beauty, or absolute good with your eyes.

Simmias: Certainly not.

Socrates: Are these things that can be touched by the senses of the body? I am not talking only about justice and beauty and good, but about such things as absolute greatness, and health, and strength, and of the essential nature of things. With our eyes we may see something that is relatively good or something that is relatively beautiful, but do we see goodness and beauty themselves with our senses? It seems that the absolute nature of goodness and beauty are not things that we perceive with our physical organs. Instead, we see we see and know such things only through the careful ordering of our intellectual vision; we grasp such things by using the mind. This intellectual vision enables us to understand the essences of such things as goodness, and beauty, and justice. Thus, it is through the rational use of our minds that we come to know about these things, not through our senses. Is this not so?

Simmias: It is certainly so.

Socrates: Yes! The person who comes to pure knowledge of these absolute realities, such as absolute justice and absolute goodness, does so by considering them with the mind alone. This kind of knowledge is only confused if you mix in the senses. Instead, the true seeker of knowledge uses the mind in its own, pure clarity, to search for the truth about each of these important realities. This seeker gets rid, as far as possible, of eyes and ears and, so to speak, of the whole body. For the truth seeker believes that these things are distractions which sidetrack the mind and hinder it from acquiring pure truth and knowledge. So the person who seeks knowledge about the great realities of such things as justice and goodness and beauty will put aside all physical distractions and focus on thinking about these things with the mind alone. Don't you think so?

Simmias: What you say has a wonderful truth in it, Socrates.

Socrates: And when true lovers of wisdom, for that is what philosophers are, think about all these things, they will conclude that by removing themselves from the influence of the body as much as possible, they have found a path of thought which seems to bring them and their arguments to a conclusion. They will see that while their minds were infected with the evils of the body, their desires were never satisfied. They desired truth, but could not attain it. For their bodies, they will realize, were a source of endless trouble, always demanding food, and always liable to sickness and disease which overtake and hinder the search for truth. The body fills us full of loves, lusts, fears, and desires of all kinds, and with endless foolishness. In fact, in this way the body takes away from a person the ability to think at all. This is why there are wars, and fighting, and divisions among people. These things come from the lusts and desires of the body. Wars are motivated by the love of money, and those who enter into the wars want money to feed the desires of the body. And in this endless quest to satisfy the body's demands, a person has no time to give to philosophy, the love of wisdom. And

The body fills us full of loves, lusts, fears, and desires of all kinds, and with endless foolishness. In fact, in this way the body takes away from a person the ability to think at all.

even if a person tries to take some time to think about wisdom and true reality, the body constantly makes demands and causes inner turmoil and confusion in the attempt to think carefully. It is the body, then, that prevents us, or anyone seeking to know the truth, from fully grasping it.

Thus, experience makes clear to us that if we want to have pure knowledge of anything, we must strictly discipline the body, setting aside its demands and desires, and consider things with the mind alone. The mind by itself, when it has separated itself from the body, can see the truth of things as they are in themselves. The soul alone can see things as they really are. But since the soul is never entirely free from the distraction of the body while we live, it is only after death that we shall attain the wisdom we desire as lovers of wisdom. Since the soul can never attain pure knowledge while it is distracted by the body, pure knowledge will be impossible unless it can be gotten after death. For it is at death, and not until death, that the soul will be entirely separate from the body and will exist by itself alone. In this present life, it seems to me, we make the nearest approach to knowledge when the body is so disciplined that the mind is distracted by the body as little as possible and when our thinking is not compromised by our physical nature. We come closest to attaining knowledge and wisdom when we keep ourselves pure until the hour when God himself is pleased to release us from our bodies. When that time comes, having removed ourselves from the body's foolishness, we shall be pure and converse with what is pure, and come to clear light and wisdom about everything, the light of truth itself.

Isn't this how true philosophers, lovers of wisdom, will see things, Simmias?

Simmias: Undoubtedly, Socrates.

Socrates: But, O my friend, if this is true, there is great reason to hope that wherever I go when I die, when I come to the end of my journey, I can look forward to attaining the very thing that I have searched for during my life: knowledge and wisdom. Therefore, I am happy and rejoice as I leave this life. This is true for any person who believes that his mind has been prepared and that he has been purified for the afterlife. Doesn't this make sense Simmias?

Simmias: Certainly.

Socrates: But the purification that prepares one for death is nothing but the separation of the soul from the body, as I was saying before. Purification in this life, as much as it is possible, is the habit of the soul gathering and collecting herself into herself from all sides out of the body. It is the soul living as if it were already separate from the body as much as it can. Purification is nothing less that the release of the soul from the chains of the body.

Simmias: Very true.

Socrates: In its most complete form, then, the purification of the soul, the separation and release of the body from the soul, is called death. Is that not what death is?

Simmias: To be sure, he said.

Socrates: But haven't we just shown that the release of the body from the soul, the separation of soul from body, is what philosophers seek in order to find true wisdom?

Simmias: Yes, that is what we have agreed.

Socrates: So we are back to where we were earlier. It would be ridiculous and contradictory for philosophers, who try to live as if the soul was already separate from the body, in other words to live as much as possible as if the body were dead to the soul, should worry and fear when they are faced with death itself. In fact, Simmias, it's not a joke. True philosophers are always engaged in dying. So death is less terrible and fearful for them than it is to anyone else. Look at it this way: If true philosophers are in every way the enemies of the body, and if they desire to be alone with the soul, it would be extremely inconsistent of them to tremble and have regrets when the desire is finally granted to them. Instead they should rejoice that at last they are going to the place where they can attain that which they have desires most in life – which is wisdom. Many persons have been willing to die to go to the world below in the hope that they would see again their earthly lover, or wife, or child, and would be able to be with them again. In the same way, the philosopher will be happy and not shrink back from the opportunity to go to the world where wisdom can be enjoyed. The philosopher will depart with joy!! Surely this is true, O my friend. The true philosopher will have a firm conviction that there and there only, can wisdom be found in her purity. And if this is true, it would be very absurd, for a philosopher to be afraid of death. But the true philosopher will, above all, face death with courage.

The true philosopher will, above all, face death with courage.

Simmias: It would, indeed.

Socrates: So if you ever see a person who shrinks back from death, and does not face it joyfully, you have proof that that person is not a lover of wisdom, but a lover of the body, and probably at the same time a lover of either money or power, or both.

Simmias: Certainly. . . .

Socrates: If a person has not studied philosophy and is not entirely pure at the time of his departure from this life, then that person is not allowed to enter the company of the gods. It is only the lover of true knowledge who is given that privilege. And this is the reason, Simmias and Cebes, why those truly committed to the love of wisdom discipline themselves to resist all fleshly lusts, and never give in to their desires. Some people seek to restrain themselves from evil desires because they fear that if they do not, their desires will drive them to poverty, or ruin their families. These people are lovers of money and of the world. Some people restrain themselves because they fear they will be dishonored and disgraced when their evil deeds are revealed. These people are lovers of fame and honor. But these are not the motivations of the lover of wisdom. True philosophers, who love wisdom, abstain from the gratification of physical desires in order to be prepared for death, when they will be welcomed into the company of the gods, where wisdom itself dwells and is to be enjoyed.

So, my friends, lovers of wisdom, who care about their souls, and do not live to caress and dress up their bodies, say farewell to all the body's pleasures and adornments. They refuse to walk in the ways of the blind. When philosophy offers them a way to be purified and to be released from evil, they choose to follow, realizing they should not resist philosophy's invitation.

Simmias: Explain what you mean, Socrates.

Socrates: I will tell you, he said. The lovers of knowledge are aware that the soul was simply fastened and glued to the body—until philosophy released her, the soul, while it was in the body, could only view real existence through the bars of a prison, not with unobstructed vision. The soul was bogged down in ignorance. Lust for physical gratification had made the soul the principal accomplice in her own captivity. This was soul's original state. But philosophy reveals how terrible this imprisonment is. It shows that the soul that gives in to physical desire is itself part of the cause of its own imprisonment. So philosophy seeks to gently release the soul, coaxing it away from dependence on physical senses and enslavement to physical desire, and inviting it to think about things that cannot be seen or heard, but are invisible and intangible, like the soul itself. The soul which accepts this invitation, therefore, puts aside physical pleasures and desires and pains and fears, as far as possible. The soul which accepts this invitation realizes that joys and sorrows and fears and desires, actually bring suffering and evil, not just the suffering that comes when joys and possessions are lost, but an even greater evil.

Cebes: What evil are you talking about Socrates?

Socrates: The great evil is this: When a person experiences intense pleasure or release from intense pain, he or she thinks this intense feeling is the best and truest. And the soul, in thinking that these feelings are good, becomes enthralled with the body itself and what it can give. And then, each pleasure experienced, and even each pain felt, is a sort of nail that nails and rivets the soul to the body, until the soul becomes like the body. The soul, as it were, becomes stuck to the body so tightly that it believes whatever the body says, and is caught up in the habits and haunts of the body itself. So when death comes the soul is not pure, but infected by the body. That soul sinks into another body and begins to grow and live another physical life. It misses out on the experience of living and interacting with the gods and with truth itself. This is a terrible evil, to miss out on the greatest good.

The Natural Purpose of Plants and Animals

—Aristotle

The Greek philosopher, Aristotle (384-322 BCE), was perhaps the earliest thinker in the European cultural tradition to systematically observe the natural world and attempt to categorize and understand various types of life on earth. In addition to writing about topics that we would now call philosophy, such as metaphysics and ethics, he studied many areas that we would now identify as scientific fields, including astronomy, anatomy, biology, and psychology. The works of Aristotle that we have today include extensive discussions of the workings of the natural world. Although many of his theories have been eclipsed by more recent science, for years Aristotle was seen as a source of knowledge about nature. He was understood as a scientist more than as a "philosopher," as we think of the term. Aristotle's impact on European and North American ways of viewing nature has been immense and continues in some respects to the present day.

In the following very brief excerpt, Aristotle reflects on the purpose of different kinds of plants and animals. Central to Aristotle's philosophy is the idea that every living thing has a natural purpose of function. To understand a thing one must understand its function. Plants have a function. Non-human animals have a function. Human beings have a function as well. What sets humans apart from other animals, in Aristotle's thought is their rationality. Thus the human function is to live up to the full potential that we have as rational beings. Our goal should be to live as rationally as possible. In fact, Aristotle believed that humans were rational and no other living things were. And since rationality was the highest capacity a living being could have, he believed that humans were above all other living things in the organizational chart of reality. If the function of human beings is to live rationally, what is the function of plants and non-human animals? What are plants for? What is their function? What are non-human animals for? What is their function? In the following very brief excerpt from Politics, Aristotle reflects on different kinds of animals and in the end suggests that ultimately all other living things exist to serve human beings.

There are many kinds of food. Because of this, there are many kinds of living things, human beings and animals. And they must all have food. Since the kind of food that each type eats is different, they have different ways of life. Some animals are social, interacting with one another, and others are solitary, living by themselves. Each type of animal lives in the way that best sustains its own way of life and meets its nutritional needs. There are carnivorous (meat-eating) animals, herbivorous (plant-eating) animals, and omnivorous (meat and plant-eating) animals. Nature has assigned each type of animals instincts (habits) that enable it to obtain the type of food it prefers. Since different species have different tastes, the same things are not naturally pleasant to all of them, and therefore the lives of carnivorous or herbivorous animals also differ from one another. Human beings have greatly different ways of life also. Shepherds are the laziest. Their lifestyle is easy and they get there food from tame animals. They follow their flocks from place to place, creating a kind of living farm. Other people provide for themselves by different sorts of hunting. Even thievery is a kind of hunting to which some humans resort in order to get what they need. Others, who live near lakes, watersheds, rivers or the sea, get their food by fishing. Still others hunt birds or wild beasts. Most eat cultivated fruits and vegetables. These are the ways of living among those whose food

---

This excerpt from Aristotle's *Politics* has been paraphrased by Roger P. Ebertz.

comes from their own labor, and who do not obtain their food through exchange or retail trade. . . . Some make their living by combining two types of labor, supplementing the lacks from one with the other. For example, they may combine shepherding and thievery, or farming and hunting. So humans combine various ways of living as necessary to have what they need to live.

Every animal seems to be given the bare necessities of life from the time it is born and when they are grown. Some animals produce enough food, such as milk, for their offspring until they are able to supply themselves. In the same manner, after animals are born, plants exist for the sake of animals, other animals exist for the sake of humans. Tame animals exist for use for food. Most wild animals exist for food for humans, or for the provision of clothing or various tools. Nature makes nothing in vain or incomplete. It would follow that Nature has made all animals for the sake of man. . . .

Nature has made all animals for the sake of man. . . .

NAME ____________________

Worksheet 2 — Ancient Views of Creation and Animals

1. After reading the passages from the Bible included in this chapter, answer the following questions:

   How does the Bible evaluate God's work at the end of the sixth day? (Genesis 1:31)

   What role did God give to humankind on the earth? (See especially Genesis 1:26–28 and Genesis 2:15)

   According the Bible, who "owns" the animals? (See especially Psalm 50:9–11)

   What parts of the created world are called upon to praise God in Psalm 148?

   According to Romans 8, what does the creation groan for and expect?

2. **Read both the Bible's and Plato's accounts of creation and make lists of similarities and differences below.**

   Similarities between the Bible's and Plato's creation accounts:

   Differences between the Bible's and Plato's creation accounts:

3. **How would you describe the relationship between the body and the soul, as described by Plato in the Timaeus?**

4. **How does Aristotle describe the relationship of humans, other animals, and plants?**

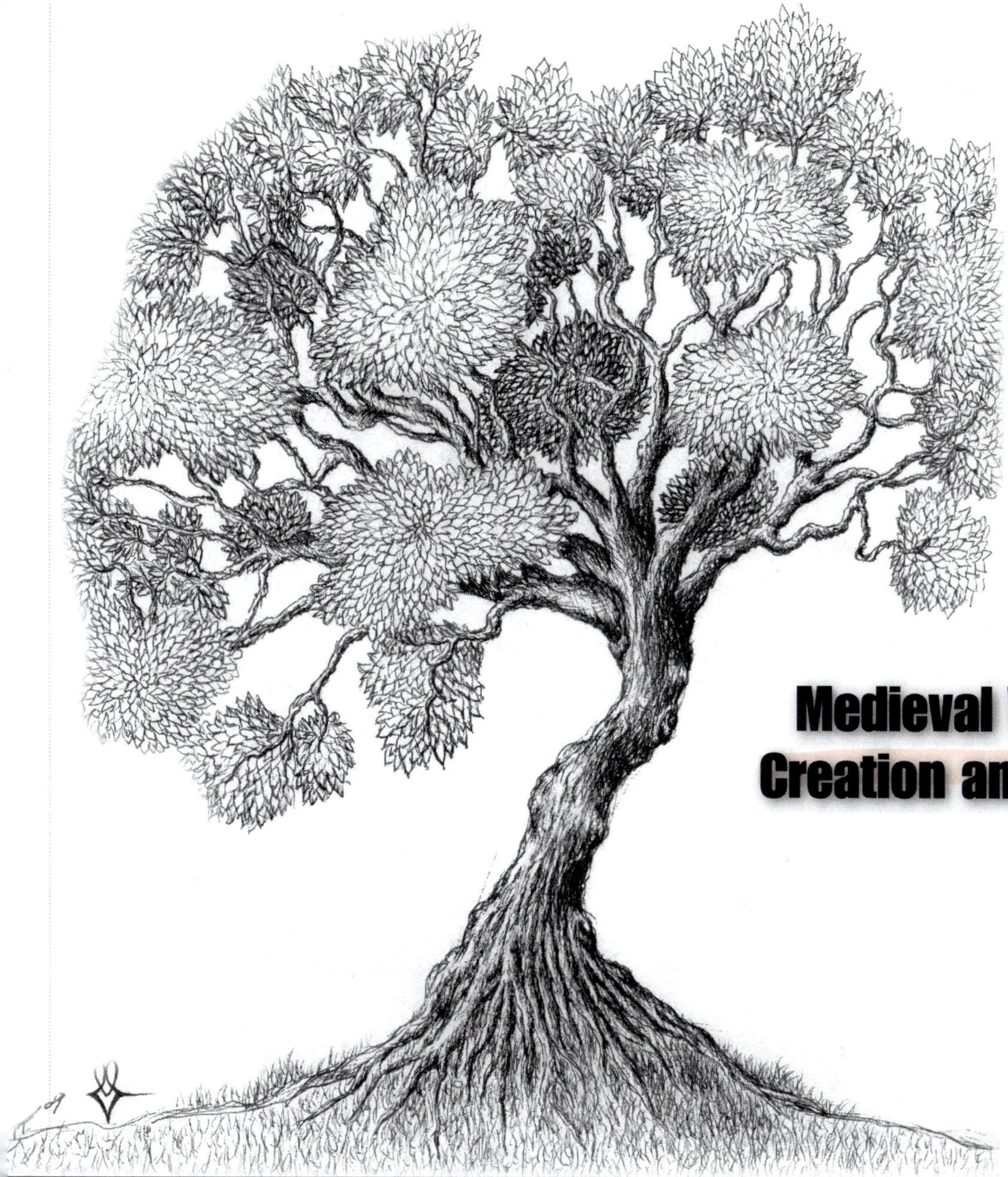

Chapter 3

Medieval Views of Creation and Animals

INTRODUCTION

In the previous chapter we have explored two major "roots" of Western thought, taking a look at several readings from ancient Greek philosophers and several passages from the Bible, representing ancient Hebrew ideas. Through the years, the Western tradition has drawn nourishment from both Greek and Hebrew roots in many ways. As we explore later periods, we will see the influences of these major roots on our ideas, even today. In this chapter, we take a look at representative thinkers writing on the natural world during the Middle Ages. To understand this period, it is crucial to see that it was a period, during the Roman Empire, in which Greek and Hebrew ideas collided and began to intertwine. Although the Romans were a mighty power, they borrowed much of their philosophy, religion, literature, and art from the Greeks. So, when the Romans conquered Palestine, home of the Jews, they brought with them Greek ideas, "hellenizing" the whole Roman Empire. It was in Roman occupied Palestine that Jesus lived and died, and in which Jesus' followers experienced the risen Christ. It was by means of Roman roads and trade routes that the news of the new Christian movement spread. Both Christians and Jews, facing intermittent persecution under various Roman emperors, scattered throughout the Roman empire.

Some Christians and Jews took a dim view of the pagan ideas they encountered around them, including the ideas of Greek and Roman philosophers. Others sought to synthesize their religion and pagan philosophy. Whether they approved or not, however, it was inevitable that Greek ideas would influence the shape of Christian and Jewish ideas. As Christianity developed, theologians like Origen and Clement of Alexandria and Augustine of Hippo began to articulate Christian ideas systematically. Reaching out for ways to express their faith, and communicate it to those around them, they found Greek and Roman concepts helpful. The Jewish philosopher, Philo of Alexandria, had already fused Greek ideas and the religion of the Hebrew Bible. Christian theologians followed suit and were particularly attracted to Platonic and neoplatonic ideas. Later, as Islam emerged, Muslim philosophers also engaged in attempts to express their faith in ways that employed the concepts of Greek philosophy. The dialogue between the theistic faiths of Judaism, Christianity, and Islam continued for a millennium.

Thus, during the Middle Ages, it is not always easy to separate Greek and Hebrew influences. The two were twisted together. Biblical concepts, such as the idea that humans are created in the image of God, were interpreted through "Greek-colored" lenses. The Bible said that humans were created in the image of God. The Greeks defined human beings as "rational animals." Bringing the two together in the Middle Ages, the notion of "the image of God" becomes closely linked with reason. Similarly, the idea of the resurrection of humans after death, accepted by Christians and some Jews, became intertwined in the minds of at least some Christians with the more Greek notion of the immortality of the soul. Thus, some saw death as a time when the immortal soul departs from the mortal body, a picture of death very similar to the picture we read suggested in Plato's Phaedo. Lasting influences of these fusions between philosophical concepts and religious ideas can be seen even today in the ways many in the West think of human nature and life after death. This fusion of Greek ideas with the major theistic religions of the West impacted human perspectives of the non-human world as well. In this chapter, we examine several excerpts that give us a window on medieval views of nature.

The Book of Creation: Three Brief Excerpts

—Augustine

Our first reading in this chapter consists of three brief selections form the work of Augustine of Hippo (354–430). Augustine was one of the early systematic theologians who attempted to think through the Christian faith in light of the ideas of his age. Before becoming a Christian, he studied several of the alternative philosophies of the day, being particularly influenced by neoplatonic thought. In the selections included here, Augustine speaks of the created world as "a great book" which points toward God. In the last of these selections, Augustine also reflects a hierarchical perspective on the relationship between plants, animals, and humans, similar to the view we saw in Aristotle.

CREATION IS A GREAT BOOK*

Some people read books in order to find God. Yet there is a great book, the very appearance of created things. Look above you; look below you! Note it; read it! God, whom you wish to find, never wrote that book with ink. Instead, He set before your eyes the things that He had made. Can you ask for a louder voice than that? Why, heaven and earth cry out to you: "God made me!"

THE WHOLE WORLD PROCLAIMS ITS MAKER†

I have gone round and have offered in His tabernacle a sacrifice of jubilation. We offer up a sacrifice of jubilation, we offer up a sacrifice of gladness, a sacrifice of rejoicing, a sacrifice of thanksgiving, which no words can express. But where do we offer it? In his own tabernacle, in Holy Church. And what is the sacrifice we offer? An overflowing and ineffable joy, beyond words, not to be expressed in speech. This is the sacrifice of jubilation. Where seek for it, how find it? By looking everywhere. *I have gone round,* says the Psalmist, *and have offered in His tabernacle a sacrifice of jubilation.* Let your mind roam through the whole creation; everywhere the created world will cry out to you: "God made me." Whatever pleases you in a work of art brings to your mind the artist who wrought it; much more, when you survey the universe, does the consideration of it evoke praise for its Maker. You look on the heavens; they are God's great work. You behold the earth; God made its numbers of seeds, its varieties of plants, its multitude of animals. Go round the heavens again and back to the earth, leave out nothing; on all sides everything cries out to you of its Author; nay, the very forms of created things are as it were the voices with which they praise their Creator. But who can fathom the whole creation? Who shall set forth its praises? Who shall worthily praise heaven and earth, the sea, and all things that are in them? And these indeed are visible things. Who shall worthily praise the angels, thrones, dominations, principalities, and powers? Who shall worthily praise that power which works actively within ourselves, quickening the body, giving movement to the members, bringing the senses into play, embracing so many things in the memory, distinguishing so many things by the intelligence; who can worthily praise it? Now if in considering these creatures of God human language

---

From The Essential Augustine,*edited by Vernon J. Bourke.*

**Sermon,* Mai 126.6; trans. V.J.B

†*On Psalm 26*, Serm. 2, 12; trans. ACW 29, 272–273.

is so at a loss, what is it to do in regard to the Creator? When words fail, can aught but triumphant music remain? *I have gone round and have offered in His tabernacle a sacrifice of jubilation.*

ASCENDING TO THE SUPREME TRUTH‡

Now that we have discussed the advantages of authority to a length that at present seems sufficient, let us see to what extent reason can succeed in ascending from the visible to the invisible and from the temporal to the eternal. We must not be passive and thoughtless in our contemplation of the beauty of the Heavens, the order of the heavenly bodies, the splendor of the sun, the alternation of day and night, the monthly phases of the moon, the four seasons of the year corresponding to the four fundamental elements, the great power of the seeds in producing figures and numbers, all things preserving their proper order and nature, each in its own kind. Our consideration of these phenomena must not be one of thoughtless, passing curiosity, but should become a step towards the undying and the everlasting. And so we must next consider what is that vital nature which perceives all these things and which, because it gives life to the body, is necessarily superior to the body. For no mass whatsoever, no matter how it glitters in the light, is to be valued highly, if it be without life. And any living substance is by the law of nature exalted above any nonliving substance.

What metaphor does Augustine use to speak of the created world? In what way does creation communicate about God? What is its message?

But since even irrational animals are admittedly alive and sentient, that power in the human soul is not noblest by which man perceives sensible objects, but that by which man makes judgments about those objects. For most brutes see more clearly and perceive things more keenly with their other bodily senses than does man. But to form judgments about things is not the part of mere sentient life, but of rational life—a life which brutes lack, and in which men excel. Now, it is certainly easy to see that the judge is nobler than what he judges. The judgment of a rational being is exercised not only on the objects of the senses, but on the senses themselves; for example: why an oar should appear broken in water, although it is intact, and why the eyes perceive it as broken. A glance of the eyes can report the fact, but by no means can it pass judgment. Thus it is evident that, just as sentient life surpasses nonlife, so rational life surpasses both.

---

‡*The True Religion,* 29.52–31.58; trans. G. R. Sheahan, S.J.,*De vera religione (Chapters 18–38)* (St. Louis University Master's Thesis, 1946), pp. 49–63.

On the Human Treatment of Animals

—Thomas Aquinas

Thomas Aquinas (1225–1274 CE) was one of the most influential thinkers in Medieval Europe. In the course of his life of less than fifty years, Thomas was instrumental in bringing about a major shift in official Roman Catholic views. Thomas ushered in a revival of Aristotelian ideas in Christian Europe. Early Christians, like Augustine, found Plato's philosophy more consistent with Christian doctrine than Aristotle's, so only a few of Aristotle's works were translated into Latin, the language of Rome. As Latin became the major language in northern Europe, fewer and fewer scholars were able to read the remainder of Aristotle's writings. In the meantime, Arab forces had conquered parts of southern Europe, spreading Islam in the region. It was primarily in the Arab world that Aristotle's works were studied and preserved. Arabic scholars translated Aristotle's works into Arabic and wrote extensive commentaries on his philosophy. Finally, in the 12th century, translations of Aristotle from Arabic to Latin appeared and through these translations Aristotle's scientific ideas were reintroduced into the mainstream of European scholarship.

Aristotle had observed a wide range of natural phenomena and had developed systematic scientific theories based on his observations, so when his ideas reemerged, it was as if a "new" science had been discovered. The controversy that broke out was not unlike more recent controversies that are associated with the names of Galileo and Darwin. When Aristotle's ideas began to become popular, Roman Catholic authorities condemned his teachings. The Church's reaction was partially the result of the fact that some Aristotelian ideas did not seem to be reconcilable with Christian doctrine. The association of Aristotelian scholarship with Islam, no doubt increased the level of anti-Aristotelian suspicion. In spite of the church's initial rejection of Aristotle, however, Thomas Aquinas sifted through the great Greek philosopher's writings, developing a major synthesis of Aristotelian science and Christian thought. Primarily through Thomas' work, the Aristotelian world picture became the official scientific perspective of the Church.

Thomas' view of nature is complex and would require extensive study to master. In this reading, I introduce only one aspect of his perspective on the natural world, namely his thoughts about the value and purpose of animals. I draw from two excerpts from Thomas' works. Because Thomas' style can be difficult to follow, I have both paraphrased and rearranged passages, adding my own comments.

The first reading is a short passage from Thomas' *Contra Gentiles* (Against the Gentiles). Thomas first presents a thesis:

> ***The God-ordained order of things is such that rational animals (human beings) are valuable in themselves, while other beings (non-human animals, plants) are valued subordinately to the rational animals.***

Thomas is going to defend the thesis that human beings are valuable in themselves and non-humans are valuable only because they are valuable for human beings. We have seen this idea in Aristotle. It is not surprising that we see it repeated here. In his defense of the thesis, Thomas presents thirteen arguments. I am including my own paraphrases of five of these arguments here:

---

*Excerpts from the works of Thomas Aquinas in this reading have been paraphrased by Roger P. Ebertz.

[1] First of all, the superiority of rational animals (humans) is seen by the difference in the way that they were made. Rational animals were created to be the master of their own actions, and thus they were made in such a way that they can provide for themselves. Other creatures (non-rational animals and plants) were created in such a way that they do not have control over their own actions. This fact shows that they must be valued, not for their own sake, but as subordinated to others. A thing which is moved by another being has the nature of a tool, an instrument, something that is valued because it is useful to someone else. In contrast, a thing which moves itself has the essential nature of an agent (one who initiates action). An instrument is not valued for its own sake, but as useful to an agent. So, while much care might be put into the construction of instruments, this is actually done for the good of the agent. But the work that is done by the agent himself or herself is done for the agent's own good. As self-moving agents, then, rational animals (human beings) are created and ordered by God as objects that are valuable for their own sake; all other created things are subordinate to rational creatures. . .

As self-moving agents, then, rational animals (human beings) are created and ordered by God as objects that are valuable for their own sake; all other created things are subordinate to rational creatures. . .

[5] Furthermore, it is obvious that for any being, each part of that being is designed so that the whole being might be complete. The whole does not exist to make the part complete, but the part exists to contribute to the completeness of the whole. (For example, an eye is a part of a human body. An eye exists to contribute to the health and well-being of the human body of which it is a part. The body as a whole does not exist for the sake of the eye.) Rational substances (humans), more than non-rational beings, are closely related to the whole. Each individual rational substance, in a way, encompasses the whole, since it can come to understand the whole of reality through the use of reason. On the other hand, every non-rational substance has only a partial share of reality. Thus it is fitting that God would provide all other substances for the care of intellectual substances (human beings).

Notice that in both of these arguments, Thomas starts with the assumption that humans are uniquely rational. In agreement with both Plato and Aristotle, as well as the dominant strain of Christian theology up to his time, Thomas believes that human beings, among earthly creatures, are the only ones that are rational. All other earthly creatures lack rational abilities. When one draws a line between rational beings and non-rational beings and places all of the elements of God's earthly creation on one side or the other, human beings end up on the side of reason, and all the rest of creation on the other. In the above arguments Thomas builds on this belief in humanity's unique rationality and argues that rationality makes humans superior to the rest of creation. In the argument numbered [1] he argues that because they are rational humans can initiate free actions. Other creatures (other animals and plants) cannot. Non-humans can only be moved by something outside of themselves or by instinct. Since humans can act freely, they are superior to other creatures. In the argument numbered [5], Thomas first asserts that the whole is always greater than the parts, since the parts of something always contribute to the completeness of the whole, not the other way around. He goes on to argue that reason enables humans to grasp the whole of reality. Thus, rational beings are more closely associated with the whole than are any of the other parts. Thus, since rational beings (humans) are associated with the whole, they are more valuable than non-rational beings. It is appropriate, he concludes, that plants and non-human animals (which are non-rational) would be created to be used by humans (who are rational).

After presenting these more philosophical arguments, Thomas seeks to support his case by appeal to the Bible.

[11] Thus there are numerous arguments for the conclusion that non-rational creatures exist for the sake of rational beings. This is why it is said in Deuteronomy (4:19), "when you look up to the sky and see the sun, the moon and the stars—all the heavenly array—be careful not to be deceived into thinking they are gods and bow to them, for your God has created them for the service of all the nations under heaven." And as it says in the Psalm (8:5–8), "You subjected all

things under their feet, all sheep and oxen, and animals of the field." And in Wisdom (12:18) it says, "You, our powerful Master, judge us with mildness and gently rule over us."

Humans are not wrong when they use animals, either by killing them or by employing them in any other way.

[12] All of these arguments will also refute the error of those who claim that it is wrong for humans to kill other animals. For animals are provided by God for human use in the natural course of things. Consequently, humans are not wrong when they use animals, either by killing them or by employing them in any other way.

This is why God said to Noah (Genesis 9:3), "Just as I gave you the plants, now all living things shall be food for you."

Finally, Thomas is aware that there are passages of the Bible which forbid cruelty to animals. Some might suggest that these texts show that animals are valuable in themselves, and not just for humans. No, Thomas argues, the Bible forbids cruelty to animals not because the animals are valuable in themselves, but because cruelty to animals may result in cruelty or harm to humans. Thus it is God's concern for human well-being that is behind these commands, not concern for animals.

[13] Indeed, if any statements are found in Sacred Scripture prohibiting cruelty against irrational animals, for instance, that one should not kill a bird who is with her young (Deut. 22:6), this command is given for some other reason. For example, it might be given to prevent humans from being cruel to other humans. If someone is cruel to an irrational animal, the person who is cruel may become habitually cruel and begin to act cruelly to other people. Second, the command might be given because the cruelty to an animal may also harm a human being, (as when cruelty to a farm animal results in a loss to the owner of the animal). . . .

The second passage we will consider is from Thomas' *Summa Theologica* (Summary of Theology). To help modern readers follow Thomas' argument, I have rearranged the order of presentation, and again paraphrased the text for ease of understanding today.

Whether it is wrong to kill any living thing.

Augustine says (City of God I. 20), "When we hear it said, 'You shall not kill," we do not take it to be referring to trees, for they have no sense. We do not take it to be referring to irrational animals, since they cannot interact with us on a human level. So it follows that the words, "You shall not kill," refer to the killing of humans, not to the killing of animals and plants.

So it is not wrong for a human being to use plants for the good of animals or to use animals for the good of human beings.

Argument: It is not wrong to use something for the purpose for which it exists. Now, reality is structured in such a way that imperfect things are for perfect things. (In in the process of birth and growth, living things move from imperfection to perfection.) Human growth begins with a living thing, then an animal, and finally a human being. The ultimate end of the process is human life. So, in the same way, plants, which have life alone are all for animals, and animals are for human beings. So it is not wrong for a human being to use plants for the good of animals or to use animals for the good of human beings, as Aristotle has said (Politics 1.3).

Now the most obvious way in which animals use plants and humans use animals is for food. And to do this, they have to take the life of that which is used for food. It follows, then that if it is right for animals to use plants, and for humans to use animals, for food, then it is not wrong to take their life in these cases. In fact, even the commandment of God says, "Look, I have given you all the herbs and all the trees to be your food, and I have given them to the non-human animals as well." And in Genesus 9:3, God says "All things that move and live shall be food for you."

Thomas summarizes the following objections to his views:

Objection 1: *It would seem that it is wrong to kill any living thing. For in the Bible, the Apostle Paul says that anyone who opposes God's ordinances will be damned (Romans 13:2). Furthermore,* *according to Psalm 146:8, 9, God, in his providence, has ordained that all living things should be preserved. It says, that God "makes grass to grow on the mountains . . . (and) gives animals their food." So, since God takes care of the animals, it would be wrong to take any animal's life.*

Objection 2: *Further, murder is a sin because it takes away life from a person. All animals and plants share in having life. So, for the same reason that it is wrong to kill a human being, it is wrong to kill irrational animals and plants.*

Objection 3: *Further, in God's law, punishments are only given for sins. But in the Old Testament God's law includes a special commandment that those who kill another man's ox or sheep receive be punished (Exodus 22:1). So the slaying of an irrational animal must be a sin.*

Thomas gives the following replies to these objections:

Reply to 1: *It is true that God wants the lives of animals and plants to be preserved. But the purpose of their preservation is not for their own good, but for the good of human beings. Accordingly, Augustine says (City of God I, 20). "The Creator has given a command that both their life and their deaths are subject to our use."*

Reply to 2: *Irrational animals and plants do not have the ability to reason by which they can set themselves in motion. They are moved by other things, and by instinct. This is a sign that they are naturally enslaved and created for the use of others.*

Reply 3: *The person who kills another person's farm animal, does not sin because he kills the animal, but because he has harmed another person's property. The crime that the Bible provides a punishment for, then, is not the crime of murder, but the crime of theft or robbery.*

What view of animals emerges from these passages? Are there ethical limits to the ways in which human beings can treat animals? We can summarize Thomas' view with three key points. First, Thomas continues a long theological and philosophical tradition which affirms that among earthly beings, only human beings are rational. All other animals lack the capacity to reason. In this, Thomas agrees with many European and American thinkers both before and after him. Second, Thomas clearly sees living things as falling within a hierarchical ordering that places the most valuable at the top and assigns value to those creatures below on the basis of their value to those at the top. Because humans are rational, they are at the top of the hierarchy, with God above and all other earthly things below. Human beings are valuable in themselves, other animals and plants are valuable because they are valuable for humans. Finally, at least in these passages, Thomas seems to find no basis for any restrictions on human treatment of animals if that treatment does not affect humans in any way. Cruelty to animals, it would seem to follow, is not wrong in itself. It is only wrong in as much as it harms human beings.

The Song of Creation's Praise

*—Francis of Assisi**

Francis of Assisi (1181–1226) was born in 1181, the son of a wealthy cloth merchant in Assisi, Umbria, now part of Italy. He was baptized with the name Giovanni, but his father, reflecting a love for France, renamed him Francesco. As a boy, Francis enjoyed partying and getting into trouble with his friends. In spite of this, to the delight of his father, Francis was becoming a successful young business person. Seeking glory, he joined the army of Assisi and went to war against the neighboring city-state of Perugia, and later Francis responded to a call for knights to join in the Fourth Crusade. On his way to join the troops, however, Francis had a dream in which God told him he was going in the wrong direction and should return home. Back in town, having returned without even seeing battle, Francis was humiliated in his home town. He struggled to discover what God was calling him to do. Eventually his call became clear. In a dramatic parting from his father, he renounced all his wealth, committing himself to living like the poor. Preaching in the public areas, Francis spoke a message of Christ's love and call to leave riches behind. Others began to follow, and eventually he founded the Order of Friars Minor for men, and then the Order of the St. Clare for women. Francis died on October 6, 1226. Less than two years later he was declared a saint by Pope Gregory IX.

Stories about Francis abound in which he preaches not only to human beings, but to birds and other animals. He is well-known for having a deep love for nature, seeing all creatures as his brothers and sisters. Thus, Francis has become the patron saint of animals, and in more recent years, of ecology. The poem here is a paraphrase in common English of one of Francis' most famous works. Translations give it differing titles, including "The Canticle of the Brother Sun," and "The Canticle of the Creatures." In this beautiful song, Francis names many of the elements of creation as his brothers and sisters and joins with them in praising the Creator of them all.

The Song of Creation's Praise

Highest, all-powerful, awesomely good Lord,
May You receive praise and glory, honor and blessing.
To You alone do such tributes belong, Most High God,
And there is no one truly fit to even speak of You.

Praise be to You, My Lord, with all of Your creation
Especially for my brother sun, who continually praises You,
Through whom you light up the day and brighten Your creation.
He shines with great beauty and splendor.
In radiance, he always points to You, O Most High God.

---

*The Song of Creation included here is a modern paraphrase of The Canticle of the Sun, by Roger P. Ebertz

Praise be to You, my Lord, for sister moon and all the stars.
You formed them clear and precious and fair.
Praise be to you, my Lord, for brother wind,
For air and clouds and weather of every kind.
By these You give Your creatures nourishment.

Praise be to You, my Lord, for sister water,
She is helpful, humble, precious and pure.
Praise be to You, my Lord, for brother fire,
By which You lighten the darkness.
He is fair, joyful, mighty and strong.

Praised be to You, my Lord, for our sister, mother earth
Who sustains and keeps us,
Bringing forth all kinds of fruits, grasses, and beautiful flowers.

Praise be to You, my Lord, for those who forgive others,
Even in the face of trouble and hardship, because of your love.
Blessed are those who keep the peace
For by You, Most High God, they will be crowned.

Praise be to You, my Lord, for our sister, physical death,
From which no man or woman can escape.
Despair and suffering will come to those who die in mortal sin
But blessings will come to those who find themselves doing Your holy will.
For the second death will not harm them.

May all creation praise and bless my Lord, and give Him thanks.
Bow down and humbly submit to Him.

Prayer for Creation

—Attributed to St. Basil the Great

The following prayer is frequently attributed to Basil the Great (c. 329–379 CE). Basil is revered in both the Roman Catholic and the Eastern Orthodox traditions. Recently, scholars have called into question the correctness of the claim that this prayer is really from Basil. Having searched his works, they have failed to find the prayer. If the prayer is truly from Basil, it comes from a much earlier period than that of St. Francis. If the prayer was not written by Basil, it is still a wonderful prayer expressing a view of animals and other creatures that is much more consistent with ideas of Francis than the views of Thomas Aquinas.

May we realize that they live not for us alone but for themselves and for thee and that they love the sweetness of life.

O God, enlarge within us the sense of fellowship with all living things, our brothers, the animals, to whom thou gavest the earth as their home in common with us.

We remember with shame that in the past we have exercised the high dominion of man with ruthless cruelty, so that the voice of the earth, which should have gone up to thee in song, has been a groan of travail.

May we realize that they live not for us alone but for themselves and for thee and that they love the sweetness of life.

NAME ____________________

Worksheet 3 Medieval Views of Creation

1. Augustine compares the world to a book; what do we read in this book?

2. In the third excerpt from Augustine in this chapter, he describes and ranks three types of earthly beings. Place then in the chart below and briefly summarize what he says about each.

   Highest:

   Middle:

   Lowest:

3. In Reading 3.2 Thomas Aquinas describes human beings as "intellectual," or "rational animals." Why does God care about these "rational animals"?

4. What does Thomas Aquinas say about the purpose of plants and animals? (For whose sake were plants given? Who were unintelligent animals given for?)

5. **In what way are Augustine and Aquinas similar in their way of describing animals?**

6. **Read 3.3 and 3.4, then write two or three sentences in which you compare the attitude toward animals in Thomas Aquinas on the one hand and Frances and the prayer for creation on the other.**

Pulling It Together

Medieval Views of Creation

QUESTIONS FOR THOUGHT AND DISCUSSION

In this chapter, you have read passages from Christian writers. What different views toward nature have you encountered in these readings?

At the beginning of this chapter, there were several selections from the Bible? Do you believe the Christian writers included in this chapter have a biblical understanding of the role of creation? Or are they mistaken in some ways?

Chapter 4

Nature as Machine in Early Modern Thought

INTRODUCTION

Scholars usually date the end of the Middle Ages with the rise of the European renaissance of the fifteenth and sixteenth centuries. "Renaissance" suggests a "rebirth," an unfortunate label, because the medieval period was not nearly as "dead" as is traditionally suggested. Nevertheless, the renaissance did bring a rebirth of humanism, of a more positive attitude toward humankind and its abilities, as well as a rediscovery of classical Greek and Roman art, literature, and philosophy. We have seen that even in the medieval period, philosophers such as Thomas Aquinas characterized human beings as uniquely rational. But the renaissance saw a new and greater confidence in human reason and other human capacities to understand and master the world. It ushered in a period in which the religious establishment, the Roman Catholic Church, faced challenges from more independent-minded thinkers.

One challenge came from within the Church itself. In 1517, Martin Luther, an Augustinian priest and university professor, tacked 95 complaints about practices within the Roman Catholic Church on the door of the church in Wittenberg, Germany. Luther's complaints focused primarily on the sale of "indulgences," the exchange of money for release from penance for sin, including time in purgatory. His arguments were based on his study of the Bible. His intention was to challenge church leaders to a debate that he believed could lead to setting things straight in the Church. But instead, his emphasis on independent thought provided oxygen to feed a fire that had begun to kindle among those who believed that the church had strayed from its Biblical roots. Challenging the authority of the Church, the reformers translated the Bible into the languages of the people and gave them liberty to read it for themselves. Subsequent debates and differences in interpretations have led to the many branches that make up Protestantism today. Although the reformation was a religious movement, it is also symptomatic of the greater independence and individualism that characterized the renaissance.

Closely following the beginning of the Reformation was another challenge to the Church's authority, this time from the natural philosophers, those thinkers we would today call "scientists." The sixteenth and seventeenth centuries have been characterized as a period of "scientific revolution." Major changes were taking place in the way scholars understood the natural world. At the risk of oversimplifying, I will highlight two major areas of change. First, the natural philosophers in this period were shifting the model by which they interpreted the lights and motions they observed in the universe (cosmos). At the end of the Middle Ages, in the fifteenth century, the philosophy of Aristotle, especially his view of the nature of the universe, and the theology of the church had become quite intertwined. The Biblical teaching that humankind was created in the image of God, enjoying a special relationship to God, seemed to fit well with the notion that the earth was at the center of the universe. God held the cosmos in existence; the earth was at the center of the cosmos; humankind was at the center of the earth. But in 1543, Nicolaus Copernicus published *On the Revolutions of the Celestial Spheres,* in which he suggested a model of the universe that placed the Sun, rather than the earth, at the center.

Copernicus was a Roman Catholic, but his book seemed to go directly against the accepted cosmological teachings of the Church. For various reasons, including the fact that a Lutheran scholar had written an introduction to Copernicus's work, suggesting that Copernicus intended the new model merely as a helpful way to calculate the motions of the stars and planets, not as a literal description, the initial reaction to the book was minimal. Another fact that perhaps minimized the reaction was Copernicus's death at about the time of the controversial work's publication. In 1632, however, Galileo Galilei published *Dialogue Concerning the Two Chief World Systems,* in which he defended heleocentrism, the view suggested by Copernicus that the sun is at the center of the universe. Galileo had been defending this view since 1610, and he had received criticism, but not official condemnation by the Church. After the publication of his ideas in his book, however, Galileo was tried by the Church, accused of heresy, and found guilty. The Roman Church at that time exercised far more power than it does in Western nations today. It used this power to force Galileo to recant and to place him under house arrest for the remainder of his life.

In spite of the Church's action toward Galileo, the heliocentric perspective gained ground among scientists. A second major change involved perception of the nature and laws of motion. It was typical in the medieval framework to separate the "celestial" (heavenly) realm, which includes the stars and planets, from the "terrestrial" (eartlhly), the realm of things on earth. Each realm had its own set of laws. But Issac Newton, studying motion on earth, devised laws that brought the two together. The new system, however, required a new way of seeing reality. Like Plato, who had described the creation of the world as the creation of an animal, Aristotle and his followers interpreted the world in terms of growth and organic life. The new science turned to mechanics for a new metaphor, seeing the cosmos not as a organism, but as a machine.

In addition to the rise of new religious and scientific ideas, the sixteenth and seventeenth centuries also saw a rise in new technologies. Sponsors were supporting scientists and mathematicians for the practical applications of their work, and among these applications were a variety of new machines. The mind of the period was captured, in particular, by the image of the mechanical clock. The parts of the new clocks were connected and fit together in intricate ways, such that one could wind them up and simply let them run on their own, following the laws of force and motion. This image, the image of the machine, became the new metaphor for human understanding of things on earth and things in the heavens. The cosmos and things on earth were seen as machines. The new science was "mechanical science." The machine metaphor was used by many prominent scientist-philosophers of the period, including Robert Boyle, Galileo, and Issac Newton.

The parts of the new clocks were connected and fit together in intricate ways, such that one could wind them up and simply let them run on their own, following the laws of force and motion.

From Organism to Machine: The Rise and Impact of the Mechanistic View of Nature

In his dialogue Timaeus, Plato has his characters describe the creation of the world as the formation of a "living creature truly endowed with soul and intelligence by the providence of God."[1] *The creator God, he says, intended "to make this world like the fairest and most perfect of intelligible beings." The world is an intelligent being and organism that has both life and a soul. By the end of the seventeenth century, the metaphor of the earth as an organism had been almost entirely replaced by the image of the earth as a machine. It would be an oversimplification to say that all ancient Greeks understood the world to be an organism, and all early modern thinkers saw the world, or the entire universe, as a machine. In every age, there are diverse views of the world. But it is clear that a shift took place in the dominant metaphor by which the natural world was conceived by the seventeenth century. In this essay, I will trace this shift by noting descriptions of the natural world in prominent natural philosophers, especially in sixteenth- and seventeenth-century Europe, and will outline a number of key implications that this shift had for the way men and women understood reality and the way they lived in the world.*

THE LIVING COSMOS IN THE MEDIEVAL WORLD PICTURE

Others have argued that the medieval period brought a desacralization of nature to Europe. I do not doubt this is true. From the third to the fifteenth centuries, the three great monotheistic religions, Judaism, Christianity, and Islam, struggled for power and influence. All three of these religions find their roots in the same basic monotheistic faith of Abraham and his children. In contrast to the peoples surrounding them, the children of Abraham were to worship only one God. All the other so-called gods were not gods at all. To worship and fear any "god" other than the one true God was idolatry. And the one true God was a transcendent God who was not associated with any particular place or any particular aspect of the natural world. The Jews were radical monotheists. The early Christians were Jews. Thus it is not surprising that as Christianity spread into what we now call Europe, it took with it a refusal to recognize the gods and spirits of the native worldviews. As pagan peoples converted to Christianity, worship of gods of forests and mountains was rejected. As others have argued, the spirits were exorcised.

In spite of this, however, medieval men and women did not think of the world as dead. It was very much alive. To grasp this aspect of medieval thought, we must take a step back and consider the dominant world-picture of the time. In the Ptolemaic-Aristotelian cosmology accepted by most in the period, the earth is at the center of the Universe. To this Greek cosmology was added the Judeo-Christian idea that human beings were created in the image of God, taking center-stage in the entire created universe. Humans were at the center, but they were not the only "living" beings. Most obviously, there were plants and other animals. But beyond this, the earth itself was seen as a living being. One common medieval idea saw humans as a microcosm of the universe. As macrocosm and microcosm, the cosmos and humanity were

---

[1] Plato, *Timaeus.* translated by Benjamin Jowett.

understood to closely reflect and participate in one another.[2] As late as the fifteenth century, for example, Leonardo da Vinci (1452–1519) describes humans and the earth as "very much alike." They are composed of the same elements. Like bones in the human body, the earth is supported with stones. "While man has within himself a pool of blood wherein the lungs as he breaths expand and contract, so the body of the earth has its ocean, which also rises and falls every six hours with the breathing of the world."[3] But even beyond the earth, the cosmos was full of life. Pico della Mirandola (1463–1494), describes the creation of the cosmos in these words,

> ***God the Father, the supreme Architect, had already built this cosmic home we behold, the most sacred temple of His Godhead, by the laws of His mysterious wisdom. The region above the heavens He adorned with Intelligences, the heavenly spheres He had quickened with eternal souls, and the excrementary and filthy parts of the lower world He filled with a multitude of animals of every kind.***[4]

We note that in his description of the universe, even the sun, the moon, and the planets are described as "Intelligences." The entire cosmos is filled with intelligent, living beings.

Finally, we should note that, even in the principles by which motion on earth was explained, one sees the influence of the image of a living being. According to the accepted scientific theory, each of the earth's four elements, earth, air, fire, and water, has a place of its own. And although the elements become stirred up, each seeks its own natural "place." The element earth, for example, seeks the center, and thus falls, whereas air rises, seeking its natural place above. Owen Barfield writes, "A stone falls to the ground—we see it seeking the centre of the earth, moved by something much more like desire than what we today call gravity."[5] A being must be alive to have desires. Even the elements were not "dead." These few examples, some from the period of renaissance that ushered in the new science, illustrate the lasting impact of the idea that the world was alive, and even intelligent. Plato had described the earth as an intelligent animal, or as a god, Medieval thinkers saw that earth, or the cosmos as a whole, as a macrocosm of human life. What is common to these views is that the earth and the cosmos that surrounds it are seen as living organisms. In spite of the death of the gods of nature, the metaphor of earth as organism continued to prevail.

The earth and the cosmos that surrounds it are seen as living organisms.

NATURE AS BOOK: A TRANSITIONAL METAPHOR

Although the organic image of nature was strong, another metaphor for the natural world had also been important through the middle ages: the metaphor of creation as a book. An early example is found in Augustine:

> ***Some people read books in order to find God. Yet there is a great book, the very appearance of created things. Look above you; look below you! Note it; read it. God, whom you wish to find, never wrote that book with ink. Instead he set before your eyes the things that He had made. Can you ask for a louder voice than that? Why, heavens and earth cry out to you: "God made me."***[6]

One finds the image of a book used throughout the history of Christianity to speak of the universe, even up to the present. But there are changes in the way the metaphor is used. And as such, I will argue that the idea of "the book of nature" serves as a transitional image, as the dominant image for the cosmos as it changes from an organism to a machine.

---

[2]Owen Barfield. *Saving the Appearances: A Study in Idolatry*, 2nd ed. (Wesleyan University Press, 1988), 51.

[3]Leonardo da Vinci. *The Notebooks*, quoted in J. J. Clarke, ed., *Voices of the Earth: An Anthology of Ideas and Arguments* (New York: George Braziller, 1994), 72.

[4]Pico della Mirandola. *Oration on the Dignity of Man*, quoted in Clarke, 73.

[5]Barfield, 77.

[6]Augustine. quoted in *The Essential Augustine*, Vernon J. Bourke, ed. (Indianapolis: Hackett Publishing Co., 1985), 123.

In Augustine's words, the role of "the book of nature" is clearly to reveal God and God's nature. In this sense, the image is rooted in the Bible itself. The Psalmist wrote, "The heavens declare the glory of God" (Psalm 19:1), and the New Testament writer Paul claims that, "since the creation of the world God's invisible qualities—his eternal power and divine nature-have been clearly seen, being understood from what has been made. . . ." (Romans 1:20). In the same context, Paul suggests that human beings are without excuse for their sin, because they could know what is right from the creation. In Christianity, then, the book of nature is a book about God and his law for humankind.

In Christianity, then, the book of nature is a book about God and his law for humankind.

The book of nature as an ethical primer seems to have continued into the Middle Ages. Peter Coates describes the use of the image as an "allegorical approach," which looked for moral lessons in nature. Nature as a whole, it seems, was seen as "a book of lessons through which God instructed people how to behave—the natural world having the inestimable advantage of being a 'book' accessible to all in a largely unlettered and illiterate society."[7] He traces this tradition back to the Stoic philosophers in Greece, but one could find precedent for moral lessons from nature in the book of Proverbs as well. Coates writes, "'Nature study' in the Middle Ages meant a search for sources of morality. The busy ant was 'a sermon to sluggards,' from whose activities one could learn the virtues of industry."[8]

In the Middle Ages, then, the "book" was a book of lessons. In contrast, sixteenth and seventeenth centuries saw the emergence of a different way of seeing the "book." Different lessons were to be learned from it. Johann Kepler (1571–1630), key mathematician and astronomer as the new science emerged, writes,

> ***If you desire venerability—nothing is more precious, nothing is more beautiful than our magnificent temple of God. If you wish to know the mysteries—nothing in Nature is, or ever has been, more recondite. It is but for one reason that my object will not satisfy everybody, for its usefulness will not be apparent to the thoughtless. I am speaking of the Book of Nature, which is so highly esteemed in the Holy Scriptures, St. Paul admonished the Heathen to reflect on God within themselves as they would on the Sun on the water or in a mirror. Why then should we Christians delight the less in this reflection, seeing that it is our proper task to honor, to revere, and to admire God in the true way? Our piety in this is the deeper the greater is our awareness of creation and its grandeur.***[9]

In one sense Kepler draws from the same biblical passages mentioned above and reflects Augustine's use of the metaphor. When one studies nature, one's mind is turned to the glory of God. But it is important that Kepler has in mind the study of the motions of the universe and the attempts to understand the motions of the planets. His nature study is not a search for moral lessons, but a search for an understanding of the working of the world. Far from believing that such an understanding will lead one away from God, Kepler believes it will in fact lead to reverence for the creator. Nevertheless, Kepler points in the direction of the scientific study of nature, not the moralizing that characterized the medieval use of the image.

Interestingly, when Kepler read the book, he read it not as a moralist, but as a scientist. We see the same theme expressed even more clearly in these words of Galileo (1564–1642):

> ***Philosophy is written in this grand book, the universe, which stands continually open to our gaze. But the book cannot be understood unless one first learns to comprehend the language and read the letters in which it is composed. It is written in the language of mathematics, and its characters are triangles, circles, and other geometric figures without which it is humanly impossible to understand a single word of it.***[10]

---

[7]Peter Coates. *Nature: Western Attitudes Since Ancient Times* (Berkeley: University of California Press, 1998), 60.

[8]Coates, 60. See Proverbs 6:6 for a biblical example of the same lesson.

[9]Johann Kepler. *Mysterium Cosmographicum*, quoted in quoted in J. J. Clarke, ed. *Voices of the Earth: An Anthology of Ideas and Arguments* (New York: George Braziller, 1994), 69.

[10]Galileo. *The Assayer*, quoted in J. J. Clarke, ed. *Voices of the Earth: An Anthology of Ideas and Arguments* (New York: George Braziller, 1994), 87.

Galileo has a clearly scientific study of the Book of Nature in mind. He has moved significantly away from the thought, suggested by Coates, that the book of nature was a book that could be understood by the unlettered and illiterate. One must know the language of mathematics and geometry to read this book. The lessons learned are not moral, but scientific.

Several important changes in the use of the metaphor have taken place. First, in the Middle Ages, the idea seems to have been that the book of nature could serve as a kind of "Bible for dummies." Written in a language everyone could understand, it nevertheless told the same story. One could learn about God and about morality from looking at nature, just as one could learn about God and about morality from studying the Bible. Clearly the Bible revealed God's truth more fully than did nature, but, nevertheless, the revelations from the two books were continuous with one another. By the modern period, thinkers are coming to see the two books as being about different subjects. Whereas the Bible is about God and morality, the book of Nature is about the physical principles of the universe. Both books have important and powerful truths to teach. But there is an excitement about the book of Nature that pervades the period. By studying this other book, the natural philosophers are discovering principles that give them important powers. When Bacon considers the book of nature, it is this knowledge that brings power to improve human live that he has in mind.

Second, and perhaps less clear, it seems to me that there is a shift away from a focus on the author of the book of nature, God, to a focus on the things that can be learned from nature. As long as nature is seen as a source of moral lessons, it is easier to understand it as collections of communications from a personal God. But once the focus turns to finding the principles by which the universe operates, the message becomes less personal and more mechanical. The lesson of the book of Nature as seen by the natural philosophers, whom we would now call scientists, was that the world was a wonderfully intricate machine. In this sense, the image of the book of nature is a transitional image. Although it is rooted in early Christian thought, it opens the door for the new mechanical world-picture as well. It is to this new world picture that we now turn.

NATURE AS MACHINE

Kepler, Galileo, and Newton studied the Book of Nature; they sought not the principles by which humans are to live their lives, but the principles by which nature operates. And as they conceptualized the world they studied, the metaphor of the machine took prominence. As early as 1370, a French mathematician and bishop, Nicole Oresme, had likened the operations of the universe to those of a clock. But the image came to the fore with Kepler, who wrote in 1605,

> ***I am occupied with the investigation of physical causes. My aim is to show that the celestial machine is to be likened not to a divine organism but rather to a clock. . . insofar as nearly all the manifold movements are carried out by means of a single, quite simple magnetic force, as is the case of a clockwork, all motions [of which are caused] by a simple weight. Moreover I show how this physical conception is to be presented through calculations and geometry."***[11]

And when Descartes (1596–1650) summarizes his own book on the workings of the cosmos, he writes, "I have described the earth and the entire visible world as if it were simply a machine. . . ."[12] Baraka

The use of the metaphor of a machine involves more than at first meets the eye. It represents deeper changes in the way the natural philosophers were studying and coming to an understanding of the world. The first of these was a new understanding of the nature of physical causes. As we noted above, Ptolemaic science, which prevailed in the medieval period, explained motions in terms much like those of human

[11]Johann Kepler, quoted in Peter Marshall. *Nature's Web: Rethinking Our Place on Earth* (New York: Paragon House, 1994), 171–2.

[12]Rene Descartes. "Principles of Philosophy," in Rene Descartes. *Philosophical Essays and Correspondence,* Roger Ariew, ed. (Indianapolis: Hackett Publishing Co., 2000), 263.

action. Earth sought to return to its proper place, motivated by something similar to desire. Based on the understanding of the universe as a macrocosm of the human body, this seems quite natural. It is desire that moves us to action, and so too a kind of desire is at work in the world. This Ptolemaic understanding of causation involves what philosophers call *final,* or teleological, causation. When we act out of desire, we act in order to accomplish something; we seek to attain that which is desired. We have a goal in mind. This goal, or "end," is the cause of our action. In a sense, then, our motion is caused by something in the future. Nature, too, in the medieval mind, moved toward its goals. With the rise of the mechanistic philosophy, the teleological account of causation was being rejected. If one thinks merely of the motion of a machine, one sees that each event is caused by something that goes before it, which in turn is caused by something which has come before that. Events are caused not by goals or ends that point to things in the future, but by events in the past. In the sixteenth century, this was a new way of thinking about causation. Although a clock has the human purpose of telling time, one can understand its physical workings by merely understanding the mechanisms that cause the motion. This is the focus of the metaphor of the machine.

Although a clock has the human purpose of telling time, one can understand its physical workings by merely understanding the mechanisms that cause the motion. This is the focus of the metaphor of the machine.

Second, natural philosophers were not content with diverse explanations of motion, but sought a single set of principles that coherently explained all motion. In the medieval world picture, the motions on earth (terrestrial motions) were understood as operating by one set of laws, and motions in the heavens (celestial motions) were understood as operating according to a different set of laws. The rise of the new mechanics, expressed in the metaphor of the machine, included the argument that one set of principles could explain all motion. When Copernicus suggested that the earth was not at the center of the universe, it was perhaps the displacement of earth from its presumed central location that raised the most controversy. But behind his suggestion lay the understanding that, in making the change, one could give a single coherent explanation for motion. In his dedication to his work, *Of the Revolutions of the Heavens,* written to the pope, Copernicus explains that as he considered the work of previous philosophers, he grew "disgusted that no more consistent scheme of the movements of the mechanisms of the universe, set up by the best and most law-abiding Architect of all things, was agreed upon by philosophers. . . ."[13] Using mechanistic language, Copernicus expresses the desire for a single theory of motion. With his discovery of the laws of motion, Isaac Newton provided the unifying theory the moderns were looking for, enabling an explanation of both terrestrial and celestial motion, by means of a single set of principles.

Third, the new science of mechanics brought with it a new language and methodology. It was based on observation, carried out through a method of analysis, and expressed mathematically. As far back as Aristotle, natural philosophers had believed that knowledge could be acquired by observing the world. But with the rise of early modern science, basing science on what could be seen with the eyes took on central importance. The philosopher Francis Bacon defended a new method of science that was to be based on careful observation. The new method also involved a method of analysis. In his *Discourse on Method,* Descartes argues that to find a solution to a problem, he set as a rule "to divide each of the difficulties I would examine into as many parts as possible and as was required in order better to resolve them."[14] Although Descartes is here speaking of questions of doubt and knowledge, the analysis of things into parts characterized the philosophy of the time in many areas of thought. Like a clock or machine that had working parts, the world as a machine had many individual parts. One could not understand the whole without a careful look at the parts. Finally, the mechanical science was expressed mathematically. We have already observed Galileo's description of mathematics and geometry as the language and grammar of the Book of Nature.

---

[13]Nichalaus Copernicus. "Dedication of the Revolutions of the Heavenly Spheres," in *The Scientific Background to Modern Philosophy: Selected Readings*, Michael R. Matthews, ed. (Indianapolis: Hackett Publishing Co., 1989), 42.

[14]Rene Descartes. "Discourse on Method," in Rene Descartes, *Philosophical Essays and Correspondence,* Roger Ariew, ed. (Indianapolis: Hackett Publishing Co., 2000), 54.

In short, the rise of the metaphor of the Book of Nature was part of significant changes in the way scholars were looking at the world. In the following sections, I will argue that the use of the machine image, and the philosophy which supported it, had a very significant influence on the way humans thought about and acted in the nonhuman world.

A NEW VIEW OF REALITY

As the mechanical philosophy of nature emerged, so too did new ways of thinking about reality. This in turn impacted the way in which human beings viewed the nonhuman world. First, it is worth mentioning that the rise of the idea that the universe is like a machine brought with it the popularity of the argument from design for the existence of God. If the universe is like a clock, and all clocks are designed by intelligent beings, there must be a supreme intelligence who created the universe. Peter Marshall writes, "From where does all the order and beauty we see in the world arise? Newton replied that every design must have a designer: 'Does it not appear from phenomena that there is a Being, incorporeal, living, intelligent, omnipresent. . . ?'"[15] More important than the argument itself, however, was the suggestion that God had created the world as a perfect great machine that could run smoothly on its own. Deism, popular in the eighteenth century, saw the world as created by God, but operating quite on its own according to physical laws. It should be noted, however, that some philosophers of the period were not convinced. Continuing to see the world in mechanical terms, they were less sure that there was a creator at all.

Second, the idea that the cosmos is a great machine brought with it the conviction that the parts of the cosmos, including living things, were machines as well. Although he believed there were aspects of human behavior that could not be explained other than by positing a human soul, Descartes believed that human bodies were machines, and that other animals were completely like machines. Thomas Hobbes (1588–1679) took the idea farther and attempts to explain even the moral and mental dimensions of human life in mechanical terms.[16] We will see later that the idea that nonhuman animals are machines, without souls or even minds, had a significant impact on how humans acted toward animals.

Third, the rise of mechanistic philosophy brought an even more fundamental change in the very understanding of what was seen as "real." We have noted that the language of the new science was mathematics. This meant that the focus was on what could be measured, what could be reduced to numbers. If reality was to be understood mathematically, what of those aspects of reality that could not be measured? As the new scientists studied the nature of physical objects, they distinguished between two types of properties of these objects. The following quotation from Galileo provides a good illustration:

> ***I say that whenever I conceive any material or corporeal substance, I immediately feel the need to think of it as bounded, and as having this or that shape; as being large or small in relation to other things, and in some specific place at any given time; as being in motion or at rest; as touching or not touching some other body; and as being one in number, or few, or many. From these conditions I cannot separate such a substance by any stretch of my imagination. But that it must be white or red, bitter or sweet, noisy or silent, and of sweet or foul odor, my mind does not feel compelled to bring in as necessary accompaniments. With the sense as our guides, reason or imagination unaided would probably never arrive at qualities like these. Hence I think that tastes, odors, colors, and so on are no more than mere names so far as the object in which we place them is concerned, and that they reside only in our consciousness. Hence if the living creature were removed, all these qualities would be wiped away and annihilated.***[17]

I think that tastes, odors, colors, and so on are no more than mere names so far as the object in which we place them is concerned, and that they reside only in our consciousness. Hence if the living creature were removed, all these qualities would be wiped away and annihilated.

[15]Peter Marshall. *Nature's Web: Rethinking Our Place On Earth* (New York: Paragon House, 1994), 176.
[16]Thomas Hobbes. *Leviathan*, "Part One: Of Man," 1568.
[17]Galileo. *The Asayer*, quoted in J. J. Clarke, ed. *Voices of the Earth: An Anthology of Ideas and Arguments* (New York: George Braziller, 1994), 87.

Galileo argues that when he looks at a physical object, there are some types of properties he perceives in the objects that he could not even imagine that the object would not have. For example, he could not imagine a physical object that was not located in some place, or did not take up a certain amount of space. Similarly, he could not imagine a physical object that was not either at rest or in motion. He lists several such properties. These properties, he believes, must clearly exist in objects themselves. They would be there whether there were any living beings to observe them or not. To use terminology that is more natural to us than to Galileo, we might say these properties of objects are "objectively real." On the other hand, Galileo claims that he perceives objects to have other properties that he does believe he can imagine an object without: color, taste, bitterness, and having an odor are examples. These properties, he suggests, are real only in the sense that we experience them; they are present only in our minds. Perhaps the clearest examples involve taste or sound. What would it mean to say that an object tastes sweet if there were no living being who was tasting it? What would it mean to say that an object was making a noise if there were no living being to hear it? "Hence, I think, that tastes, odors, colors, and so on. . . reside only in the consciousness." In today's terms, we would say these properties are at best "subjectively real." They do not exist in the world apart from some mind that gives rise to the appropriate "image."

The philosophers of the period came to label these two types of properties as "primary qualities" and "secondary" qualities. The distinction was not new in the early modern period. It had been taught by the ancient Greek atomists. But it was rediscovered and accepted by a wide range of seventeenth-century philosophers. The important point to note in our connection, however, is this. The primary qualities of objects were clearly seen as real. Secondary qualities, it would seem, were real only in a derivative sense. They were real only in the minds of perceivers. Interestingly, it was the primary qualities that the natural philosophers could quantify. Height and width and depth could be measured. The velocity of an object could be measured. These quantifiable concepts could be translated into the language of mathematics and calculated using theories of motion. It is these measurable qualities that are objectively real, in the perspective of the new scientists. On the other hand, qualities that could not be measured were passed over. One cannot measure a color or a smell. These properties of things, they believed, do not really exist in things themselves, but in the mind of the perceiver. Descartes writes,

> ***Thus it is one and the same matter that exists throughout the universe; its one distinctive characteristic everywhere is extension. All the properties we can clearly and distinctly perceive in it are reducible to divisibility and capacity for varying motions. From this there follows the potentiality of all states. Motion is the activity whereby a body travels from one place to another.***[18]

What is real? As the new science developed, a clear view emerged. Only that which is quantifiably measureable is real. And when objects are so reduced, all that ends up being real are things that take up space, that can be divided and can move, all made of the same stuff, whatever that is. Descartes made an exception for human beings, arguing that humans had souls. Some continued to believe there was a God. The natural world, however, was a world of bare matter. Peter Marshall quotes a famous twentieth-century philosopher to sum it up: "In the words of philosopher A. N. Whitehead, nature became 'a dull affair, soundless, scentless, colourless; merely the hurrying of material, endlessly, meaninglessly.'"[19]

THE MECHANICAL PHILOSOPHY AND EUROPEAN ATTITUDES TOWARD NATURE

The mechanistic philosophy that arose in the early modern period involved a new view of reality. Although some viewed the cosmos as created by an intelligent designer, the cosmos itself, and the things that existed within it, were seen as machines. Reality consisted of matter in motion, that which was truly real, could be counted and calculated. How did the mechanical perspective affect human attitudes and actions toward

---

[18]Rene Descartes. "Principles of Philosophy," quoted in J. J. Clarke, ed. *Voices of the Earth: An Anthology of Ideas and Arguments* (New York: George Braziller, 1994), 89.

[19]Peter Marshall. *Nature's Web: Rethinking Our Place on Earth* (New York: Paragon, 1994), 178–179.

nature? In this section, we will explore this question. The first attitude change that I will note is this: as humans came to see the world in more and more mechanistic terms, they came to see themselves as separated from nature. Own Barfield writes, "To the extent therefore that the phenomena are experienced as machine, they are believed to exist independently of man, not to be participated [in] and therefore not to be in the nature representations." As we saw, the medieval mindset had included the idea that humans were a microcosm of the cosmos, in one sense representing it and participating in it. As the organism metaphor faded, so did the human sense of connection with nature. One philosopher who exacerbated this trend was Rene Descartes. Although Descartes adopted the mechanistic philosophy, when it came to humans he argued that we are not just mechanical bodies, but bodies with souls. It is the soul that we can be most certain of, he argues, and suggests that it is the soul that defines our being. Thus, Cartesianism draws a clear divide between humans and nature. Nature is machine, but humans are defined by their soul. In various ways, then, the new philosophy brought a general sense of the separation of humankind from the world of nature.

Secondly, the rise of the mechanical philosophy fueled a sense of human power over nature.The renaissance had brought a renewed appreciation of human abilities, but with the discovery of laws of nature, humans began to see that the new understanding brought power. The most notable proponent of the use of the new knowledge as power over nature is Francis Bacon. Using theological language, Bacon began with Christian notion that humankind had fallen in sin and suffered the consequences of evil in the world. But through the new scientific understanding that God was revealing through the book of nature, humans had a chance to regain the place God had intended them to have, conquering the evils of nature. Humankind's role was not to live within nature, but to conquer, to make it better for the human good. As Bacon writes,

The rise of the mechanical philosophy fueled a sense of human power over nature.

> ***Now the empire of man over things depends wholly on the arts and sciences. For we cannot command nature except by obeying her. . . . Only let the human race recover that right over nature which belongs to it by divine bequest, and let power be give it; the exercise thereof will be governed by sound reason and true religion.[20]***

It is clear that for Bacon, the goal of science is mastery over nature.

With the new philosophy, humankind came to see nature as a collection of resources for human use.

Third, with the new philosophy, humankind came to see nature as a collection of resources for human use. The quotation from Bacon we have just considered suggests that the purpose of the power of knowledge is improving the human condition. In the following passage, Bacon suggests that, through the human knowledge, we can indeed make nature better.

> ***We make (by art) in the same orchards and gardens, trees and flowers to come earlier and later than their seasons, and to come up and bear more speedily than by their natural course they do. We make them also by art greater much than their nature, and their fruit greater and sweeter and of differing taste, smell, colour, figure, for their nature.[21]***

As humankind regains its proper mastery over nature, nature can be manipulated and changed, evils can be overcome and good things made even better. In this perspective, humankind becomes God's instrument in restoring the world to what it was meant to be, and the world is a treasure chest of resources for human use.

Fourth, the mechanistic philosophy included a troubling attitude toward nonhuman animals. As Peter Marshall summarizes, "The more the powers of man were elevated, the more animals became degraded. They were considered beings created by God entirely for human use, and of no intrinsic value."[22] Marshall recounts the story of Flemish anatomist Andreas Vesalius (1514–64), who performed experiments on

---

[20]Francis Bacon, quoted in J. J. Clarke, ed. *Voices of the Earth: An Anthology of Ideas and Arguments* (New York: George Braziller, 1994), 85.
[21]Francis Bacon, quoted in Peter Coates, *Nature: Western Attitudes Since Ancient Times* (Berkeley: University of California Press, 1998), 73.
[22]Peter Marshall. *Nature's Web: Rethinking Our Place on Earth* (New York: Paragon, 1994), 169.

animals without anesthetic, in order to help his student observe the workings of the nervous systems in animals. Far from trying to avoid pain to animals, Vesalius chooses his animal subjects in part because of the intensity with which they cry out. "It is better to choose a sow on account of the voice," he writes. "For a dog, after being bound for some time, no matter what you do to it, finally neither barks nor howls, so you are sometimes unable to observe the weakening of the voice."[23] He describes his experiments with some detail, taking particular care to explain the animals' reactions, with no sense that the suffering of the animals is to be avoided. Vesalius was not alone, either in believing that animals were to be used for humans, nor in believing that animal suffering was of no moral concern. Rene Descartes, based on his comparison of nonhuman animals to machines, argued that they did not have souls, and thus did not experience pain as humans do. Thus, he too defended the dissection of living animals without anesthetics to gain knowledge for humankind.

Fifth, and finally, I would suggest that the rise of the mechanical philosophy in the early modern period subtly changed the way in which human saw nature, robbing nature of much that is of value. As we have seen, in the move to base its theories on mathematics, the new science focused on bare matter and motion. Color, smell, and taste were mere subjective realities, present only in minds. What, then, becomes of the beauty of the sounds and smells one experiences as one walks through a forest, or of the beauty one is struck by as one observes autumn leaves? From the perspective of modern science, a view that became rooted in Western thought in the seventeenth century and continues to affect our thinking today, color, odor, and beauty are not aspects of reality at all, but merely subjective reactions.

IS EARLY MODERN SCIENCE TO BLAME?

I have argued that in the seventeenth and eighteenth centuries the metaphor of the organism came to be replaced with a new metaphor for the cosmos, that of a machine. The image of nature as a book served as a transitional metaphor between the two. I have outlined a number of changes that occurred in the way Europeans saw the reality and their attitudes toward the world of nature. These views and attitudes were not held by everyone. I have not, for example, noted the rejection of the quantifiable, mechanical philosophy by the Romantic Movement. Nor have I noted various complaints against the misuse of animals that were voiced even at the time. As I indicated at the outset, no period can be accurately described in simple terms. Nevertheless, I believe it is clear that a shift took place from seeing nature as organism to nature as machine. I think it is also clear that this change had important effects on how human beings lived in their world.

Is the metaphor of the machine, and the mechanical philosophy it represents, to blame for our environmental crisis? Should we blame the early modern scientists for global warming? I don't think so. The causes of our current ecological problems are complex. No metaphor, no single period in history, no school of thought is responsible for our situation today. But understanding the metaphors by which humans have conceptualized the world, and by which we continue to do so today, can help us understand why humans have acted the way they have. It can help us think through the perspective we will form of the natural world.

---

[23] Andreas Vesalius, quoted in Marshall, 170.

What Sets Humans Apart?

—Rene Descartes

In the previous reading, we saw that with the rise of modern science, the image of the world as a living thing was replaced by the image of the world as a machine. One philosopher and scientist who accepted and applied the machine metaphor was Rene Descartes (1596–1650 CE). Descartes put considerable time and thought into understanding various functions of the human body, including human sight, muscular movement, and circulation. In his attempts to explain these functions, Descartes found the model of a machine useful. The bodies of animals, both human and nonhuman, could be explained mechanically. They consisted of parts, intricately arranged, that operated according to the laws of nature. If one were to construct a monkey-like automata (which we might perhaps call a robot), this being would really be no different from a real monkey. The automata and the "real" monkey would both be machines. Neither has reasoning ability, nor has a soul; both operate according to natural laws. Human bodies, too, are machines. But unlike other animals, Descartes believed that humans also had immaterial souls. The following reading is an excerpt from a letter by Descartes in which he summarizes the ideas of a book he had written about the nature of the physical world. In the passage here, Descartes discusses the differences between animals and humans. These differences, Descartes argues, justify us in concluding that while human bodies are machines, humans also have immaterial souls which give us special reasoning abilities that animals do not have. In setting human beings apart in this way, Descartes' thought has the effect of devaluing other living things. Because it is the soul and its rational capacities that give value to humans, Descartes's argument implies that other animals and other natural entities are in no sense rational and not of value in themselves.

I had provided a sufficiently detailed explanation for all these things in the treatise that I had previously intended lo publish.[1] And then I had shown what the constitution of the nerves and muscles of the human body must be in order to make the animal spirits within them have the force to move its members, as when one observes that heads, shortly after being severed, still move about and bite the earth, even though they are no longer alive. I had also shown what changes must take place in the brain in order to cause wakefulness, sleep, and dreams; how light, sounds, odors, tastes, heat, and all the other qualities of external objects can imprint various ideas there through the mediation of the senses; how hunger, thirst, and the other internal passions can also send their ideas there; what part of them needs to be taken there for the common sense, where these ideas are received: for the memory; which preserves them; and for the imagination, which can change them in various ways and compose new ones out of them, and, by the same means, distributing the animal spirits into the muscles, make the members of this body move in as many different ways (and in a manner appropriate to the objects that present themselves to the senses and to the internal passions that are in the body) as our own bodies can, without their being guided by the will. This will in no way seem strange to those who are cognizant of how many different automata of moving machines the ingenuity of men can make, without using, in doing so, but a very small number of parts, in comparison

---

[1]This is a reference to The World, of which the Treatise on Man was a part.

with the great multitude of bones, muscles, nerves, arteries, veins, and all the other parts which are in the body of each animal. For they will regard this body as a machine which, having been made by the hands of God, is incomparably better ordered and has within itself movements far more wondrous than any of those that can be invented by men.

I paused here in particular in order to show that, if there were such machines having the organs and the shape of a monkey or of some other animal that lacked reason, we would have no way of recognizing that they were not entirely of the same nature as these animals; whereas, if there were any such machines that bore a resemblance to our bodies and imitated our actions as far as this is practically feasible, we would always have two very certain means of recognizing that they were not at all, for that reason, true men. The first is that they could never use words or other signs, or put them together as we do in order to declare our thoughts to others. For one can well conceive of a machine being so made that it utters words, and even that it utters words appropriate to the bodily actions that will cause some change in its organs (such as, if one touches it in a certain place. it asks what one wants to say to it. or, if in another place, it cries out that one is hurting it. and the like). But it could not arrange its words differently so as to respond to the sense of all that will be said in its presence, as even the dullest men can do. The second means is that, although they might perform many tasks very well or perhaps better than any of us, such machines would inevitably fail in other tasks; by this means one would discover that they were acting, not through knowledge, but only through the disposition of their organs. For while reason is a universal instrument that can be of help in all sorts of circumstances, these organs require some particular disposition for each particular action; consequently, it is for all practical purposes impossible for there to be enough different organs in a machine to make it act in all the contingencies of life in the same way as our reason makes us act.

I paused here in particular in order to show that, if there were such machines having the organs and the shape of a monkey or of some other animal that lacked reason, we would have no way of recognizing that they were not entirely of the same nature as these animals; whereas, if there were any such machines that bore a resemblance to our bodies and imitated our actions as far as this is practically feasible, we would always have two very certain means of recognizing that they were not at all, for that reason, true men.

Imagine that we could make very life-like robot dogs. In what ways do you think these dogs would be similar and different from real dogs? Now I imagine that we could create life-like robots that similated humans? In what ways would these robots be similar and different from real humans? Finally, think again about the real dog. Do you think the real dog would be more like the real human or the human-like robot? Explain your answer.

Now by these two means one can also know the difference between men and beasts. For it is rather remarkable that there are no men so dull and so stupid (excluding not even the insane), that they are incapable of arranging various words together and of composing from them a discourse by means of which they might make their thoughts understood; and that, on the other hand, there is no other animal at all, however perfect and pedigreed it may be, that does the like. This does not happen because they lack the organs, for one sees that magpies and parrots can utter words just as we can, and yet they cannot speak as we do, that is to say, by testifying to the fact that they are thinking about what they are saying; on the other hand, men born deaf and dumb, who are deprived just as much as, or more than, beasts of the organs that aid others in speaking, are wont to invent for themselves various signs by means of which they make themselves understood to those who, being with them on a regular basis, have the time to learn their language. And this attests not merely to the fact that the beasts have less reason than men but that they have none at all. For it is obvious it does not need much to know how to speak; and since we notice as much inequality among animals of the same species as among men, and that some are easier to train than others, it is unbelievable that a monkey or a parrot that is the most perfect of its species would not equal in this respect one of the most stupid children or at least a child with a disordered brain, if their soul were not of a nature entirely different from our own. And we should not confuse words with the natural movements that attest to the passions and can be imitated by machines as well as by animals. Nor should we think, as did some of the ancients, that beasts speak, although we do not understand their language; for if that were true, since they have many organs corresponding to our own, they could make themselves as well understood by us as they are by their fellow creatures. It is also a very remarkable phenomenon that, although there are many animals that show more skill than we do in some of their actions, we nevertheless see that they show none at all in many other actions. Consequently, the fact that they do something better than we do does not prove that they have any intelligence; for were that the case, they would have more of it than any of us and would excel us in everything. But rather it proves that they have no intelligence at all, and that it is nature that acts in them, according to the disposition of their organs—just as we see that a clock composed exclusively of wheels and springs can count the hours and measure time more accurately than we can with all our carefulness.

After that, I described the rational soul and showed that it can in no way be derived from the potentiality of matter, as can the other things I have spoken of, but rather that it must be expressly created; and how it is not enough for it lo be lodged in the human body like a pilot in his ship, unless perhaps in order to move its members, but rather that it must be more closely joined and united to the body in order to have, in addition to this, feelings and appetites similar to our own, and thus to constitute a true man. As to the rest, I elaborated here a little on the subject of the soul because it is of the greatest importance; for, after the error of those who deny the existence of God (which I think I have sufficiently refuted), there is none at all that puts weak minds at a greater distance from the straight path of virtue than to imagine that the soul of beasts is of the same nature as ours, and that, as a consequence, we have nothing lo fear or to hope for after this life any more than do flies and ants. On the other hand, when one knows how different they are. one understands much better the arguments which prove that our soul is of a nature entirely independent of the body, and consequently that it is not subject to die with it. Then, since we do not see any other causes at all for its destruction, we are naturally led to judge from this that it is immortal.

NAME ______________________

Worksheet 4 — Nature as Machine in Early Modern Thought

1. **In the first essay in this chapter, three metaphors for nature are discussed. Briefly describe what each metaphor suggests:**

   What does it suggest to say that the world, or the cosmos, is an organism?

   What does it suggest to say that the world is a book?

   What does it suggest to say that the world is a machine?

2. **Make a list of three items in each category:**

Organisms:	Machines:

3. **Make a list of features that characterize organisms and machines:**

Organisms are:	**Machines are:**

4. **List, and explain in a sentence, three changes in attitudes and ideas about the cosmos that took place when Western thinkers began thinking about the cosmos as a machine rather than as an organism.**

   a.

   b.

   c.

5. **What two features did Descartes believe would always distinguish humans from human-like machines?**

6. **Do you believe it would ever be possible to create a machine that was equal to a machine? Why or why not?**

Pulling It Together

Nature as Machine in Early Modern Thought

QUESTION

In this chapter, it is argued that philosophers shifted from seeing the the earth as an organism to seeing the earth as a machine. Which of these two images do you find most useful in describing the nature of the earth? Explain.

EXERCISE

Ask your friends whether they think the earth is more like an single living organism, like a plant or an animal, or like a machine. Ask them why and write down their answers. Report these answers to the class for discussion.

Chapter

5

17th and 18th Century European-American Encounter with Wilderness

INTRODUCTION

In this chapter we cross the ocean from Europe to America and consider five writers whose works represent important channels of influence in American attitudes toward nature. We begin with a short reading by William Bradford (1590–1657). Bradford was a member of the party of "Pilgrims" who sailed to American on the Mayflower, landing at Cape Cod in 1620.

The remaining readings in this chapter represent more positive views of wilderness that flourished not far from Plymouth in geographical distance, but that were separated by a little over 200 years. One school of thought, called Transcendentalism, is represented here by two of its most famous members: Ralph Waldo Emerson (1803–1882) and Henry David Thoreau (1817–1862). Emerson and Thoreau were both natives of Concord, MA, a small village 50 miles northwest of Plymouth. In the time between the settlement of Plymouth by the Pilgrims and the lives of Emerson and Thoreau, a lot had happened. America had fought its war of independence from England, the Eastern seaboard had been settled by Europeans, and various forms of Christianity had become well established in the new nation. In Massachusetts, the dominant form of religion was Puritan Christianity, primarily in the form of Congregationalism.

Reacting, perhaps, to the climate of strict religious observance and orthodoxy of earlier Puritans, many men and women in the time of Emerson and Thoreau were questioning and challenging orthodox Christian beliefs. In the eighteenth century the Unitarian movement had broken from Congregationalism, rejecting the Christian doctrine of the Trinity. Emerson's father had been a Unitarian minister, and Emerson was trained to be a minister, but he left the ministry after finding himself in disagreement with various practices of the church. Like others in their time, Emerson and Thoreau explored not only Christian ideas, but those of other major religions as well. Emerson and Thoreau were also influenced by Romanticism. Romanticism in Europe had rejected the overly rationalistic philosophy of the Enlightenment and the purely mechanical understanding of the natural world that had come to dominate with the Scientific Revolution. By focusing purely on reason and analysis, one could totally miss the insights and inspiration the natural world had to offer. The Romantics emphasized both emotions and wild nature as key sources of understanding. Transcendentalism preached a similar message in America. The transcendentalists did not see nature or "wilderness" as something to be conquered and civilized, but as the source of strength and insight. Wild places were places in which one could experience of the sublime, the divine, which transcended the reality of everyday life. Wilderness, in fact, is a source of salvation—holy land.

In addition to readings by Emerson and Thoreau, this chapter also includes excerpts from the writings of two other nature writers of the period, Susan Fenimore Cooper and John Burroughs.

Arriving at Cape Cod

—William Bradford

William Bradford (1590-1657), who landed in America as one 7f the "Pilgrims" on the Mayflower in 1620, was the primary leader of the early Plymouth colony, serving as its governor five times between from 1621 to 1657. Of Plymouth Plantation, written between 1620 and 1651, was Bradford's account of the voyage of the Pilgrims and their experience. America. In this chapter, he recounts their early days at Plymouth Colony, in what is now Massachusetts. Although Bradford was himself aboard the ship, he writes his account in the third person.

Prior to the place in the book where our selection begins, Bradford has described how the Pilgrims, and others travelling with them, set out in two ships. One of those ships, however, was found not to be seaworthy, and they were forced to turn back. Eventually, those who still wanted to make the voyage were put on one ship, the Mayflower, and they set out.

THE ACCOUNT OF THE VOYAGE ACROSS THE SEA AND ARRIVAL AT CAPE COD

They were now all crowded together in one ship. They put to sea again with a good wind, which continued for several days. This was an encouragement to them. But, as is usual with these voyages, many on board became sick. I don't want to leave out something that God did while they were on the journey. There was a very nasty young man, one of the ship's crew, who was extremely proud and haughty, made more so by the fact that he was strong and healthy. This man was always criticizing the poor people that were sick and swearing at them daily. He didn't hesitate to let them know that he hoped to throw half of them overboard before they got to their destination. He told them to be satisfied with what they had. If anyone would gently correct him, he would curse and swear all the more bitterly. But it pleased God to inflict this young man with a bad sickness himself, so that he became desperately ill and died. Thus, he was himself thrown overboard. So his curses landed on his own head and all his fellow crewmen were astonished and noted that God had justly punished him.

As they started their voyage, the weather was good and the winds favorable. But after a time, they encountered many cross winds and fierce storms which shook the ship. Her upper works began to leak and one of the main beams in the middle of the ship bowed and cracked. They began to fear that the ship could not survive the long trip. Some of the passengers heard the crew muttering their fears that the ship would fail, and so their leaders had serious talks with the master and officers to discuss whether they should turn around and return rather than face certain death at sea. Even the sailors themselves had differing opinions on what should be done. The sailors wanted to keep their jobs and get their wages but they also were concerned that the cargo would be lost. After taking into consideration all the opinions, the ship masters decided the ship was strong below the water and devised a way to repair the main beam. They could caulk the upper decks as best they could to repair the leaks. All in all, they thought the ship would be okay if they did not push it too hard with the sails. So they prayed for God's care and continued on their journey.

This excerpt from William Bradford has been paraphrased by Roger P. Ebertz.

In some of the storms, the winds were so strong and the waves so high they could not raise the sails, and were forced to wait out the storms. In one of them, as they were sitting out a mighty storm, a merry young man named John Howland was thrown into the sea. But God saved him when he caught hold of a rope which hung overboard. He held on to the rope, even though he was under water, until he was hauled up to the surface and brought into a lifeboat. And so he was saved. Although the experience made him ill, he survived and became an active member and participant in both the church and the community. In all of the journey, there was only one passenger that died, a man by the name of William Butten. William was a boy, a servant of Samuel Fuller. He died as they were approaching land at the end of the journey.

I will leave out many details, so as to be brief. After a perilous voyage, they finally arrived at the place that is called Cape Cod. They were extremely glad to see land, but after some deliberation among themselves and with the master of the ship, they decided to travel south to find a place near the Hudson River where they could land and settle. After sailing for half a day in that direction, however, they hit a dangerous area of the coastline and roaring waves, and became entangled in great danger. So they abandoned their travels south and decided to land at the Cape, happy to get out of danger before night came. And so, by God's providence, they landed at Cape Cod. The next day, they entered the Cape itself and rode along in its safe and peaceful waters.

And so, by God's providence, they landed at Cape Cod.

A word about Cape Cod: It was first named this by Captain Gosnold and his company in the year of 1602. Later it was called Cape James by Captain Smith. But sailors continue to call it by the earlier name. The Cape is surrounded by a peninsula, the tip of which they called Point Care. It is here that the dangerous shoreline begins. Sometimes it is called Tucker's Terror. The French and Dutch, to this day, call it Malabar because of the dangers in that area and the lives they have lost there.

Once they arrived in a good place to anchor the ship and come on land, the pilgrims fell down on their knees and thanked the God of heaven who had brought them over the vast and dangerous ocean, saved them from all the dangers and miseries of the journey, and brought them to solid land once again. It is no wonder that they were so joyful after such a trip. But I cannot help but pause in my story and consider the poor situation of these people. Having just crossed the vast ocean, and experienced a sea of troubles even before they began, they had no friends to welcome them. There were no restaurants or hotels to entertain or refresh their weather-beaten bodies. There were no homes, let alone towns or cities, to go to for help. It is recorded in the Bible that the Apostle Paul, when he was in a shipwreck, experienced the kindness of barbarians. But this was not so with these travelers. The barbarians that they encountered here were more ready to fill their sides with arrows than anything else. Their bleak situation was made worse by the fact that it was winter when they arrived. They knew that winters in this new land were harsh, subject to cruel and fierce storms. In such conditions it would be difficult to travel in familiar territory, even more difficult along an unknown coast. When they looked around all they could see was a hideous and desolate wilderness, full of wild beasts and wild men — and how many of them there might be they did not know. Even when they climbed up to the top of Mt Pisgah and looked around for a better place to live, whatever direction they looked (except when they looked up to heaven) they could see nothing that gave them comfort. Summer was over, and everything looked bare and bleak. The whole country, full of woods and thickets, had a wild and savage appearance. If they looked behind them, there was the mighty ocean which now separated them from all the civilized parts of the world. It might be thought that they could take comfort in the ship and the supplies aboard, but every day the master and the crew reminded them that they must find a place to settle soon. The ship would have to make a return journey and would need provisions for the trip. Some of the sailors muttered that if the settlers did not find a place to settle soon, the crew would unload their goods and leave them on the shore. The pilgrims could not rely on much help from back home or be encouraged by the love and support their fellow believers might provide. For, although their friends in Leyden were supportive and loved them dearly, they were poor themselves and in no situation to be able to help.

When they looked around all they could see was a hideous and desolate wilderness, full of wild beasts and wild men.

All that could sustain them now was the Spirit of God and His grace. The children of these people may rightly say, "Our fathers were Englishmen who came over this great ocean, and were about to die in the wilderness; but they cried to the Lord and He heard their voice and saw their trouble," etc.[1] "Let them therefore praise the Lord, because he is good; and His mercies endure forever." "Let those who have been saved by the Lord show how He has delivered them from the hand of the oppressor. When they wandered in the desert wilderness out of the way, and found no city in which to dwell, both hungry and thirsty, they were discouraged and overwhelmed. Let them tell how kind and loving the Lord was to them and tell of His wonderful works among the people of the earth."[2]

---

[1]Bradford is adapting Deuteronomy 26: 5-8 to draw a parallel between the situation of the Pilgrims and the situation of the ancient Hebrews described in this passage.
[2]Bradford quotes from Psalm 107.

On Nature

—Ralph Waldo Emerson

NATURE

To go into solitude, a man needs to retire as much from his chamber as from society. I am not solitary whilst I read and write, though nobody is with me. But if a man would be alone, let him look at the stars. The rays that come from those heavenly worlds, will separate between him and what he touches. One might think the atmosphere was made transparent with this design, to give man, in the heavenly bodies, the perpetual presence of the sublime. Seen in the streets of cities, how great they are! If the stars should appear one night in a thousand years, how would men believe and adore; and preserve for many generations the remembrance of the city of God which had been shown! But every night come out these envoys of beauty, and light the universe with their admonishing smile.

The stars awaken a certain reverence, because though always present, they are inaccessible; but all natural objects make a kindred impression, when the mind is open to their influence. Nature never wears a mean appearance. Neither does the wisest man extort her secret, and lose his curiosity by finding out all her perfection. Nature never became a toy to a wise spirit. The flowers, the animals, the mountains, reflected the wisdom of his best hour, as much as they had delighted the simplicity of his childhood.

One might think the atmosphere was made transparent with this design, to give man, in the heavenly bodies, the perpetual presence of the sublime.

When we speak of nature in this manner, we have a distinct but most poetical sense in the mind. We mean the integrity of impression made by manifold natural objects. It is this which distinguishes the stick of timber of the wood-cutter, from the tree of the poet. The charming landscape which I saw this morning, is indubitably made up of some twenty or thirty farms. Miller owns this field, Locke that, and Manning the woodland beyond. But none of them owns the landscape. There is a property in the horizon which no man has but he whose eye†can integrate all the parts, that is, the poet. This is the best part of these men's farms, yet to this their warranty-deeds give no title.

To speak truly, few adult persons can see nature. Most persons do not see the sun. At least they have a very superficial seeing. The sun illuminates only the eye of the man, but shines into the eye and the heart of the child. The lover of nature is he whose inward and outward senses are still truly adjusted to each other; who has retained the spirit of infancy even into the era of manhood. His intercourse with heaven and earth, becomes part of his daily food. In the presence of nature, a wild delight runs through the man, in spite of real sorrows. Nature says,—he is my creature, and maugre all his impertinent griefs, he shall be glad with me. Not the sun or the summer alone, but every hour and season yields its tribute of delight; for every hour and change corresponds to and authorizes a different state of the mind, from breathless noon to grimmest midnight.

Nature is a setting that fits equally well a comic or a mourning piece. In good health, the air is a cordial of incredible virtue. Crossing a bare common, in snow puddles, at twilight, under a clouded sky, without having in my thoughts any occurrence of special good fortune, I have enjoyed a perfect exhilaration. I am glad to the brink of fear. In the woods too, a man casts off his years, as the snake his slough, and at what period soever of life, is always a child. In the woods, is perpetual youth. Within these plantations of God, a decorum and sanctity reign, a perennial festival is dressed, and the guest sees not how he should tire of them in a thousand years. In the woods, we return to reason and faith. There I feel that nothing can befall me in life,—no disgrace, no calamity, (leaving me my eyes,) which nature cannot repair. Standing on the bare ground,—my head bathedby the blithe air, and uplifted into infinite space,—all mean egotism vanishes. I become a transparent eye-ball; I am nothing; I see all; the currents of the Universal Being circulate through me; I am part or particle of God. The name of the nearest friend sounds then foreign and accidental: to be brothers, to be acquaintances,—master or servant, is then a trifle and a disturbance. I am the lover of uncontained and immortal beauty. In the wilderness, I find something more dear and connate than in streets or villages. In the tranquil landscape, and especially in the distant line of the horizon, man beholds somewhat as beautiful as his own nature.

Emerson writes "To Speak truly, few adult persons can see nature." What do you think one means? What does it mean to truly see nature?

Standing on the bare ground,—my head bathed by the blithe air, and uplifted into infinite space,—all mean egotism vanishes. I become a transparent eye-ball; I am nothing; I see all; the currents of the Universal Being circulate through me; I am part or particle of God.

The greatest delight which the fields and woods minister, is the suggestion of an occult relation between man and the vegetable. I am not alone and unacknowledged. They nod to me, and I to them. The waving of the boughs in the storm, is new to me and old. It takes me by surprise, and yet is not unknown. Its effect is like that of a higher thought or a better emotion coming over me, when I deemed I was thinking justly or doing right.

Yet it is certain that the power to produce this delight, does not reside in nature, but in man, or in a harmony of both. It is necessary to use these pleasures with great temperance. For, nature is not always tricked in holiday attire, but the same scene which yesterday breathed perfume and glittered as for the frolic of the nymphs, is overspread with melancholy today. Nature always wears the colors of the spirit. To a man laboring under calamity, the heat of his own fire hath sadness in it. Then, there is a kind of contempt of the landscape felt by him who has just lost by death a dear friend. The sky is less grand as it shuts down over less worth in the population.

What does Emerson say happens to a person who goes into the woods? How is a person's perspective transformed? In what ways is it like a religious experience?

From *Rural Hours*

—*Susan Fenimore Cooper*

Susan Fenimore Cooper (1813–1894) represents the American "nature writing" tradition of mid-nineteenth century. Although Cooper's work is less well known, it is comparable in many ways to that of Henry David Thoreau, whose Walden was published in 1954, four years after Rural Hours, from which these selections are taken. In Rural Hours, Cooper reflects on her experiences living and walking in rural New York in 1948. The work was published in 1950. Susan Fenimore Cooper was the daughter of the well-known author, James Fenimore Cooper. She was very close to her father, serving as his secretary and assistant. In her work, Cooper combines her Christian faith with precise observations of the natural world. After her father's death in 1951, Susan devoted herself to numerous charitable causes, especially the founding of an orphanage and hospital. In these selections, we see one European-Americas growing appreciation of the centuries-old forests that are quickly disappearing as the "new world" is developed. We also get a glimpse of European-American interaction with America's native people at the time.

Tuesday, 17th.–Rambled about Mill Island and the woods beyond.

We are told that for some years after the village was commenced, Mill Island was a favorite resort of the Indians, who, at that time, came frequently in parties to the new settlement, remaining here for months together. The island was then covered with wood, and they seem to have chosen it for their camp, in preference to other situations. Possibly it may have been a place of resort to their fishing and hunting parties when the country was a wilderness. Now they come very seldom, and singly, or in families, craving permission to build a shanty of boughs or boards, in order to ply their trade of basket-makers. They no longer encamp on the island itself, for the oak by the bridge is almost the only tree standing on it, and they still love the woods; but three out of four families who have been here during the last ten years, have chosen the neighboring groves for their halting-place.

There are already many parts of this country where an Indian is never seen. There are thousands and hundreds of thousands of the white population who have never laid eyes upon a red man. But this ground lies within the former bounds of the Six Nations, and a remnant of the great tribes of the Iroquois still linger about their old haunts, and occasionally cross our path. The first group that we chance to see strike us strangely, appearing as they do in the midst of a civilized community with the characteristics of their wild race still clinging to them; and when it is remembered that the land over which they now wander as strangers, in the midst of an alien race, was so lately their own–the heritage of their fathers–it is impossible to behold them without a feeling of peculiar interest.

Standing at the window, one summer/s afternoon, our attention was suddenly fixed by three singular figures approaching the house. More than one member of our household had never yet seen an Indian, and unaware that any were in the neighborhood, a second glance was necessary to convince us that these visitors must belong to the red race, whom we had long been so anxious to see. They came slowly towards the door, walking singly and silently, wrapped in blankets, bareheaded and barefooted. Without knocking or speaking, they entered the house with a noiseless step, and stood silently near the open door. We gave them a friendly greeting, and they proved to be women of the Oneida tribe, belonging to a family who had encamped in the woods the day before, with the purpose of selling their baskets in the village. Meek in countenance, with delicate forms and low voices, they had far more of the peculiarities of the red race about them than one would look for in a tribe long accustomed to intercourse with the whites, and a portion of whom have become more than half civilized. Only one of the three could speak English, and she seemed to do so with effort and reluctance. They were dressed in gowns of blue calico, rudely cut, coarsely stitched together, and so short as to show their broadcloth leggings worked with beads. Their heads were entirely bare, their straight, black hair hanging loose about their shoulders, and although it was midsummer at the time, they were closely wrapped in coarse white blankets. We asked their names. "Wallee"–"Awa"–"Cootlee"–was the answer. Of what tribe? "Oneida," was the reply, in a voice low and melancholy as the note of the whip-poor-will, giving the soft Italian sound to the vowels, and four syllables to the word. They were delicately made, of the usual height of American women, and their features were good, without being pretty. About their necks, arms, and ankles, they wore strings of cheap ornaments, pewter medals, and coarse glass beads, with the addition of a few scraps of tin, the refuse of some tin-shop passed on their way. One, the grandmother, was a Christian; the other two were Pagans. There was something startling and very painful in hearing these poor creatures within our own community, and under our own roof, declaring themselves heathens! They paid very little attention to the objects about them, until the youngest of the three observed a small Chinese basket on a table near her. She rose silently, took the basket in her hand, examined it carefully, made a single exclamation of pleasure, and then exchanged a few words with her companions in their own wild but musical tongue. They all seemed struck with this specimen of Chinese ingenuity. They asked, as usual, for bread and cold meat, and a supply was cheerfully given them, with the addition of some cake, about which they appeared to care very little. In the mean time a messenger had been sent to one of the shops of the village, where toys and knicknacks for children were sold, and he returned with a handful of copper rings and brooches, pewter medals, and bits of bright ribbons, which were presented to our guests; the simple creatures looking much gratified, as well as surprised, although their thanks were brief, and they still kept up the true Indian etiquette of mastering all emotion. They were, indeed, very silent and unwilling to talk, so that it was not easy to gather much information from them; but their whole appearance was so much more Indian than we had been prepared for, while their manners were so gentle and womanly, so free from anything coarse or rude in the midst of their untutored ignorance, that we were much pleased with the visit. Later in the day we went to their camp, as they always call their halting-place; here we found several children and two men of the family. . . .

Since the visit of the Oneida squaws, several other parties have been in the village. The very next season a family of three generations made their appearance at the door, claiming an hereditary acquaintance with the master of the house. They were much less wild than our first visitors, having discarded the blanket entirely, and speaking English very well. The leader and patriarch of the party bore a Dutch name, given him, probably, by some of his friends on the Mohawk Flats; and he was, moreover, entitled to write Reverend before it, being a Methodist minister–the Rev. Mr. Kunkerpott. He was notwithstanding a full-blooded Indian, with the regular copper-colored complexion and high cheekbones; the outline of his face was decidedly Roman, and his long, gray hair had a wave which is rare among his people; the mouth, where the savage expression is usually most strongly marked, was small, with a kindly expression about it. Altogether he was a strange mixture of the Methodist preacher and the Indian patriarch. His son was much more savage than himself in appearance–a silent, cold-looking man; and the grandson, a boy of ten or twelve, was one of the most uncouth, impish-looking creatures we ever beheld. . . .

It is easy to wish these poor people well; but surely something more may justly be required of us–of those who have taken their country and their place on the earth. The time seems at last to have come when their own eyes are opening to the real good of civilization, the advantages of knowledge, the blessings of Christianity. Let us acknowledge the strong claim they have upon us, not in word only, but in deed also. The native intellect of the red men who peopled this part of America surpassed that of many other races laboring under the curses of savage life; they have shown bravery, fortitude, religious feeling, eloquence, imagination, quickness of intellect, with much dignity of manner; and if we are true to our duty, now at the moment when they are making of their own accord a movement in the path of improvement, perhaps the day may not be distant when men of Indian blood may be numbered among the wise and the good, laboring in behalf of our common country.

It is painful, indeed, to remember how little has yet been done for the Indian during the three centuries since he and the white man first met on the Atlantic coast. But such is only the common course of things; a savage race is almost invariable corrupted rather than improved by its earliest contact with a civilized people; they suffer from the vices of civilization before they learn justly to comprehend its merits. It is with nations as with individuals–amelioration is a slow process, corruption a rapid one. . . .

It is painful, indeed, to remember how little has yet been done for the Indian during the three centuries since he and the white man first met on the Atlantic coast.

Monday, 23d.–Just at the point where the village street becomes a road and turns to climb the hillside, there stands a group of pines, a remnant of the old forest. There are many trees like these among the woods; far and near such may be seen rising from the hills, now tossing their arms in the stormy winds, now drawn in still and dark relief against the glowing evening sky. Their gaunt, upright forms standing about the hill-tops, and the ragged gray stumps of those which have fallen, dotting the smooth fields, make up the sterner touches in a scene whose general aspect is smiling. But although these old trees are common upon the wooded heights, yet the group on the skirts of the valley stands alone among the fields of the valley; their nearer brethren have all been swept away, and these are left in isolated company, differing in character from all about them, a monument of the past.

It is upon a narrow belt of land, a highway and a corn-field on one side, a brook and an orchard on the other, that these trees are rooted; a strip of woodland connected with the forest on the hills above, and suddenly cut off where it approaches the first buildings of the village. There they stand, silent spectators of the wonderful changes that have come over the valley. Hundreds of winters have passed since the cones which contained the seed of that grove fell from the parent tree; centuries have elapsed since their heads emerged from the topmost wave of the sea of verdure to meet the sunshine, and yet it is but yesterday that their shadows first fell, in full length, upon the sod at their feet.

There they stand, silent spectators of the wonderful changes that have come over the valley.

Sixty years since, those trees belonged to a wilderness; the bear, the wolf, and the panther brushed their trunks; the ungainly moose and the agile deer browsed at their feet; the savage hunter crept stealthily about their roots, and painted braves passed noiselessly on the war-path beneath their shade. How many successive generations of the red man have trod the soil they overshadowed, and then sat down in their narrow graves–how many herds of wild creatures have chased each other through that wood, and left their bones to bleach among the fern and moss, there is no human voice can tell. We only know that the summer winds, when they filled the canvas of Columbus and Cabot, three hundred years ago, came sweeping over these forest pines, murmuring then as we hear them murmur to-day.

There is no record to teach us even the name of the first white man who saw this sequestered valley, with its limpid lake; it was probably some bold hunter from the Mohawk, chasing the deer, or in quest of the beaver. But while towns were rising on the St. Lawrence and upon the sea-board, this inland region lay still unexplored; long after trading-houses had been opened, and fields had been tilled, and battles had been fought to the north, south, east, ay, and even at many points westward, those pines stood in the heart of a silent wilderness. This, little lake lay embedded in a forest until after the great struggle of the Revolution was over. A few months after the war was brought to an honorable close, Washington made a journey of observation among the inland waters of this part of the country; writing to a friend in France, he names this little lake, the source of a river, which, four degrees farther south, flows into the Chesapeake in near neighborhood with his own Potomac. As he passed along through a half-wild region, where the few marks of civilization then existing bore the blight of war, he conceived the outline of many of those improvements which have since been carried out by others, and have yielded so rich a revenue of prosperity. It is a pleasing reflection to those who live here, that while many important places in the country were never honored by his presence, Washington has trod the soil about our lake. But even at that late day, when the great and good man came, the mountains were still clothed in wood to the water/s edge, and mingled with giant oaks and ashes, those tall pines waved above the valley.

At length, nearly three long centuries after the Genoese had crossed the ocean, the white man came to plant a home on this spot, and it was then the great change began; the axe and the saw, the forge and the wheel, were busy from dawn to dusk, cows and swine fed in thickets whence the wild beasts had fled, while the ox and the horse drew away in chains the fallen trunks of the forest. The tenants of the wilderness shrunk deeper within its bounds with every changing moon; the wild creatures fled away within the receding shades of the forest, and the red man followed on their track;

The tenants of the wilderness shrunk deeper within its bounds with every changing moon; the wild creatures fled away within the receding shades of the forest, and the red man followed on their track.

his day of power was gone, his hour of pitiless revenge had passed, and the last echoes of the war-whoop were dying away forever among these hills, when the pale-faces laid their hearth-stones by the lake shore. The red man, who for thousands of years had been lord of the land, no longer treads the soil; he exists here only in uncertain memories, and in forgotten graves.

Such has been the change of the last half century. Those who from childhood have known the cheerful dwellings of the village, the broad and fertile farms, the well beaten roads, such as they are to-day, can hardly credit that this has all been done so recently by a band of men, some of whom, white-headed and leaning on their staves, are still among us. Yet such is the simple truth. This village lies just on the borders of the tract of country which was opened and peopled immediately after the Revolution; it was among the earliest of those little colonies from the sea-board which struck into the wilderness at that favorable moment, and whose rapid growth and progress in civilization have become a byword. Other places, indeed, have far surpassed this quiet borough; Rochester, Buffalo, and others of a later date, have become great cities, while this remains a rural village; still, whenever we pause to recall what has been done in this secluded valley during the lifetime of one generation, we must needs be struck with new astonishment. And throughout every act of the work, those old pines were there. Unchanged themselves, they stand surrounded by objects over all of which a great change has passed. The open valley, the half-shorn hills, the paths, the flocks, the buildings, the woods in their second growth, even the waters in the different images they reflect on their bosom, the very race of men who come and go, all are different from what they were; and those calm old trees seem to heave the sigh of companionless age, as their coned heads rock slowly in the winds. . . .

The grove upon the skirts of the village numbers, perhaps, some forty trees, varying in their girth from five or six to twelve feet; and in height, from a hundred and twenty to a hundred and sixty feet. Owing to their unscreened position and their height, these trees may be clearly distinguished for miles, whether from the lake, the hills, or the roads about the country–a landmark overtopping the humble church-spires, and every object raised by man within the bounds of the valley. Their rude simplicity of outline, the erect, unbending trunks, their stern, changeless character, and their scanty drapery of foliage, unconsciously lead one to fancy them an image of some band of savage chiefs, emerging in a long, dark line from the glen in their rear, and gazing in wonder upon their former hunting-ground in its altered aspect.

It needs but a few short minutes to bring one of these trees to the ground; the rudest boor passing along the highway may easily do the deed; but how many years must pass ere its equal stand on the same spot! Let us pause to count the days, the months, the years; let us number the generations that must come and go, the centuries that must roll onward, ere the seed sown from this year/s cones shall produce a wood like that before us. The stout arm so ready to raise the axe to-day, must grow weak with age, it must drop into the grave; its bone and sinew must crumble into dust long before another tree, tall and great as those, shall have grown from the cone in our hand. Nay, more, all the united strength of sinew, added to all the powers of mind, and all the force of will, of millions of men, can do no more toward the work than the poor ability of a single arm; these are of the deeds which time alone can perform. But allowing even that hundreds of years hence other trees were at length to succeed these with the same dignity of height and age, no other younger wood can ever claim the same connection as this, with

It needs but a few short minutes to bring one of these trees to the ground; the rudest boor passing along the highway may easily do the deed; but how many years must pass ere its equal stand on the same spot!

a state of things now passed away forever; they cannot have that wild, stern character of the aged forest pines. This little town itself must fall to decay and ruin; its streets must become choked with bushes and brambles; the farms of the valley must be anew buried within the shades of a wilderness; the wild deer and the wolf and the bear must return from beyond the great lakes; the bones of the savage men buried under our feet must arise and move again in the chase, ere trees like those, with the spirit of the forest in every line, can stand on the same ground in wild dignity of form like those old pines now looking down upon our homes. . . .

Saturday, 28th.–Passed the afternoon in the woods.

What a noble gift to man are the forests! What a debt of gratitude and admiration we owe for their utility and their beauty!

How pleasantly the shadows of the wood fall upon our heads, when we turn from the glitter and turmoil of the world of man! The winds of heaven seem to linger amid these balmy branches, and the sunshine falls like a blessing upon the green leaves; the wild breath of the forest, fragrant with bark and berry, fans the brow with grateful freshness; and the beautiful wood-light, neither garish nor gloomy, full of calm and peaceful influences, sheds repose over the spirit. The view is limited, and the objects about us are uniform in character; yet within the bosom of the woods the mind readily lays aside its daily littleness, and opens to higher thoughts, in silent consciousness that it stands alone with the works of God. The humble moss beneath our feet, the sweet flowers, the varied shrubs, the great trees, and the sky gleaming above in sacred blue, are each the handiwork of God. They were all called into being by the will of the Creator, as we now behold them, full of wisdom and goodness. Every object here has a deeper merit than our wonder can fathom; each has a beauty beyond our full perception; the dullest insect crawling about these roots lives by the power of the Almighty; and the discolored shreds of last year/s leaves wither away upon the lowly herbs in a blessing of fertility. But it is the great trees, stretching their arms above us in a thousand forms of grace and strength, it is more especially the trees which fill the mind with wonder and praise.

Within the bosom of the woods the mind readily lays aside its daily littleness, and opens to higher thoughts, in silent consciousness that it stands alone with the works of God.

It is more especially the trees which fill the mind with wonder and praise.

Of the infinite variety of fruits which spring from the bosom of the earth, the trees of the wood are the greatest in dignity. Of all the works of the creation which know the changes of life and death, the trees of the forest have the longest existence. Of all the objects which crown the gray earth, the woods preserve unchanged, throughout the greatest reach of time, their native character: the works of man are every varying their aspect; his towns and his fields alike reflect the unstable opinions, the fickle wills and fancies of each passing generation; but the forests on his borders remain to-day the same they were ages of years since. Old as the everlasting hills, during thousands of seasons they have put forth and laid down their verdure in calm obedience to the decree which first bade them cover the ruins of the Deluge.

But, although the forests are great and old, yet the ancient trees within their bounds must each bend individually beneath the doom of every earthly existence; they have their allotted period when the mosses of Time gather upon their branches; when, touched by decay, they break and crumble to dust. Like man, they are decked in living beauty; like man, they fall a prey to death; and while we admire their duration, so far beyond our own brief years, we also acknowledge that especial interest which can only belong to the graces of life and to the desolation of death. We raise our eyes, and we see collected in one company vigorous trunks, the oak, the ash, the pine, firm in the strength of maturity; by their side stand a young group, elm, and birch, and maple, their supple branches playing in the breezes, gay and fresh as youth itself; and yonder, rising in unheeded gloom, we behold a skeleton trunk, an old fir, every branch broken, every leaf fallen,–dull, still, sad, like the finger of Death.

It is the peculiar nature of the forest, that life and death may ever be found within its bounds, in immediate presence of each other; both with ceaseless, noiseless advances, aiming at the mastery; and if the influences of the first be most general, those of the last are the most striking. Spring, with all her wealth of life and joy, finds within the forest many a tree unconscious of her approach; a thousand young plants springing up about the fallen trunk, the shaggy roots, seek to soften the gloomy wreck with a semblance of the verdure it bore of old; but ere they have thrown their fresh and graceful wreaths over the mouldering wood, half their own tribe wither and die with the year. We owe to this perpetual presence of death an impression, calm, solemn, almost religious in character, a chastening influence, beyond what we find in the open fields. But this subdued spirit is far from gloomy or oppressive, since it never fails to be relieved by the cheerful animation of living beauty. Sweet flowers grow beside the fallen trees, among the shattered branches, the season through; and the freedom of the woods, the unchecked growth, the careless position of every tree, are favorable to a thousand wild beauties, and fantastic forms, opening to the mind a play of fancy which is in itself cheering and enlivening, like the bright sunbeams which checker with golden light the shadowy groves.

From "Walking"

—Henry David Thoreau

Henry David Thoreau (1817–1862) was born in Concord, Massachusetts in 1817, attended Harvard University, and returned to Concord, at first as a school teacher, but later as a writer and student of nature who supported himself by making pencils, surveying land and performing other odd jobs. Thoreau became more influential after death than he was in his own day. He spoke out against slavery, criticized the cultivation and dividing up of wild lands, and protested the encroachment of government into the lives of citizens. His ideas on civil disobedience as a form of protest against the government influenced both Mahatma Gandhi (1869–1948) in India and Martin Luther King, Jr. (1929–1968) in the United States. He is perhaps best known for his writings on nature. In 1845, Thoreau built a small cabin on the banks of Walden Pond, near Concord, on land owned by his friend, Ralph Waldo Emerson, where he lived for two years. In 1954, Thoreau published his account of his Walden Pond experience in the book, ***Walden****. "The mass of men lead lives of quiet desperation," he writes early in the book. ". . . A stereotyped but unconscious despair is concealed even under what are called the games and amusements of mankind." In* ***Walden****, and elsewhere, Thoreau points away for the desperation within which many of us are still trapped today toward a more contented and satisfying life. Men and women live hectic lives, desperately seeking the "good life" through work, money and things. Thoreau writes of simplicity of lifestyle, wildness and nature as keys to life. The following reading is an edited version of "Walking," an essay he began as a lecture 1851 and developed over the years. It was finally published in the* ***Atlantic Monthly*** *after his death in 1862. "[I]n Wildness is the preservation of the World," he writes at the heart of this essay. "Every tree sends its fibers forth in search of the Wild. The cities import it at any price. Men plow and sail for it. From the forest and wilderness come the tonics and barks which brace mankind."*

I wish to speak a word for Nature, for absolute freedom and wildness, as contrasted with a freedom and culture merely civil—to regard man as an inhabitant, or a part and parcel of Nature, rather than a member of society. I wish to make an extreme statement, if so I may make an emphatic one, for there are enough champions of civilization: the minister and the school committee and every one of you will take care of that.

I have met with but one or two persons in the course of my life who understood the art of Walking, that is, of taking walks—who had a genius, so to speak, for SAUNTERING, which word is beautifully derived "from idle people who roved about the country, in the Middle Ages, and asked charity, under pretense of going a la Sainte Terre," to the Holy Land, till the children exclaimed, "There goes a Sainte-Terrer," a Saunterer, a Holy-Lander. They who never go to the Holy Land in their walks, as they pretend, are indeed mere idlers and vagabonds; but they who do go there are saunterers in the good sense, such as I mean . . . For every walk is a sort of crusade, preached by some Peter the Hermit in us, to go forth and reconquer this Holy Land from the hands of the Infidels.

. . . If you are ready to leave father and mother, and brother and sister, and wife and child and friends, and never see them again—if you have paid your debts, and made your will, and settled all your affairs, and are a free man—then you are ready for a walk. . . .

. . . the walking of which I speak has nothing in it akin to taking exercise, as it is called, as the sick take medicine at stated hours—as the Swinging of dumb-bells or chairs; but is itself the enterprise and adventure of the day. If you would get exercise, go in search of the springs of life. Think of a man/s swinging dumbbells for his health, when those springs are bubbling up in far-off pastures unsought by him!

Moreover, you must walk like a camel, which is said to be the only beast which ruminates when walking. When a traveler asked Wordsworth/s servant to show him her master/s study, she answered, "Here is his library, but his study is out of doors." . . .

What sort of walks is Thoreau describing here?

When we walk, we naturally go to the fields and woods: what would become of us, if we walked only in a garden or a mall? Even some sects of philosophers have felt the necessity of importing the woods to themselves, since they did not go to the woods. "They planted groves and walks of Platanes," where they took*subdiales ambulationes* in porticos open to the air. Of course it is of no use to direct our steps to the woods, if they do not carry us thither. I am alarmed when it happens that I have walked a mile into the woods bodily, without getting there in spirit. In my afternoon walk I would fain forget all my morning occupations and my obligations to Society. But it sometimes happens that I cannot easily shake off the village. The thought of some work will run in my head and I am not where my body is—I am out of my senses. In my walks I would fain return to my senses. What business have I in the woods, if I am thinking of something out of the woods? I suspect myself, and cannot help a shudder when I find myself so implicated even in what are called good works—for this may sometimes happen. . . .

Nowadays almost all man's improvements, so called, as the building of houses and the cutting down of the forest and of all large trees, simply deform the landscape, and make it more and more tame and cheap. A people who would begin by burning the fences and let the forest stand! I saw the fences half consumed, their ends lost in the middle of the prairie, and some worldly miser with a surveyor looking after his bounds, while heaven had taken place around him, and he did not see the angels going to and fro, but was looking for an old post-hole in the midst of paradise. I looked again, and saw him standing in the middle of a boggy Stygian fen, surrounded by devils, and he had found his bounds without a doubt, three little stones, where a stake had been driven, and looking nearer, I saw that the Prince of Darkness was his surveyor.

I saw that the Prince of Darkness was his surveyor.

I can easily walk ten, fifteen, twenty, any number of miles, commencing at my own door, without going by any house, without crossing a road except where the fox and the mink do: first along by the river, and then the brook, and then the meadow and the woodside. There are square miles in my vicinity which have no inhabitant. From many a hill I can see civilization and the abodes of man afar. The farmers and their works are scarcely more obvious than woodchucks and their burrows. Man and his affairs, church and state and school, trade and commerce, and manufactures and agriculture even politics, the most alarming of them all—I am pleased to see how little space they occupy in the landscape. Politics is but a narrow field, and that still narrower highway yonder leads to it. I sometimes direct the traveler thither. If you would go to the political world, follow the great road—follow that market-man, keep his dust in your eyes, and it will lead you straight to it; for it, too, has its place merely, and does not occupy all space. I pass from it as from a bean field into the forest, and it is forgotten. In one half-hour I can walk off to some portion of the earth/s surface where a man does not stand from one year/s end to another, and there, consequently, politics are not, for they are but as the cigar-smoke of a man. . . .

At present, in this vicinity, the best part of the land is not private property; the landscape is not owned, and the walker enjoys comparative freedom. But possibly the day will come when it will be partitioned off into so-called pleasure-grounds, in which a few will take a narrow and exclusive pleasure only—when fences shall be multiplied, and man-traps and other engines invented to confine men to the PUBLIC road, and walking over the surface of God/s earth shall be construed to mean trespassing on some gentleman/s grounds. To enjoy a thing exclusively is commonly to exclude yourself from the true enjoyment of it.Let us improve our opportunities, then, before the evil days come.

To enjoy a thing exclusively is commonly to exclude yourself from the true enjoyment of it.

What is it that makes it so hard sometimes to determine whither we will walk? I believe that there is a subtle magnetism in Nature, which, if we unconsciously yield to it, will direct us aright. It is not indifferent to us which way we walk. . . .

When I go out of the house for a walk, uncertain as yet whither I will bend my steps, and submit myself to my instinct to decide for me, I find, strange and whimsical as it may seem, that I finally and inevitably settle southwest, toward some particular wood or meadow or deserted pasture or hill in that direction. My needle is slow to settle,—varies a few degrees, and does not always point due southwest, it is true, and it has good authority for this variation, but it always settles between west and south-southwest. The future lies that way to me, and the earth seems more unexhausted and richer on that side. . . . Eastward I go only by force; but westward I go free. . . .

The West of which I speak is but another name for the Wild; and what I have been preparing to say is, that in Wildness is the preservation of the World. Every tree sends its fibers forth in search of the Wild. The cities import it at any price. Men plow and sail for it. From the forest and wilderness come the tonics and barks which brace mankind. Our ancestors were savages. The story of Romulus and Remus being suckled by a wolf is not a meaningless fable. The founders of every state which has risen to eminence have drawn their nourishment and vigor from a similar wild source. It was because the children of the Empire were not suckled by the wolf that they were conquered and displaced by the children of the northern forests who were.

Life consists with wildness. The most alive is the wildest. I believe in the forest, and in the meadow, and in the night in which the corn grows. We require an infusion of hemlock, spruce or arbor vitae in our tea. There is a difference between eating and drinking for strength and from mere gluttony. The Hottentots eagerly devour the marrow of the koodoo and other antelopes raw, as a matter of course. Some of our northern Indians eat raw the marrow of the Arctic reindeer. . . .

Life consists with wildness. The most alive is the wildest.

Not yet subdued to man, its presence refreshes him. One who pressed forward incessantly and never rested from his labors, who grew fast and made infinite demands on life, would always find himself in a new country or wilderness, and surrounded by the raw material of life. He would be climbing over the prostrate stems of primitive forest trees.

Hope and the future for me are not in lawns and cultivated fields, not in towns and cities, but in the impervious and quaking swamps. When, formerly, I have analyzed my partiality for some farm which I had contemplated purchasing, I have frequently found that I was attracted solely by a few square rods of impermeable and unfathomable bog—a natural sink in one corner of it. That was the jewel which dazzled me. I derive more of my subsistence from the swamps which surround my native town than from the cultivated gardens in the village. . . .

Yes, though you may think me perverse, if it were proposed to me to dwell in the neighborhood of the most beautiful garden that ever human art contrived, or else of a Dismal Swamp, I should certainly decide for the swamp. How vain, then, have been all your labors, citizens, for me!. . . My spirits infallibly rise in proportion to the outward dreariness. Give me the ocean, the desert, or the wilderness! In the desert, pure air and solitude compensate for want of moisture and fertility. . . .

When I would recreate myself, I seek the darkest woods the thickest and most interminable and, to the citizen, most dismal, swamp. I enter a swamp as a sacred place,— a sanctum sanctorum. There is the strength, the marrow, of Nature. The wildwood covers the virgin mould,—and the same soil is good for men and for trees. A man/s health requires as many acres of meadow to his prospect as his farm does loads of muck. There are the strong meats on which he feeds. A town is saved, not more by the righteous men in it than by the woods and swamps that surround it. A township where one primitive forest waves above while another primitive forest rots below—such a town is fitted to raise not only corn and potatoes, but poets and philosophers for the coming ages. In such a soil grew Homer and Confucius and the rest, and out of such a wilderness comes the Reformer eating locusts and wild honey.

To preserve wild animals implies generally the creation of a forest for them to dwell in or resort to. So it is with man. . . . In short, all good things are wild and free. There is something in a strain of music, whether produced by an instrument or by the human voice—take the sound of a bugle in a summer night, for instance—which by its wildness, to speak without satire, reminds me of the cries emitted by wild beasts in their native forests. It is so much of their wildness as I can understand. Give me for my friends and neighbors wild men, not tame ones. The wildness of the savage is but a faint symbol of the awful ferity with which good men and lovers meet.

I love even to see the domestic animals reassert their native rights—any evidence that they have not wholly lost their original wild habits and vigor; as when my neighbor/s cow breaks out of her pasture early in the spring and boldly swims the river, a cold, gray tide, twenty-five or thirty rods wide, swollen by the melted snow. It is the buffalo crossing the Mississippi. This exploit confers some dignity on the herd in my eyes—already dignified. The seeds of instinct are preserved under the thick hides of cattle and horses, like seeds in the bowels of the earth, an indefinite period.

Any sportiveness in cattle is unexpected. I saw one day a herd of a dozen bullocks and cows running about and frisking in unwieldy sport, like huge rats, even like kittens. They shook their heads, raised their tails, and rushed up and down a hill, and I perceived by their horns, as well as by their activity, their relation to the deer tribe. But, alas! a sudden loud WHOA! would have damped their ardor at once, reduced them from venison to beef, and stiffened their sides and sinews like the locomotive. Who but the Evil One has cried "Whoa!" to mankind? . . .

Here is this vast, savage, hovering mother of ours, Nature, lying all around, with such beauty, and such affection for her children, as the leopard; and yet we are so early weaned from her breast to society, to that culture which is exclusively an interaction of man on man—a sort of breeding in and in, which produces at most a merely English nobility, a civilization destined to have a speedy limit.

In society, in the best institutions of men, it is easy to detect a certain precocity. When we should still be growing children, we are already little men. Give me a culture which imports much muck from the meadows, and deepens the soil—not that which trusts to heating manures, and improved implements and modes of culture only! . . .

While almost all men feel an attraction drawing them to society, few are attracted strongly to Nature. In their reaction to Nature men appear to me for the most part, notwithstanding their arts, lower than the animals. It is not often a beautiful relation, as in the case of the animals. How little appreciation of the beauty of the land- scape there is among us! We have to be told that the Greeks called the world Beauty, or Order, but we do not see clearly why they did so, and we esteem it at best only a curious philological fact.

For my part, I feel that with regard to Nature I live a sort of border life, on the confines of a world into which I make occasional and transient forays only, and my patriotism and allegiance to the state into whose territories I seem to retreat are those of a moss-trooper. Unto a life which I call natural I would gladly follow even a will-o/-the-wisp through bogs and sloughs unimaginable, but no moon nor firefly has shown me the causeway to it. Nature is a personality so vast and universal that we have never seen one of her features. The walker in the familiar fields which stretch around my native town sometimes finds himself in another land than is described in their owners/ deeds, as it were in some faraway field on the confines of the actual Concord, where her jurisdiction ceases, and the idea which the word Concord suggests ceases to be suggested. These farms which I have myself surveyed, these bounds which I have set up, appear dimly still as through a mist; but they have no chemistry to fix them; they fade from the surface of the glass, and the picture which the painter painted stands out dimly from beneath. The world with which we are commonly acquainted leaves no trace, and it will have no anniversary.

I took a walk on Spaulding/s Farm the other afternoon. I saw the setting sun lighting up the opposite side of a stately pine wood. Its golden rays straggled into the aisles of the wood as into some noble hall. I was impressed as if some ancient and altogether admirable and shining family had settled there in that part of the land called Concord, unknown to me—to whom the sun was servant—who had not gone into society in the village—who had not been called on. I saw their park, their pleasure-ground, beyond through the wood, in Spaulding/s cranberry-meadow. The pines furnished them with gables as they grew. Their house was not obvious to vision; the trees grew through it. I do not know whether I heard the sounds of a suppressed hilarity or not. They seemed to recline on the sunbeams. They have sons and daughters. They are quite well. The farmer/s cart-path, which leads directly through their hall, does not in the least put them out, as the muddy bottom of a pool is sometimes seen through the reflected skies. They never heard of Spaulding, and do not know that he is their neighbor—notwithstanding I heard him whistle as he drove his team through the house. Nothing can equal the serenity of their lives. Their coat-of-arms is simply a lichen. I saw it painted on the pines and oaks. Their attics were in the tops of the trees. They are of no politics. There was no noise of labor. I did not perceive that they were weaving or spinning. Yet I did detect, when the wind lulled and hearing was done away, the finest imaginable sweet musical hum,—as of a distant hive in May, which perchance was the sound of their thinking. They had no idle thoughts, and no one without could see their work, for their industry was not as in knots and excrescences embayed.

But I find it difficult to remember them. They fade irrevocably out of my mind even now while I speak, and endeavor to recall them and recollect myself. It is only after a long and serious effort to recollect my best thoughts that I become again aware of their cohabitancy. If it were not for such families as this, I think I should move out of Concord.

We are accustomed to say in New England that few and fewer pigeons visit us every year. Our forests furnish no mast for them. So, it would seem, few and fewer thoughts visit each growing man from year to year, for the grove in our minds is laid waste—sold to feed unnecessary fires of ambition, or sent to mill—and there is scarcely a twig left for them to perch on. They no longer build nor breed with us. In some more genial season, perchance, a faint shadow flits across the landscape of the mind, cast by the WINGS of some thought in its vernal or autumnal migration, but, looking up, we are unable to detect the substance of the thought itself. Our winged thoughts are turned to poultry. They no longer soar. . . .

We hug the earth—how rarely we mount! Methinks we might elevate ourselves a little more. We might climb a tree, at least. I found my account in climbing a tree once. It was a tall white pine, on the top of a hill; and though I got well pitched, I was well paid for it, for I discovered new mountains in the horizon which I had never seen before—so much more of the earth and the heavens. I might have walked about the foot of the tree for threescore years and ten, and yet I certainly should never have seen them. But, above all, I discovered around me—it was near the end of June—on the ends of the topmost branches only, a few minute and delicate red cone-like blossoms, the fertile flower of the white pine looking heavenward. I carried straightway to the village the topmost spire, and showed it to stranger jurymen who walked the streets—for it was court week—and to farmers and lumber-dealers and woodchoppers and hunters, and not one had ever seen the like before, but they wondered as at a star dropped down. Tell of ancient architects finishing their works on the tops of columns as perfectly as on the lower and more visible parts! Nature has from the first expanded the minute blossoms of the forest only toward the heavens, above men/s heads and unobserved by them. We see only the flowers that are under our feet in the meadows. The pines have developed their delicate blossoms on the highest twigs of the wood every summer for ages, as well over the heads of Nature/s red children as of her white ones; yet scarcely a farmer or hunter in the land has ever seen them.

Above all, we cannot afford not to live in the present. He is blessed over all mortals who loses no moment of the passing life in remembering the past. Unless our philosophy hears the cock crow in every barnyard within our horizon, it is belated. That sound commonly reminds us that we are growing rusty and antique in our employments and habits of thoughts. His philosophy comes down to a more recent time than ours. There is something suggested by it that is a newer testament,—the gospel according to this moment. He has not fallen astern; he has got up early and kept up early, and to be where he is is to be in season, in the foremost rank of time. It is an expression of the health and soundness of Nature, a brag for all the world,—healthiness as of a spring burst forth, a new fountain of the Muses, to celebrate this last instant of time.

Where he lives no fugitive slave laws are passed. Who has not betrayed his master many times since last he heard that note?

Above all, we cannot afford not to live in the present.

The merit of this bird/s strain is in its freedom from all plaintiveness. The singer can easily move us to tears or to laughter, but where is he who can excite in us a pure morning joy? When, in doleful dumps, breaking the awful stillness of our wooden sidewalk on a Sunday, or, perchance, a watcher in the house of mourning, I hear a cockerel crow far or near, I think to myself, "There is one of us well, at any rate,"—and with a sudden gush return to my senses.

We had a remarkable sunset one day last November. I was walking in a meadow, the source of a small brook, when the sun at last, just before setting, after a cold, gray day, reached a clear stratum in the horizon, and the softest, brightest morning sunlight fell on the dry grass and on the stems of the trees in the opposite horizon and on the leaves of the shrub oaks on the hillside, while our shadows stretched long over the meadow east- ward, as if we were the only motes in its beams. It was such a light as we could not have imagined a moment before, and the air also was so warm and serene that nothing was wanting to make a paradise of that meadow. When we reflected that this was not a solitary phenomenon, never to happen again, but that it would happen forever and ever, an infinite number of evenings, and cheer and reassure the latest child that walked there, it was more glorious still.

The sun sets on some retired meadow, where no house is visible, with all the glory and splendor that it lavishes on cities, and perchance as it has never set before—where there is but a solitary marsh hawk to have his wings gilded by it, or only a musquash looks out from his cabin, and there is some little black-veined brook in the midst of the marsh, just beginning to meander, winding slowly round a decaying stump. We walked in so pure and bright a light, gilding the withered grass and leaves, so softly and serenely bright, I thought I had never bathed in such a golden flood, without a ripple or a murmur to it. The west side of every wood and rising ground gleamed like the boundary of Elysium, and the sun on our backs seemed like a gentle herdsman driving us home at evening.

So we saunter toward the Holy Land, till one day the sun shall shine more brightly than ever he has done, shall perchance shine into our minds and hearts, and light up our whole lives with a great awakening light.

So we saunter toward the Holy Land, till one day the sun shall shine more brightly than ever he has done, shall perchance shine into our minds and hearts, and light up our whole lives with a great awakening light, as warm and serene and golden as on a bankside in autumn.

From *Leaf and Tendril*

—John Burroughs

Continuing in the tradition of writers like Susan Fenimore Cooper and Henry David Thoreau, John Burroughs (1837–1921) was a well-known writer of "nature essays." Writing at the beginning of the twentieth century, his work was well-liked, expressing a growing American attachment to nature and wilderness. Burroughs accompanied Theodore Roosevelt, John Muir and Henry Ford on explorations of America's undeveloped areas, authoring several accounts of his trips with Roosevelt. Like earlier nature writers, Burrough's works include careful observations of the natural world combined with deeper reflection of nature's beauty and value. Burroughs argued in many of his essays that a full explanation of the world around us must include both an evolutionary account of life's development and a reverence for divine direction within this process. Two excerpts are included here. The first brief excerpt is from an essay entitled, "The Art of Seeing." The second is a more extended excerpt from the essay, "The Grist of the Gods." Both are found in Leaf and Tendril, a collection of Burrough's essays, published in 1908.

FROM "THE ART OF SEEING"

If I were to name the three most precious resources of life, I should say books, friends, and nature; and the greatest of these, at least the most constant and always at hand, is nature. Nature we have always with us, an inexhaustible storehouse of that which moves the heart, appeals to the mind, and fires the imagination,—health to the body, a stimulus to the intellect, and joy to the soul. To the scientist Nature is a storehouse of facts, laws, processes; to the artist she is a storehouse of pictures; to the poet she is a storehouse of images, fancies, a source of inspiration; to the moralist she is a storehouse of precepts and parables; to all she may be a source of knowledge and joy.

If I were to name the three most precious resources of life, I should say books, friends, and nature; and the greatest of these, at least the most constant and always at hand, is nature.

There is nothing in which people differ more than in their powers of observation. Some are only half alive to what is going on around them. Others, again, are keenly alive: their intelligence, their powers of recognition, are in full force in eye and ear at all times. They see and hear everything, whether it directly concerns them or not. They never pass unseen a familiar face on the street; they are never oblivious of any interesting feature or sound or object in the earth or sky about them. Their power of attention is always on the alert, not by conscious effort, but by natural habit and disposition. Their perceptive faculties may be said to be always on duty. They turn to the outward world a more highly sensitized mind than other people. The things that pass before them are caught and individualized instantly. If they visit new countries, they see the characteristic features of the people and scenery at once. The impression is never blurred or confused. Their powers of observation suggest the sight and scent of wild animals; only, whereas it is fear that sharpens the one, it is love and curiosity that sharpens the other.

FROM "THE GRIST OF THE GODS"

About all we have in mind when we think of the earth is this thin pellicle of soil with which the granite framework of the globe is clothed—a red and brown film of pulverized and oxidized rock, scarcely thicker, relatively, than the paint or enamel which some women put on their cheeks, and which the rains often wash away as a tear washes off the paint and powder. But it is the main thing to us. Out of it we came and unto it we return. "Earth to earth, and dust to dust." The dust becomes warm and animated for a little while, takes on form and color, stalks about recuperating itself from its parent dust underfoot, and then fades and is resolved into the original earth elements. We are built up out of the ground quite as literally as the trees are, but not quite so immediately. The vegetable is between us and the soil, but our dependence is none the less real. "As common as dust" is one of our sayings, but the common, the universal, is always our mainstay in this world. When we see the dust turned into fruit and flowers and grain by that intangible thing called vegetable life, or into the bodies of men and women by the equally mysterious agency of animal life, we think better of it. The trembling gold of the pond-lily's heart, and its petals like carved snow, are no more a transformation of a little black muck and ooze by the chemistry of the sunbeam than our bodies and minds, too, are a transformation of the soil underfoot.

We are built up out of the ground quite as literally as the trees are, but not quite so immediately.

We are rooted to the air through our lungs and to the soil through our stomachs. We are walking trees and floating plants. The soil which in one form we spurn with our feet, and in another take into our mouths and into our blood—what a composite product it is! It is the grist out of which our bread of life is made, the grist which the mills of the gods, the slow patient gods of Erosion, have been so long grinding—grinding probably more millions of years than we have any idea of. The original stuff, the pulverized granite, was probably not very nourishing, but the fruitful hand of time has made it so. It is the kind of grist that improves with the keeping, and the more the meal-worms have worked in it, the better the bread. Indeed, until it has been eaten and digested by our faithful servitors the vegetables, it does not make the loaf that is our staff of life. The more death has gone into it, the more life comes out of it; the more it is a cemetery, the more it becomes a nursery; the more the rocks perish, the more the fields flourish.

This story of the soil appeals to the imagination. To have a bit of earth to plant, to hoe, to delve in, is a rare privilege. If one stops to consider, one cannot turn it with his spade without emotion. We look back with the mind/s eye through the vista of geologic time and we see islands and continents of barren, jagged rocks, not a grain of soil anywhere. We look again and behold a world of rounded hills and fertile valleys and plains, depth of soil where before were frowning rocks. The hand of time with its potent fingers of heat, frost, cloud, and air has passed slowly over the scene, and the miracle is done. The rocks turn to herbage, the fetid gases to the breath of flowers. The mountain melts down into a harvest field; volcanic scoria changes into garden mould; where towered a cliff now basks a green slope; where the strata yawned now bubbles a fountain; where the earth trembled, verdure now undulates. Your lawn and your meadow are built up of the ruins of the foreworld. The leanness of granite and gneiss has become the fat of the land. What transformation and promotion!—the decrepitude of the hills becoming the strength of the plains, the decay of the heights resulting in the renewal of the valleys!

Many of our hills are but the stumps of mountains which the hand of time has cut down. Hence we may say that if God made the mountains, time made the hills.

What adds to the wonder of the earth/s grist is that the millstones that -did the work and are still doing it are the gentle forces that career above our heads—the sunbeam, the cloud, the air, the frost. The rain/s gentle fall, the air/s velvet touch, the sun/s noiseless rays, the frost/s exquisite crystals, these combined are the agents that crush the rocks and pulverize the mountains, and transform continents of sterile granite into a world of fertile soils. It is as if baby fingers did the work of giant powder and dynamite. Give the clouds and the sunbeams time enough, and the Alps and the Andes disappear before them, or are transformed into plains where com may grow and cattle graze. The snow falls as softly as down and lies almost as lightly, yet the crags crumble beneath it; compacted by gravity, out of it grew the tremendous ice sheet that ground off the mountain summits, that scooped out lakes and valleys, and modeled our northern landscapes as the sculptor his clay image.

Not only are the mills of the gods grinding here, but the great cosmic mill in the sidereal heavens is grinding also, and some of its dust reaches our planet. Cosmic dust is apparently falling on the earth at all times. It is found in the heart of hailstones and in Alpine snows, and helps make up the mud of the ocean floors.

During the unthinkable time of the revolution of the earth around the sun, the amount of cosmic matter that has fallen upon its surface from out the depths of space must be enormous. It certainly must enter largely into the composition of the soil and of the sedimentary rocks. Celestial dirt we may truly call it, stardust, in which we plant our potatoes and grain and out of which Adam was made, and every son of man since Adam—the divine soil in very fact, the garden of the Eternal, contributed to by the heavens above and all the vital forces below, incorruptible, forever purifying itself, clothing the rocky framework of the globe as with flesh and blood, making the earth truly a mother with a teeming fruitful womb, and her hills veritable mammary glands. The iron in the fruit and vegetables we eat, which thence goes into our blood, may, not very long. ago, have formed a part of the cosmic dust that drifted for untold ages along the highways of planets and suns.

The soil underfoot, or that we turn with our plow, how it thrills with life or the potencies of life!

The soil underfoot, or that we turn with our plow, how it thrills with life or the potencies of life! What a fresh, good odor it exhales when we turn it with our spade or plow in spring! It is good. No wonder children and horses like to eat it!

How inert and dead it looks, yet what silent, potent fermentations are going on there—millions and trillions of minute organisms ready to further your scheme of agriculture or horticulture. Plant your wheat or your com in it, and behold the miracle of a birth of a plant or a tree. How it pushes up, fed and stimulated by the soil, through the agency of heat and moisture! It makes visible to the eye the life that is latent or held in suspense there in the cool, impassive ground. The acorn, the chestnut, the maple keys, have but to lie on the surface of the moist earth to feel its power and send down rootlets to meet it. . . .

But let not care and humdrum deaden us to the wonders and the mysteries amid which we live, nor to the splendors and the glories. We need not translate ourselves in imagination to some other sphere or state of being to find the marvelous, the divine, the transcendent; we need not postpone our day of wonder and appreciation to some future time and condition. The true inwardness of this gross visible world, hanging like an apple on the bough of the great cosmic tree, and swelling with all the juices and potencies of life, transcends anything we have dreamed of super-terrestrial abodes. It is because of these things, because of the vitality, spirituality, oneness, and immanence of the universe as revealed by science, its condition of transcending time and space, without youth and without age, neither beginning nor ending, neither material nor spiritual, but forever passing from one into the other, that I was early and deeply impressed by Walt Whitman/s lines: —

"There was never any more inception than there is now. Nor any more youth or age than there is now; And will never be any more perfection than there is now, Nor any more heaven or hell than there is now."

And I may add, nor any more creation than there is now, nor any more miracles, or glories, or wonders, or immortality, or judgment days, than there are now. And we shall never be nearer God and spiritual and transcendent things than we are now. The babe in its mother/s womb is not nearer its mother than we are to the invisible sustaining and mothering powers of the universe, and to its spiritual entities, every moment of our lives.

We shall never be nearer God and spiritual and transcendent things than we are now.

The doors and windows of the universe are all open; the screens are all transparent. We are not barred or shut off; there is nothing foreign or unlike; we find our own in the stars as in the ground underfoot; this clod may become a man; yon shooting star may help redden his blood.

Whatever is upon the earth is of the earth; it came out of the divine soil, beamed upon by the fructifying heavens, the soul of man not less than his body.

I never see the spring flowers rising from the mould, or the pond-lilies born of the black ooze, that matter does not become transparent and reveal to me the working of the same celestial powers that fashioned the first man from the common dust.

Man/s mind is no more a stranger to the earth than is his body. Is not the clod wise? Is not the chemistry underfoot intelligent? Do not the roots of the trees find their way ? Do not the birds know their times and seasons? Are not all things about us filled to overflowing with mind-stuff? The cosmic mind is the earth mind, and the earth mind is man/s mind, freed but narrowed, with vision but with erring reason, conscious but troubled, and—shall we say?—human but immortal.

NAME ______________________________

Worksheet 5

17th and 18th Century European—American Encounters with Wilderness

1. Draw a picture of what you think the Pilgrims saw as they looked from the Mayflower toward the shore of Massachusetts.

2. How do you think they felt?

3. What did the Pilgrims and other Puritans believe that God wanted them to do in the "new world" to which they had come?

4. In their essays, both Ralph Waldo Emerson and Henry David Thoreau write that we benefit greatly from walking in natural or wild places. What benefits do they believe such interaction with nature will bring? List as many benefits as you can, with page references:

5. Emerson writes, "To speak truly, few adult persons can see nature." Both Emerson and Thoreau suggest that there are things that can be seen, yet that many miss, in the wild. What do they believe you can see? Again, make a list with page references:

6. **Thoreau likens a walk in the wild to a journey to "the holy land." What do you think he means by this?**

7. **Susan Fenimore Cooper and John Burroughs were nature writers in the same general period as Emerson and Thoreau. What similarities do you see in all of the writers/ ideas?**

Pulling It Together

17th and 18th Century European-American Encounters with Wilderness

QUESTIONS FOR THOUGHT AND DISCUSSION

1. William Bradford and other puritans seem to have had a very different view of the wilderness than Transcendentalists Ralph Waldo Emerson and Henry David Thoreau. What do you think might have caused these differences? Think about the dates in which they wrote. What important events had taken place in New England between these dates?
2. Do you ever go for long walks in wild places? Why? Are these walks simply a physical activity or do they have an impact on your mind and soul?

EXERCISES

1. In a two page essay, first explain the attitude toward nature and wildness found in the writings of Ralph Waldo Emerson and Henry David Thoreau. In particular, what ideas do you see in both of these authors. Then, compare and contrast the perspective on nature and wildness found in the transcendentalists with that found in Bradford's description of the new world.
2. Take a walk in an undeveloped area. (If it is not safe to go alone, go with a friend with the agreement that you will not spend the time talking to each other, but walking quietly.) Leave your music player and phone at home. Try to spend as much time being quiet and observing as many sights and sounds as you can. As you walk, think about your relationship to the natural world. When you return, sit down and write down what you observed and thought during your walk.

Chapter

6

Nature in 19th and 20th Century America

INTRODUCTION

As Europeans continued the push across the North American continent, the earlier settlers were more in tune with the Puritan attitude that the new world was a place to be conquered and civilized than the Transcendentalist idea that "wildness" was a source of salvation. The American settlers lived difficult lives, struggling to provide for themselves in the midst of nature's powerful forces. Perhaps it is to be expected, then, that they should see their task as making this new land "habitable," sometimes by clearing woodlands, cultivating prairies, planting their crops and "improving" their land. The Homestead Act, signed into law in 1862 by Abraham Lincoln, liberalized a practice that had begun earlier by which persons could acquire title to a tract of "unowned" land by making improvements upon it. And the new land seemed endless. At the end of the nineteenth and into the twentieth century, however, there was a dawning realization that at least some of America's natural places should be protected. In 1872, Yellowstone National Park was established by the U.S. government, and in the 1890's the national Forest Reserves were set aside. In 1908, 69,000 tourists visited Yellowstone Park. By 1915, the number had grown to 335,000. The Boy Scouts of America was founded in

1910, followed by the Girl Scouts two years later. Although not specifically environmentalist organizations, they have had tremendous influence on the attitudes of young men and women by introducing them to nature through a variety of camping and outdoor experiences and through their educational programs. By 1916, there were 13 national parks under the authority of the newly formed National Park Service.

In the 1930s the Dust Bowl added an additional concern. Whereas the movements mentioned above focused on preserving forests and wilderness areas, the Dust Bowl helped Americans see that even farmland must be properly conserved. Brought on by a drought and years of deep tilling of fertile prairie lands, the lower Midwest experienced a crisis as crop yields decreased, and topsoil literally blew away in the wind. Americans realized that efforts had to be made to protect farmland from the depletion of fertility by constant farming and from the erosion of soil by winds and water. Farmers received lessons from experts on effective conservation practices, and states and counties established conservation agencies to assist in establishing more this practices.

When we survey the history of American concerns for the natural environment in the twentieth century, three perspectives emerge. One perspective was that of the business interests, who were developing new lands as the country moved westward. Business representatives often objected to the government's growing role in land policies and saw government intervention chiefly as a detriment to business, upon which America was built. Among those inclined to advocate protection of the lands, two differing perspectives emerged. On the one hand, the **Conservationists** argued that America had been blessed with abundant resources to use in its support of human life and pursuit of well-being. It was in part because of these resources that America had flourished and become prosperous. But these resources were limited, so if we desired to continue to prosper as a nation, we must conserve the supplies of coal, wood, and other resources, upon which our well-being depended. In short, we must conserve and use them wisely. On the other hand, **Preservationists** believed that nature, and particularly certain "natural places," such as areas of undeveloped wilderness, had value quite apart from their monetary value to us as resources for gaining material prosperity. Whereas Conservationists used economic arguments, Preservationists appealed to aesthetic and even ethical considerations to defend the preservation of forests and other natural areas. At times, the two schools of environmental fought together for conservation measures. But at other times they found themselves at odds.

In this chapter, I have included three readings to illustrate these two "environmental" approaches to nature that influenced American attitudes in the early twentieth century: Conservationism and Preservationism. In the first reading, Gifford Pinchot (1865–1946), argues for the importance of conserving America's resources. After attending Yale University, Pinchot went to France to study forestry because there were no universities in the United States that offered such a program. From 1905 to 1910, Pinchot served as the Chief of the National Forest Service under President Theodore Roosevelt. Under Pinchot's leadership, America's conservation policies, especially with regard to forests, were shaped in accordance with the Conservationist philosophy which emphasized using the forests in the way that would benefit humans most. The second and third readings are written by John Muir (1838–1914), perhaps the most famous Preservationist. In writings reminiscent of Transcendentalists, Muir describes the beauty of the Western wilderness, focusing not only on the wonder of nature, but recounting experiences of the sublime in the natural world. Like the Transcendentalists, Muir's approach to nature is at times almost "spiritual," although not in a traditional Christian sense. In 1892, Muir founded the Sierra Club, an organization that continues to fight for protection of natural places. Through his writings and political activism, Muir played a very significant role in the preservation of the Yosemite Valley and Sequoia National Park in California. Interestingly, the third reading in this chapter not only illustrates Muir's approach to nature, but represents a political battle between Muir and his fellow Preservationists, and Pinchot, and the Conservationists. Pinchot had advocated a project damming the Hetch Hetchy Valley to create a reservoir for the city of San Francisco. In his essay, Muir argues against the building of the dam on the basis of the beauty of nature that would be destroyed by the project. Muir, who was a friend of President Theodore Roosevelt, pled with the administration not to go through with the project. In the end, Muir lost the battle.

The next President, Woodrow Wilson, signed the bill in 1913 that gave the okay for the dam. The defeat was one of Muir's greatest disappointments in life.

These early twentieth-century thinkers represent ways of thinking about nature that are clearly still present with us today. Our attitudes toward nature are a "mixed bag." Business interests continue to argue that government should not create policies restricting how property owners use their land and other resources. In other words, the Western business interest that was present in the early twentieth century is still represented in our debates about environmental issues today. Conservationists and Preservationist thinking is still with us as well. Conservationists still argue that we must work hard to protect our natural resources for the good of humankind. Preservationists still argue that natural places are valuable in themselves and should be preserved even at the expense of human interests. Sometimes Conservationists and Preservationists are able to join forces, but at other times they may find themselves at odds. In short, even within the mind of an environmentally minded person today, there may well be a mix of "right use," Conservationist, and Preservationist perspectives.

From *The Fight for Conservation*

—Gifford Pinchot

Gifford Pinchot (1865–1946) was an American politician and conservationist. Pinchot attended Yale University. After graduation, Pinchot travelled to Europe for further study, becoming America's first scientifically trained forester. On his return, Pinchot authored numerous books on forestry and conservation. From 1905 to 1910 he served as the founding director of the United Stated Forest Service. He served the cause of conservationism in America under the supportive Presidency of Theodore Roosevelt. In 1910, however, he was fired by President William Howard Taft, who sided with the business interests, against the conservation movement. Pinchot returned to his home state of Pennsylvania, where he was active in that state's forestry service, eventually serving as Pennsylvania governor for two terms (1923–1927, 1931–1935). Although Pinchot fought hard for conservation of natural resources, he often sided against the preservation of natural areas for their intrinsic value, a stance that brought him face to face with preservationist John Muir. This excerpt, take from the first chapter of The Fight for Conservation, *illustrates the types of arguments Pinchot advanced in the name of conservation.*

The most prosperous nation of to-day is the United States. Our unexampled wealth and well-being are directly due to the superb natural resources of our country, and to the use which has been made of them by our citizens, both in the present and in the past. We are prosperous because our forefathers bequeathed to us a land of marvelous resources still unexhausted. Shall we conserve those resources, and in our turn transmit them, still unexhausted, to our descendants? Unless we do, those who come after us will have to pay the price of misery, degradation, and failure for the progress and prosperity of our day. When the natural resources of any nation become exhausted, disaster and decay in every department of national life follow as a matter of course. Therefore the conservation of natural resources is the basis, and the only permanent basis, of national success. There are other conditions, but this one lies at the foundation.

When the natural resources of any nation become exhausted, disaster and decay in every department of national life follow as a matter of course.

Perhaps the most striking characteristic of the American people is their superb practical optimism; that marvelous hopefulness which keeps the individual efficiently at work. This hopefulness of the American is, however, as short-sighted as it is intense. As a rule, it does not look ahead beyond the next decade or score of years, and fails wholly to reckon with the real future of the Nation. I do not think I have often heard a forecast of the growth of our population that extended beyond a total of two hundred millions, and that only as a distant and shadowy goal. The point of view which this fact illustrates is neither true nor far-sighted. We shall reach a population of two hundred millions in the very near future, as time is counted in the lives of nations, and there is nothing more certain than that this country of ours will some day support double or triple or five times that number of prosperous people if only we can bring ourselves so to handle our natural resources in the present as not to lay an embargo on the prosperous growth of the future.

We, the American people, have come into the possession of nearly four million square miles of the richest portion of the earth. It is ours to use and conserve for ourselves and our descendants, or to destroy. The fundamental question which confronts us is, What shall we do with it?

That question cannot be answered without first considering the condition of our natural resources and what is being done with them to-day. As a people, we have been in the habit of declaring certain of our resources to be inexhaustible. To no other resource more frequently than coal has this stupidly false adjective been applied. Yet our coal supplies are so far from being inexhaustible that if the increasing rate of consumption shown by the figures of the last seventy-five years continues to prevail, our supplies of anthracite coal will last but fifty years and of bituminous coal less than two hundred years. From the point of view of national life, this means the exhaustion of one of the most important factors in our civilization within the immediate future. Not a few coal fields have already been exhausted, as in portions of Iowa and Missouri. Yet, in the face of these known facts, we continue to treat our coal as though there could never be an end of it. The established coal-mining practice at the present date does not take out more than one-half the coal, leaving the less easily mined or lower grade material to be made permanently inaccessible by the caving in of the abandoned workings. The loss to the Nation from this form of waste is prodigious and inexcusable.

The loss to the Nation from this form of waste is prodigious and inexcusable.

The waste in use is not less appalling. But five per cent, of the potential power residing in the coal actually mined is saved and used. For example, only about five per cent, of the power of the one hundred and fifty million tons annually burned on the railways of the United States is actually used in traction; ninety-five per cent, is expended unproductively or is lost. In the best incandescent electric lighting plants but one-fifth of one per cent, of the potential value of the coal is converted into light.

Many oil and gas fields, as in Pennsylvania, West Virginia, and the Mississippi Valley, have already failed, yet vast amounts of gas continue to be poured into the air and great quantities of oil into the streams. Cases are known in which great volumes of oil were systematically burned in order to get rid of it.

The prodigal squandering of our mineral fuels proceeds unchecked in the face of the fact that such resources as these, once used or wasted, can never be replaced. If waste like this were not chiefly thoughtless, it might well be characterized as the deliberate destruction of the Nation's future. Many fields of iron ore have already been exhausted, and in still more, as in the coal mines, only the higher grades have been taken from the mines, leaving the least valuable beds to be exploited at increased cost or not at all. Similar waste in the case of other minerals is less serious only because they are less indispensable to our civilization than coal and iron. Mention should be made of the annual loss of millions of dollars worth of by-products from coke, blast, and other furnaces now thrown into the air, often not merely without benefit but to the serious injury of the community. In other countries these by-products are saved and used.

If waste like this were not chiefly thoughtless, it might well be characterized as the deliberate destruction of the Nation's future.

We are in the habit of speaking of the solid earth and the eternal hills as though they, at least, were free from the vicissitudes of time and certain to furnish perpetual support for prosperous human life. This conclusion is as false as the term "inexhaustible" applied to other natural resources. The waste of soil is among the most dangerous of all wastes now in progress in the United States. In 1896, Professor Shaler, than whom no one has spoken with greater authority on this subject, estimated that in the upland regions of the states south of Pennsylvania three thousand square miles of soil had been destroyed as the result of forest denudation, and that destruction was then proceeding at the rate of one hundred square miles of fertile soil per year. No seeing man can travel through the United States without being struck with the enormous and unnecessary loss of fertility by easily preventable soil wash. The soil so lost, as in the case of many other wastes, becomes itself a source of damage and expense, and must be removed from the channels of our navigable streams at an enormous annual cost. The Mississippi River alone is estimated to transport yearly four hundred million tons of sediment, or about twice the amount of material to be excavated from the Panama Canal. This material is the most fertile portion of our richest fields, transformed from a blessing to a curse by unrestricted erosion. . . .

The lowest estimate reached by the Forest Service of the timber now standing in the United States is 1,400 billion feet, board measure; the highest, 2,500 billion. The present annual consumption is approximately 100 billion feet, while the annual growth is but a third of the consumption, or from 30 to 40 billion feet. If we accept the larger estimate of the standing timber, 2,500 billion feet, and the larger estimate of the annual growth, 40 billion feet, and apply the present rate of consumption, the result shows a probable duration of our supplies of timber of little more than a single generation.

Estimates of this kind are almost inevitably misleading. For example, it is certain that the rate of consumption of timber will increase enormously in the future, as it has in the past, so long as supplies remain to draw upon. Exact knowledge of many other factors is needed before closely accurate results can be obtained. The figures cited are, however, sufficiently reliable to make it certain that the United States has already crossed the verge of a timber famine so severe that its blighting effects will be felt in every household in the land. The rise in the price of lumber which marked the opening of the present century is the beginning of a vastly greater and more rapid rise which is to come. We must necessarily begin to suffer from the scarcity of timber long before our supplies are completely exhausted.

It is well to remember that there is no foreign source from which we can draw cheap and abundant supplies of timber to meet a demand per capita so large as to be without parallel in the world, and that the suffering which will result from the progressive failure of our timber has been but faintly foreshadowed by temporary scarcities of coal.

What will happen when the forests fail? In the first place, the business of lumbering will disappear. It is now the fourth greatest industry in the United States. All forms of building industries will suffer with it, and the occupants of houses, offices, and stores must pay the added cost. Mining will become vastly more expensive; and with the rise in the cost of mining there must follow a corresponding rise in the price of coal, iron, and other minerals. The railways, which have as yet failed entirely to develop a satisfactory substitute for the wooden tie (and must, in the opinion of their best engineers, continue to fail), will be profoundly affected, and the cost of transportation will suffer a corresponding increase. Water power for lighting, manufacturing, and transportation, and the movement of freight and passengers by inland waterways, will be affected still more directly than the steam railways. The cultivation of the soil, with or without irrigation, will be hampered by the increased cost of agricultural tools, fencing, and the wood needed for other purposes about the farm. Irrigated agriculture will suffer most of all, for the destruction of the forests means the loss of the waters as surely as night follows day. With the rise in the cost of producing food, the cost of food itself will rise. Commerce in general will necessarily be affected by the difficulties of the primary industries upon which it depends. In a word, when the forests fail, the daily life of the average citizen will inevitably feel the pinch on every side. And the forests have already begun to fail, as the direct result of the suicidal policy of forest destruction which the people of the United States have allowed themselves to pursue.

What is a "resource"? In chapter 1, several ways of seeing nature were described. Look back and note what is said about seeing nature as a material resource. What does this image suggest? What reasons does Pinchot give for preserving the resources that have made America prosperous?

It is true that about twenty per cent, of the less valuable timber land in the United States remains in the possession of the people in the National Forests, and that it is being cared for and conserved to supply the needs of the present and to mitigate the suffering of the near future. But it needs no argument to prove that this comparatively small area will be insufficient to meet the demand which is now exhausting an area four times as great, or to prevent the suffering I have described. Measures of greater vigor are imperatively required.

The conception that water is, on the whole, the most important natural resource has gained firm hold in the irrigated West, and is making rapid progress in the humid East. Water, not land, is the primary value in the Western country, and its conservation and use to irrigate land is the first condition of prosperity. The use of our streams for irrigation and for domestic and manufacturing uses is comparatively well developed. Their use for power is less developed, while their use for transportation has only begun. The conservation of the inland waterways of the

United States for these great purposes constitutes, perhaps, the largest single task which now confronts the Nation. The maintenance and increase of agriculture, the supply of clear water for domestic and manufacturing uses, the development of electrical power, transportation, and lighting, and the creation of a system of inland transportation by water whereby to regulate freight-rates by rail and to move the bulkier commodities cheaply from place to place, is a task upon the successful accomplishment of which the future of the Nation depends in a peculiar degree. We are accustomed, and rightly accustomed, to take pride in the vigorous and healthful growth of the United States, and in its vast promise for the future. Yet we are making no preparation to realize what we so easily foresee and glibly predict. The vast possibilities of our great future will become realities only if we make ourselves, in a sense, responsible for that future. The planned and orderly development and conservation of our natural resources is the first duty of the United States. It is the only form of insurance that will certainly protect us against the disasters that lack of foresight has in the past repeatedly brought down on nations since passed away.

The planned and orderly development and conservation of our natural resources is the first duty of the United States.

"Hetch Hetchy Valley," from *Yosemite*

—John Muir

John Muir (1838–1914) is well-known as an American preservationist. A Scottish-born immigrant, who grew up in Wisconsin, Muir moved to the Western United States, where he hiked the mountains and documented the wonders and beauty of the western U. S. forests. A prolific writer, Muir passionately defended the preservation of the wilderness areas, especially in the California mountains. He championed the bill that created Yosemite and Sequoia National Parks in 1890. In 1982 he founded the Sierra Club, which has become one of the most active environmentalist organizations, even today. Unlike conservationists, like Gifford Pinchot, Muir argued that America's wild places should be preserved not for the sake of future use as resources, but for the sake of preserving the beauty and wonder of nature. Thus, Muir's arguments invoke much different images than do Pinchot's.

In the essay from which the following excerpt is taken, Muir argues against a project that had been proposed by the government to dam up a valley in the Yosemite National Park to create a water supply for the city of San Francisco. Muir describes the beauty of the valley, outlines the many plants that can be enjoyed there, and argues for the value of experiences that will never again be enjoyed if the valley is dammed up. Toward the end of the reading, Muir alludes to the Bible, comparing the destruction of Hetch Hetchy Valley to the destruction of the garden of Eden by the devil, and likening those who would build the dam to the merchants that Jesus threw out of the temple. It is clear that for Muir, the wilderness is not just a collection of resources for human use. It is a sacred temple or church, a place where one encounters the divine. To destroy Hetch Hetchy Valley is not just to waste resources. It is to destroy the holiest of temples, a temple that has been built by God. The debate for and against the Hetch Hetchy dam pitted conservationist Gifford Pinchot, who argued for the dam, against John Muir and preservationists, who argued against it. In 1913 Muir lost the battle when Woodrow Wilson signed the bill that authorized the building of the dam.

Yosemite is so wonderful that we are apt to regard it as an exceptional creation, the only valley of its kind in the world; but Nature is not so poor as to have only one of anything. Several other yosemites have been discovered in the Sierra that occupy the same relative positions on the Range and were formed by the same forces in the same kind of granite. One of these, the Hetch Hetchy Valley, is in the Yosemite National Park about twenty miles from Yosemite and is easily accessible to all sorts of travelers by a road and trail that leaves the Big Oak Flat road at Bronson Meadows a few miles below Crane Flat, and to mountaineers by way of Yosemite Creek basin and the head of the middle fork of the Tuolumne. . . .

Standing boldly out from the south wall is a strikingly picturesque rock called by the Indians, Kolana, the outermost of a group 2300 feet high, corresponding with the Cathedral Rocks of Yosemite both in relative position and form. On the opposite side of the Valley, facing Kolana, there is a counterpart of the El Capitan that rises sheer and plain to a height of 1800 feet, and over its massive brow flows a stream which makes the most graceful fall I have ever seen. From the edge of the cliff to the top of an earthquake talus it is perfectly free in the air for a thousand feet before it is broken into cascades among talus boulders. It is in all its glory in June, when the snow is melting fast, but fades and vanishes toward the end of summer. The only fall I know with which it may fairly be compared is the Yosemite Bridal Veil; but it excels even that favorite fall both in height and airy-fairy beauty and behavior. . . .

So fine a fall might well seem sufficient to glorify any valley; but here, as in Yosemite, Nature seems in nowise moderate, for a short distance to the eastward of Tueeulala booms and thunders the great Hetch Hetchy Fall, Wapama, so near that you have both of them in full view from the same standpoint. It is the counterpart of the Yosemite Fall, but has a much greater volume of water, is about 1700 feet in height, and appears to be nearly vertical, though considerably inclined, and is dashed into huge out bounding bosses of foam on projecting shelves and knobs. No two falls could be more unlike—Tueeulala out in the open sunshine descending like thistledown; Wapama in a jagged, shadowy gorge roaring and plundering, pounding its way like an earthquake avalanche. . . .

The floor of the Valley is about three and a half miles long, and from a fourth to half a mile wide. The lower portion is mostly a level meadow about a mile long, with the trees restricted to the sides and the riverbanks, and partially separated from the main, upper, forested portion bya low bar of glacier-polished granite across which the river breaks in rapids.

The principal trees are the yellow and sugar pines, digger pine, incense cedar, Douglas spruce, silver fir, the California and golden-cup oaks, balsam cottonwood, Nuttall's flowering dogwood, alder, maple, laurel, tumion, etc. The most abundant and influential are the great yellow or silver pines like those of Yosemite, the tallest over two hundred feet in height, and the oaks assembled in magnificent groves with massive rugged trunks four to six feet in diameter, and broad, shady, wide-spreading heads. The shrubs forming conspicuous flowery clumps and tangles are manzanita, azalea, spiræa, brier-rose, several species of ceanothus, calycanthus, philadelphus, wild cherry, etc.; with abundance of showy and fragrant herbaceous plants growing about them or out in the open in beds by themselves—lilies, Mariposa tulips, brodiaeas, orchids, iris, spraguea, draperia, collomia, collinsia, castilleja, nemophila, larkspur, columbine, goldenrods, sunflowers, mints of many species, honeysuckle, etc. Many fine ferns dwell here also, especially the beautiful and interesting rock-ferns—pellaea, and cheilanthes of several species—fringing and rosetting dry rock-piles and ledges; woodwardia and asplenium on damp spots with fronds six or seven feet high; the delicate maiden-hair in mossy nooks by the falls, and the sturdy, broad-shouldered pteris covering nearly all the dry ground beneath the oaks and pines.

Hetch Hetchy Valley, far from being a plain, common, rock-bound meadow, as many who have not seen it seem to suppose, is a grand landscape garden, one of Nature's rarest and most precious mountain temples.

It appears, Hetch Hetchy Valley, far from being a plain, common, rockbound meadow, as many who have not seen it seem to suppose, is a grand landscape garden, one of Nature's rarest and most precious mountain temples. therefore, that As in Yosemite, the sublime rocks of its walls seem to glow with life, whether leaning back in repose or standing erect in thoughtful attitudes, giving welcome to storms and calms alike, their brows in the sky, their feet set in the groves and gay flowery meadows, while birds, bees, and butterflies help the river and waterfalls to stir all the air into music—things frail and fleeting and types of permanence meeting here and blending, just as they do in Yosemite, to draw her lovers into close and confiding communion with her. Sad to say, this most precious and sublime feature of the Yosemite National Park, one of the greatest of all our natural resources for the uplifting joy and peace and health of the people, is in danger of being dammed and made into a reservoir to help supply San Francisco with water and light, thus flooding it from wall to wall and burying its gardens and groves one or two hundred feet deep. This grossly destructive commercial scheme has long been planned and urged (though water as pure and abundant can be got from outside of the people's park, in a dozen different places), because of the comparative cheapness of the dam and of the territory which it is sought to divert from the great uses to which it was dedicated in the Act of 1890 establishing the Yosemite National Park.

This most precious and sublime feature of the Yosemite National Park, one of the greatest of all our natural resources for the uplifting joy and peace and health of the people, is in danger of being dammed and made into a reservoir to help supply San Francisco with water and light,

The making of gardens and parks goes on with civilization all over the world, and they increase both in size and number as their value is recognized. Everybody needs beauty as well as bread, places to play in and pray in, where Nature may heal and cheer and give strength to body and soul alike. This natural beauty-hunger is made manifest in the little window-sill gardens of the poor, though perhaps only a geranium slip in a broken cup, as well as in the carefully tended rose and lily gardens of the rich, the thousands of spacious city parks and botanical gardens, and in our magnificent National parks—the Yellowstone, Yosemite, Sequoia, etc.–Nature's sublime wonderlands, the admiration and joy of the world. Nevertheless, like anything else worth while, from the very beginning, however well guarded, they have always been subject to attack by despoiling gainseekers and mischief-makers of every degree from Satan to Senators, eagerly trying to make everything immediately and selfishly commercial, with schemes disguised in smug-smiling philanthropy, industriously, shampiously crying, "Conservation, conservation, panutilization," that man and beast may be fed and the dear Nation made great. Thus long ago a few enterprising merchants utilized the Jerusalem temple as a place of business instead of a place of prayer, changing money, buying and selling cattle and sheep and doves; and earlier still, the first forest reservation, including only one tree, was likewise despoiled. Ever since the establishment of the Yosemite National Park, strife has been going on around its borders and I suppose this will go on as part of the universal battle between right and wrong, however much its boundaries may be shorn, or its wild beauty destroyed. . . .

That any one would try to destroy such a place seems incredible; but sad experience shows that there are people good enough and bad enough for anything. The proponents of the dam scheme bring forward a lot of bad arguments to prove that the only righteous thing to do with the people's parks is to destroy them bit by bit as they are able. Their arguments are curiously like those of the devil, devised for the destruction of the first garden–so much of the very best Eden fruit going to waste; so much of the best Tuolumne water and Tuolumne scenery going to waste. Few of their statements are even partly true, and all are misleading.

Their arguments are curiously like those of the devil, devised for the destruction of the first garden.

Thus, Hetch Hetchy, they say, is a "low-lying meadow." On the contrary, it is a high-lying natural landscape garden, as the photographic illustrations show.

"It is a common minor feature, like thousands of others." On the contrary it is a very uncommon feature; after Yosemite, the rarest and in many ways the most important in the National Park.

"Damming and submerging it 175 feet deep would enhance its beauty by forming a crystal-clear lake." Landscape gardens, places of recreation and worship, are never made beautiful by destroying and burying them. The beautiful sham lake, forsooth, should be only an eyesore, a dismal blot on the landscape, like many others to be seen in the Sierra. For, instead of keeping it at the same level all the year, allowing Nature centuries of time to make new shores, it would, of course, be full only a month or two in the spring, when the snow is melting fast; then it would be gradually drained, exposing the slimy sides of the basin and shallower parts of the bottom, with the gathered drift and waste, death and decay of the upper basins, caught here instead of being swept on to decent natural burial along the banks of the river or in the sea. Thus the Hetch Hetchy dam-lake would be only a rough imitation of a natural lake for a few of the spring months, an open sepulcher for the others."

These temple destroyers, devotees of ravaging commercialism, seem to have a perfect contempt for Nature, and, instead of lifting their eyes to the God of the mountains, lift them to the Almighty Dollar.

"Hetch Hetchy water is the purest of all to be found in the Sierra, unpolluted, and forever unpollutable." On the contrary, excepting that of the Merced below Yosemite, it is less pure than that of most of the other Sierra streams, because of the sewerage of camp grounds draining into it, especially of the Big Tuolumne Meadows camp ground, occupied by hundreds of tourists and mountaineers, with their animals, for months every summer, soon to be followed by thousands from all the world.

These temple destroyers, devotees of ravaging commercialism, seem to have a perfect contempt for Nature, and, instead of lifting their eyes to the God of the mountains, lift them to the Almighty Dollar.

In what ways does he describe Hetch Hetchy's value? Why does Muir believe it would be wrong to dam it?

Dam Hetch Hetchy! As well dam for water-tanks the people's cathedrals and churches, for no holier temple has ever been consecrated by the heart of man.

Near View of the High Sierras

—John Muir

Early one bright morning in the middle of Indian summer, while the glacier meadows were still crisp with frost crystals, I set out from the foot of Mount Lyell, on my way down to Yosemite Valley, to replenish my exhausted store of bread and tea. I had spent the past summer, as many preceding ones, exploring the glaciers that lie on the head waters of the San Joaquin, Tuolumne, Merced, and Owen's rivers; measuring and studying their movements, trends, crevasses, moraines, etc., and the part they had played during the period of their greater extension in the creation and development of the landscapes of this alpine wonderland. The time for this kind of work was nearly over for the year, and I began to look forward with delight to the approaching winter with its wondrous storms, when I would be warmly snow-bound in my Yosemite cabin with plenty of bread and books; but a tinge of regret came on when I considered that possibly I might not see this favorite region again until the next summer, excepting distant views from the heights about the Yosemite walls.

To artists, few portions of the High Sierra are, strictly speaking, picturesque. The whole massive uplift of the range is one great picture, not clearly divisible into smaller ones; differing much in this respect from the older, and what may be called, riper mountains of the Coast Range. All the landscapes of the Sierra, as we have seen, were born again, remodeled from base to summit by the developing ice-floods of the last glacial winter. But all those new landscapes were not brought forth simultaneously; some of the highest, where the ice lingered longest, are tens of centuries younger than those of the warmer regions below them. In general, the younger the mountain-landscapes,—younger, I mean, with reference to the time of their emergence from the ice of the glacial period,—the less separable are they into artistic bits capable of being made into warm, sympathetic, lovable pictures with appreciable humanity in them.

Here, however, on the head waters of the Tuolumne, is a group of wild peaks on which the geologist may say that the sun has but just begun to shine, which is yet in a high degree picturesque, and in its main features so regular and evenly balanced as almost to appear conventional—one somber cluster of snow-laden peaks with gray pine-fringed granite bosses braided around its base, the whole surging free into the sky from the head of a magnificent valley, whose lofty walls are beveled away on both sides so as to embrace it all without admitting anything not strictly belonging to it. The foreground was now aflame with autumn colors, brown and purple and gold, ripe in the mellow sunshine; contrasting brightly with the deep, cobalt blue of the sky, and the black and gray, and pure, spiritual white of the rocks and glaciers. Down through the midst, the young Tuolumne was seen pouring from its crystal fountains, now resting in glassy pools as if changing back again into ice, now leaping in white cascades as if turning to snow; gliding right and left between granite bosses, then sweeping on through the smooth, meadowy levels of the valley, swaying pensively from side to side with calm, stately gestures past dipping willows and sedges, and around groves of arrowy pine; and throughout its whole eventful course, whether flowing fast or slow, singing loud or low, ever filling the landscape with spiritual animation, and manifesting the grandeur of its sources in every movement and tone.

Pursuing my lonely way down the valley, I turned again and again to gaze on the glorious picture, throwing up my arms to inclose it as in a frame. After long ages of growth in the darkness beneath the glaciers, through sunshine and storms, it seemed now to be ready and waiting for the elected artist, like yellow wheat for the reaper; and I could not help wishing that I might carry colors and brushes with me on my travels, and learn to paint. In the mean time I had to be content with photographs on my mind and sketches in my note-books. At length, after I had rounded a precipitous headland that puts out from the west wall of the valley, every peak vanished from sight, and I pushed rapidly along the frozen meadows, over the divide between the waters of the Merced and Tuolumne, and down through the forests that clothe the slopes of Cloud's Rest, arriving in Yosemite in due time—which, with me, is _any_ time. And, strange to say, among the first people I met here were two artists who, with letters of introduction, were awaiting my return. They inquired whether in the course of my explorations in the adjacent mountains I had ever come upon a landscape suitable for a large painting; whereupon I began a description of the one that had so lately excited my admiration. Then, as I went on further and further into details, their faces began to glow, and I offered to guide them to it, while they declared that they would gladly follow, far or near, whithersoever I could spare the time to lead them.

Since storms might come breaking down through the fine weather at any time, burying the colors in snow, and cutting off the artists' retreat, I advised getting ready at once.

I led them out of the valley by the Vernal and Nevada Falls, thence over the main dividing ridge to the Big Tuolumne Meadows, by the old Mono trail, and thence along the upper Tuolumne River to its head. This was my companions' first excursion into the High Sierra, and as I was almost always alone in my mountaineering, the way that the fresh beauty was reflected in their faces made for me a novel and interesting study. They naturally were affected most of all by the colors—the intense azure of the sky, the purplish grays of the granite, the red and browns of dry meadows, and the translucent purple and crimson of huckleberry bogs; the flaming yellow of aspen groves, the silvery flashing of the streams, and the bright green and blue of the glacier lakes. But the general expression of the scenery—rocky and savage—seemed sadly disappointing; and as they threaded the forest from ridge to ridge, eagerly scanning the landscapes as they were unfolded, they said: "All this is huge and sublime, but we see nothing as yet at all available for effective pictures. Art is long, and art is limited, you know; and here are foregrounds, middle-grounds, backgrounds, all alike; bare rock-waves, woods, groves, diminutive flecks of meadow, and strips of glittering water." "Never mind," I replied, "only bide a wee, and I will show you something you will like."

At length, toward the end of the second day, the Sierra Crown began to come into view, and when we had fairly rounded the projecting headland before mentioned, the whole picture stood revealed in the flush of the alpenglow. Their enthusiasm was excited beyond bounds, and the more impulsive of the two, a young Scotchman, dashed ahead, shouting and gesticulating and tossing his arms in the air like a madman. Here, at last, was a typical alpine landscape.

After feasting awhile on the view, I proceeded to make camp in a sheltered grove a little way back from the meadow, where pine-boughs could be obtained for beds, and where there was plenty of dry wood for fires, while the artists ran here and there, along the river-bends and up the sides of the cañon, choosing foregrounds for sketches. After dark, when our tea was made and a rousing fire had been built, we began to make our plans. They decided to remain several days, at the least, while I concluded to make an excursion in the mean time to the untouched summit of Ritter.

It was now about the middle of October, the springtime of snow-flowers. The first winter-clouds had already bloomed, and the peaks were strewn with fresh crystals, without, however, affecting the climbing to any dangerous extent. And as the weather was still profoundly calm, and the distance to the foot of the mountain only a little more than a day, I felt that I was running no great risk of being storm-bound.

Mount Ritter is king of the mountains of the middle portion of the High Sierra, as Shasta of the north and Whitney of the south sections. Moreover, as far as I know, it had never been climbed. I had explored the adjacent wilderness summer after summer, but my studies thus far had never drawn me to the top of it. Its height above sea-level is about 13,300 feet, and it is fenced round by steeply inclined glaciers, and cañons of tremendous depth and ruggedness, which render it almost inaccessible. But difficulties of this kind only exhilarate the mountaineer.

Next morning, the artists went heartily to their work and I to mine. Former experiences had given good reason to know that passionate storms, invisible as yet, might be brooding in the calm sun-gold; therefore, before bidding farewell, I warned the artists not to be alarmed should I fail to appear before a week or ten days, and advised them, in case a snow-storm should set in, to keep up big fires and shelter themselves as best they could, and on no account to become frightened and attempt to seek their way back to Yosemite alone through the drifts.

My general plan was simply this: to scale the cañon, wall, cross over to the eastern flank of the range, and then make my way southward to the northern spurs of Mount Ritter in compliance with the intervening topography; for to push on directly southward from camp through the innumerable peaks and pinnacles that adorn this portion of the axis of the range, however interesting, would take too much time, besides being extremely difficult and dangerous at this time of year.

All my first day was pure pleasure; simply mountaineering indulgence, crossing the dry pathways of the ancient glaciers, tracing happy streams, and learning the habits of the birds and marmots in the groves and rocks. Before I had gone a mile from camp, I came to the foot of a white cascade that beats its way down a rugged gorge in the cañon wall, from a height of about nine hundred feet, and pours its throbbing waters into the Tuolumne. I was acquainted with its fountains, which, fortunately, lay in my course. What a fine traveling companion it proved to be, what songs it sang, and how passionately it told the mountain's own joy! Gladly I climbed along its dashing border, absorbing its divine music, and bathing from time to time in waftings of irised spray. Climbing higher, higher, now beauty came streaming on the sight: painted meadows, late-blooming gardens, peaks of rare architecture, lakes here and there, shining like silver, and glimpses of the forested middle region and the yellow lowlands far in the west. Beyond the range I saw the so-called Mono Desert, lying dreamily silent in thick purple light—a desert of heavy sun-glare beheld from a desert of ice-burnished granite. Here the waters divide, shouting in glorious enthusiasm, and falling eastward to vanish in the volcanic sands and dry sky of the Great Basin, or westward to the Great Valley of California, and thence through the Bay of San Francisco and the Golden Gate to the sea.

All my first day was pure pleasure; simply mountaineering indulgence, crossing the dry pathways of the ancient glaciers, tracing happy streams, and learning the habits of the birds and marmots in the groves and rocks.

Passing a little way down over the summit until I had reached an elevation of about 10,000 feet, I pushed on southward toward a group of savage peaks that stand guard about Ritter on the north and west, groping my way, and dealing instinctively with every obstacle as it presented itself. Here a huge gorge would be found cutting across my path, along the dizzy edge of which I scrambled until some less precipitous point was discovered where I might safely venture to the bottom and then, selecting some feasible portion of the opposite wall, reascend with the same slow caution. Massive, flat-topped spurs alternate with the gorges, plunging abruptly from the shoulders of the snowy peaks, and planting their feet in the warm desert. These were everywhere marked and adorned with characteristic sculptures of the ancient glaciers that swept over this entire region like one vast ice-wind, and the polished surfaces produced by the ponderous flood are still so perfectly preserved that in many places the sunlight reflected from them is about as trying to the eyes as sheets of snow.

God's glacial-mills grind slowly, but they have been kept in motion long enough in California to grind sufficient soil for a glorious abundance of life, though most of the grist has been carried to the lowlands, leaving these high regions comparatively lean and bare; while the post-glacial agents of erosion have not yet furnished sufficient available food over the general surface for more than a few tufts of the hardiest plants, chiefly carices and eriogonae. And it is interesting to learn in this connection that the sparseness and repressed character of the vegetation at this height is caused more by want of soil than by harshness of climate; for, here and there, in sheltered hollows (countersunk beneath the general surface) into which a few rods of well-ground moraine chips have been dumped, we find groves of spruce and pine thirty to forty feet high, trimmed around the edges with willow and huckleberry bushes, and oftentimes still further by an outer ring of tall grasses, bright with lupines, larkspurs, and showy columbines, suggesting a climate by no means repressingly severe. All the streams, too, and the pools at this elevation are furnished with little gardens wherever soil can be made to lie, which, though making scarce any show at a distance, constitute charming surprises to the appreciative observer. In these bits of leanness a few birds find grateful homes.

Having no acquaintance with man, they fear no ill, and flock curiously about the stranger, almost allowing themselves to be taken in the hand. In so wild and so beautiful a region was spent my first day, every sight and sound inspiring, leading one far out of himself, yet feeding and building up his individuality.

Now came the solemn, silent evening. Long, blue, spiky shadows crept out across the snow-fields, while a rosy glow, at first scarce discernible, gradually deepened and suffused every mountain-top, flushing the glaciers and the harsh crags above them. This was the alpenglow, to me one of the most impressive of all the terrestrial manifestations of God. At the touch of this divine light, the mountains seemed to kindle to a rapt, religious consciousness, and stood hushed and waiting like devout worshipers. Just before the alpenglow began to fade, two crimson clouds came streaming across the summit like wings of flame, rendering the sublime scene yet more impressive; then came darkness and the stars.

Icy Ritter was still miles away, but I could proceed no farther that night. I found a good campground on the rim of a glacier basin about 11,000 feet above the sea. A small lake nestles in the bottom of it, from which I got water for my tea, and a storm-beaten thicket near by furnished abundance of resiny fire-wood. Somber peaks, hacked and shattered, circled half-way around the horizon, wearing a savage aspect in the gloaming, and a waterfall chanted solemnly across the lake on its way down from the foot of a glacier. The fall and the lake and the glacier were almost equally bare; while the scraggy pines anchored in the rock-fissures were so dwarfed and shorn by storm-winds that you might walk over their tops. In tone and aspect the scene was one of the most desolate I ever beheld. But the darkest scriptures of the mountains are illumined with bright passages of love that never fail to make themselves felt when one is alone.

I made my bed in a nook of the pine-thicket, where the branches were pressed and crinkled overhead like a roof, and bent down around the sides. These are the best bedchambers the high mountains afford—snug as squirrel-nests, well ventilated, full of spicy odors, and with plenty of wind-played needles to sing one asleep. I little expected company, but, creeping in through a low side-door, I found five or six birds nestling among the tassels. The night-wind began to blow soon after dark; at first only a gentle breathing, but increasing toward midnight to a rough gale that fell upon my leafy roof in ragged surges like a cascade, bearing wild sounds from the crags overhead. The waterfall sang in chorus, filling the old ice-fountain with its solemn roar, and seeming to increase in power as the night advanced—fit voice for such a landscape. I had to creep out many times to the fire during the night, for it was biting cold and I had no blankets. Gladly I welcomed the morning star.

The dawn in the dry, wavering air of the desert was glorious. Everything encouraged my undertaking and betokened success. There was no cloud in the sky, no storm-tone in the wind. Breakfast of bread and tea was soon made. I fastened a hard, durable crust to my belt by way of provision, in case I should be compelled to pass a night on the mountain-top; then, securing the remainder of my little stock against wolves and wood-rats, I set forth free and hopeful.

How glorious a greeting the sun gives the mountains! To behold this alone is worth the pains of any excursion a thousand times over. The highest peaks burned like islands in a sea of liquid shade. Then the lower peaks and spires caught the glow, and long lances of light, streaming through many a notch and pass, fell thick on the frozen meadows. The majestic form of Ritter was full in sight, and I pushed rapidly on over rounded rock-bosses and pavements, my iron-shod shoes making a clanking sound, suddenly hushed now and then in rugs of bryanthus, and sedgy lake-margins soft as moss. Here, too, in this so-called "land of desolation," I met cassiope, growing in fringes among the battered rocks. Her blossoms had faded long ago, but they were still clinging with happy memories to the evergreen sprays, and still so beautiful as to thrill every fiber of one's being. Winter and summer, you may hear her voice, the low, sweet melody of her purple bells. No evangel among all the mountain plants speaks Nature's love more plainly than cassiope. Where she dwells, the redemption of the coldest solitude is complete. The very rocks and glaciers seem to feel her presence, and become imbued with her own fountain sweetness. All things were warming and awakening. Frozen rills began to flow, the marmots came out of their nests in boulder-piles and climbed sunny rocks to bask, and the dun-headed sparrows were flitting about seeking their breakfasts. The lakes seen from every ridge-top were brilliantly rippled and spangled, shimmering like the thickets of the low

Dwarf Pines. The rocks, too, seemed responsive to the vital heat—rock-crystals and snow-crystals thrilling alike. I strode on exhilarated, as if never more to feel fatigue, limbs moving of themselves, every sense unfolding like the thawing flowers, to take part in the new day harmony.

All along my course thus far, excepting when down in the cañons, the landscapes were mostly open to me, and expansive, at least on one side. On the left were the purple plains of Mono, reposing dreamily and warm; on the right, the near peaks springing keenly into the thin sky with more and more impressive sublimity. But these larger views were at length lost. Rugged spurs, and moraines, and huge, projecting buttresses began to shut me in. Every feature became more rigidly alpine, without, however, producing any chilling effect; for going to the mountains is like going home. We always find that the strangest objects in these fountain wilds are in some degree familiar, and we look upon them with a vague sense of having seen them before.

On the southern shore of a frozen lake, I encountered an extensive field of hard, granular snow, up which I scampered in fine tone, intending to follow it to its head, and cross the rocky spur against which it leans, hoping thus to come direct upon the base of the main Ritter peak. The surface was pitted with oval hollows, made by stones and drifted pine-needles that had melted themselves into the mass by the radiation of absorbed sun-heat. These afforded good footholds, but the surface curved more and more steeply at the head, and the pits became shallower and less abundant, until I found myself in danger of being shed off like avalanching snow. I persisted, however, creeping on all fours, and shuffling up the smoothest places on my back, as I had often done on burnished granite, until, after slipping several times, I was compelled to retrace my course to the bottom, and make my way around the west end of the lake, and thence up to the summit of the divide between the head waters of Rush Creek and the northernmost tributaries of the San Joaquin.

Arriving on the summit of this dividing crest, one of the most exciting pieces of pure wilderness was disclosed that I ever discovered in all my mountaineering. There, immediately in front, loomed the majestic mass of Mount Ritter, with a glacier swooping down its face nearly to my feet, then curving westward and pouring its frozen flood into a dark blue lake, whose shores were bound with precipices of crystalline snow; while a deep chasm drawn between the divide and the glacier separated the massive picture from everything else. I could see only the one sublime mountain, the one glacier, the one lake; the whole veiled with one blue shadow—rock, ice, and water close together without a single leaf or sign of life. After gazing spellbound, I began instinctively to scrutinize every notch and gorge and weathered buttress of the mountain, with reference to making the ascent. The entire front above the glacier appeared as one tremendous precipice, slightly receding at the top, and bristling with spires and pinnacles set above one another in formidable array. Massive lichen-stained battlements stood forward here and there, hacked at the top with angular notches, and separated by frosty gullies and recesses that have been veiled in shadow ever since their creation; while to right and left, as far as I could see, were huge, crumbling buttresses, offering no hope to the climber. The head of the glacier sends up a few finger-like branches through narrow _couloirs_; but these seemed too steep and short to be available, especially as I had no ax with which to cut steps, and the numerous narrow-throated gullies down which stones and snow are avalanched seemed hopelessly steep, besides being interrupted by vertical cliffs; while the whole front was rendered still more terribly forbidding by the chill shadow and the gloomy blackness of the rocks.

Descending the divide in a hesitating mood, I picked my way across the yawning chasm at the foot, and climbed out upon the glacier. There were no meadows now to cheer with their brave colors, nor could I hear the dun-headed sparrows, whose cheery notes so often relieve the silence of our highest mountains. The only sounds were the gurgling of small rills down in the veins and crevasses of the glacier, and now and then the rattling report of falling stones, with the echoes they shot out into the crisp air.

I could not distinctly hope to reach the summit from this side, yet I moved on across the glacier as if driven by fate. Contending with myself, the season is too far spent, I said, and even should I be successful, I might be storm-bound on the mountain; and in the cloud-darkness, with the cliffs and crevasses covered with snow, how could I escape? No; I must wait till next summer. I would only approach the mountain now, and inspect it, creep about its flanks, learn what I could of its history, holding myself ready to flee on the approach of the first storm-cloud. But we little know until tried how much of the uncontrollable there is in us, urging across glaciers and torrents, and up dangerous heights, let the judgment forbid as it may.

I succeeded in gaining the foot of the cliff on the eastern extremity of the glacier, and there discovered the mouth of a narrow avalanche gully, through which I began to climb, intending to follow it as far as possible, and at least obtain some fine wild views for my pains. Its general course is oblique to the plane of the mountain-face, and the metamorphic slates of which the mountain is built are cut by cleavage planes in such a way that they weather off in angular blocks, giving rise to irregular steps that greatly facilitate climbing on the sheer places. I thus made my way into a wilderness of crumbling spires and battlements, built together in bewildering combinations, and glazed in many places with a thin coating of ice, which I had to hammer off with stones. The situation was becoming gradually more perilous; but, having passed several dangerous spots, I dared not think of descending; for, so steep was the entire ascent, one would inevitably fall to the glacier in case a single misstep were made. Knowing, therefore, the tried danger beneath, I became all the more anxious concerning the developments to be made above, and began to be conscious of a vague foreboding of what actually befell; not that I was given to fear, but rather because my instincts, usually so positive and true, seemed vitiated in some way, and were leading me astray. At length, after attaining an elevation of about 12,800 feet, I found myself at the foot of a sheer drop in the bed of the avalanche channel I was tracing, which seemed absolutely to bar further progress. It was only about forty-five or fifty feet high, and somewhat roughened by fissures and projections; but these seemed so slight and insecure, as footholds, that I tried hard to avoid the precipice altogether, by scaling the wall of the channel on either side. But, though less steep, the walls were smoother than the obstructing rock, and repeated efforts only showed that I must either go right ahead or turn back. The tried dangers beneath seemed even greater than that of the cliff in front; therefore, after scanning its face again and again, I began to scale it, picking my holds with intense caution. After gaining a point about halfway to the top, I was suddenly brought to a dead stop, with arms outspread, clinging close to the face of the rock, unable to move hand or foot either up or down. My doom appeared fixed. I _must_ fall. There would be a moment of bewilderment, and then a lifeless rumble down the one general precipice to the glacier below.

After gaining a point about halfway to the top, I was suddenly brought to a dead stop, with arms outspread, clinging close to the face of the rock, unable to move hand or foot either up or down. My doom appeared fixed. I _must_ fall. There would be a moment of bewilderment, and then a lifeless rumble down the one general precipice to the glacier below.

When this final danger flashed upon me, I became nerve-shaken for the first time since setting foot on the mountains, and my mind seemed to fill with a stifling smoke. But this terrible eclipse lasted only a moment, when life blazed forth again with preternatural clearness. I seemed suddenly to become possessed of a new sense. The other self, bygone experiences, Instinct, or Guardian Angel,—call it what you will,—came forward and assumed control. Then my trembling muscles became firm again, every rift and flaw in the rock was seen as through a microscope, and my limbs moved with a positiveness and precision with which I seemed to have nothing at all to do. Had I been borne aloft upon wings, my deliverance could not have been more complete.

I seemed suddenly to become possessed of a new sense.

Above this memorable spot, the face of the mountain is still more savagely hacked and torn. It is a maze of yawning chasms and gullies, in the angles of which rise beetling crags and piles of detached boulders that seem to have been gotten ready to be launched below. But the strange influx of strength I had received seemed inexhaustible. I found a way without effort, and soon stood upon the topmost crag in the blessed light.

How truly glorious the landscape circled around this noble summit!—giant mountains, valleys innumerable, glaciers and meadows, rivers and lakes, with the wide blue sky bent tenderly over them all. But in my first hour of freedom from that terrible shadow, the sunlight in which I was laving seemed all in all.

Looking southward along the axis of the range, the eye is first caught by a row of exceedingly sharp and slender spires, which rise openly to a height of about a thousand feet, above a series of short, residual glaciers that lean back against their bases; their fantastic sculpture and the unrelieved sharpness with which they spring out of the ice rendering them peculiarly wild and striking. These are "The Minarets." Beyond them you behold a sublime wilderness of mountains, their snowy summits towering together in crowded abundance, peak beyond peak, swelling higher, higher as they sweep on southward, until the culminating point of the range is reached on Mount Whitney, near the head of the Kern River, at an elevation of nearly 14,700 feet above the level of the sea.

Westward, the general flank of the range is seen flowing sublimely away from the sharp summits, in smooth undulations; a sea of huge gray granite waves dotted with lakes and meadows, and fluted with stupendous cañons that grow steadily deeper as they recede in the distance. Below this gray region lies the dark forest zone, broken here and there by upswelling ridges and domes; and yet beyond lies a yellow, hazy belt, marking the broad plain of the San Joaquin, bounded on its farther side by the blue mountains of the coast.

Turning now to the northward, there in the immediate foreground is the glorious Sierra Crown, with Cathedral Peak, a temple of marvelous architecture, a few degrees to the left of it; the gray, massive form of Mammoth Mountain to the right; while Mounts Ord, Gibbs, Dana, Conness, Tower Peak, Castle Peak, Silver Mountain, and a host of noble companions, as yet nameless, make a sublime show along the axis of the range.

Eastward, the whole region seems a land of desolation covered with beautiful light. The torrid volcanic basin of Mono, with its one bare lake fourteen miles long; Owen's Valley and the broad lava table-land at its head, dotted with craters, and the massive Inyo Range, rivaling even the Sierra in height; these are spread, map-like, beneath you, with countless ranges beyond, passing and overlapping one another and fading on the glowing horizon.

At a distance of less than 3000 feet below the summit of Mount Ritter you may find tributaries of the San Joaquin and Owen's rivers, bursting forth from the ice and snow of the glaciers that load its flanks; while a little to the north of here are found the highest affluents of the Tuolumne and Merced. Thus, the fountains of four of the principal rivers of California are within a radius of four or five miles.

Lakes are seen gleaming in all sorts of places,—round, or oval, or square, like very mirrors; others narrow and sinuous, drawn close around the peaks like silver zones, the highest reflecting only rocks, snow, and the sky. But neither these nor the glaciers, nor the bits of brown meadow and moorland that occur here and there, are large enough to make any marked impression upon the mighty wilderness of mountains. The eye, rejoicing in its freedom, roves about the vast expanse, yet returns again and again to the fountain peaks. Perhaps some one of the multitude excites special attention, some gigantic castle with turret and battlement, or some Gothic cathedral more abundantly spired than Milan's. But, generally, when looking for the first time from an all-embracing standpoint like this, the inexperienced observer is oppressed by the incomprehensible grandeur, variety, and abundance of the mountains rising shoulder to shoulder beyond the reach of vision; and it is only after they have been studied one by one, long and lovingly, that their far-reaching harmonies become manifest. Then, penetrate the wilderness where you may, the main telling features, to which all the surrounding topography is subordinate, are quickly perceived, and the most complicated clusters of peaks stand revealed harmoniously correlated and fashioned like works of art—eloquent monuments of the ancient ice-rivers that brought them into relief from the general mass of the range. The cañons, too, some of them a mile deep, mazing wildly through the mighty host of mountains, however lawless and ungovernable at first sight they appear, are at length recognized as the necessary effects of causes which followed each other in harmonious sequence—Nature's poems carved on tables of stone—the simplest and most emphatic of her glacial compositions.

Could we have been here to observe during the glacial period, we should have overlooked a wrinkled ocean of ice as continuous as that now covering the landscapes of Greenland; filling every valley and cañon with only the tops of the fountain peaks rising darkly above the rock-encumbered ice-waves like islets in a stormy sea—those islets the only hints of the glorious landscapes now smiling in the sun. Standing here in the deep, brooding silence all the wilderness seems motionless, as if the work of creation were done. But in the midst of this outer steadfastness we know there is incessant motion and change. Ever and anon, avalanches are falling from yonder peaks. These cliff-bound glaciers, seemingly wedged and immovable, are flowing like water and grinding the rocks beneath them. The lakes are lapping their granite shores and wearing them away, and every one of these rills and young rivers is fretting the air into music, and carrying the mountains to the plains. Here are the roots of all the life of the valleys, and here more simply than elsewhere is the eternal flux of nature manifested. Ice changing to water, lakes to meadows, and mountains to plains. And while we thus contemplate Nature's methods of landscape creation, and, reading the records she has carved on the rocks, reconstruct, however imperfectly, the landscapes of the past, we also learn that as these we now behold have succeeded those of the pre-glacial age, so they in turn are withering and vanishing to be succeeded by others yet unborn.

But in the midst of these fine lessons and landscapes, I had to remember that the sun was wheeling far to the west, while a new way down the mountain had to be discovered to some point on the timber line where I could have a fire; for I had not even burdened myself with a coat. I first scanned the western spurs, hoping some way might appear through which I might reach the northern glacier, and cross its snout; or pass around the lake into which it flows, and thus strike my morning track. This route was soon sufficiently unfolded to show that, if practicable at all, it would require so much time that reaching camp that night would be out of the question. I therefore scrambled back eastward, descending the southern slopes obliquely at the same time. Here the crags seemed less formidable, and the head of a glacier that flows northeast came in sight, which I determined to follow as far as possible, hoping thus to make my way to the foot of the peak on the east side, and thence across the intervening cañons and ridges to camp.

The inclination of the glacier is quite moderate at the head, and, as the sun had softened the _névé_, I made safe and rapid progress, running and sliding, and keeping up a sharp outlook for crevasses. About half a mile from the head, there is an ice-cascade, where the glacier pours over a sharp declivity and is shattered into massive blocks separated by deep, blue fissures. To thread my way through the slippery mazes of this crevassed portion seemed impossible, and I endeavored to avoid it by climbing off to the shoulder of the mountain. But the slopes rapidly steepened and at length fell away in sheer precipices, compelling a return to the ice. Fortunately, the day had been warm enough to loosen the ice-crystals so as to admit of hollows being dug in the rotten portions of the blocks, thus enabling me to pick my way with far less difficulty than I had anticipated. Continuing down over the snout, and along the left lateral moraine, was only a confident saunter, showing that the ascent of the mountain by way of this glacier is easy, provided one is armed with an ax to cut steps here and there.

The lower end of the glacier was beautifully waved and barred by the outcropping edges of the bedded ice-layers which represent the annual snowfalls, and to some extent the irregularities of structure caused by the weathering of the walls of crevasses, and by separate snowfalls which have been followed by rain, hail, thawing and freezing, etc. Small rills were gliding and swirling over the melting surface with a smooth, oily appearance, in channels of pure ice—their quick, compliant movements contrasting most impressively with the rigid, invisible flow of the glacier itself, on whose back they all were riding.

Night drew near before I reached the eastern base of the mountain, and my camp lay many a rugged mile to the north; but ultimate success was assured. It was now only a matter of endurance and ordinary mountain-craft. The sunset was, if possible, yet more beautiful than that of the day before. The Mono landscape seemed to be fairly saturated with warm, purple light. The peaks marshaled along the summit were in shadow, but through every notch and pass streamed vivid sun-fire, soothing and irradiating their rough, black angles, while companies of small, luminous clouds hovered above them like very angels of light.

Darkness came on, but I found my way by the trends of the cañons and the peaks projected against the sky. All excitement died with the light, and then I was weary. But the joyful sound of the waterfall across the lake was heard at last, and soon the stars were seen reflected in the lake itself. Taking my bearings from these, I discovered the little pine thicket in which my nest was, and then I had a rest such as only a tired mountaineer may enjoy. After lying loose and lost for awhile, I made a sunrise fire, went down to the lake, dashed water on my head, and dipped a cupful for tea. The revival brought about by bread and tea was as complete as the exhaustion from excessive enjoyment and toil. Then I crept beneath the pine-tassels to bed. The wind was frosty and the fire burned low, but my sleep was none the less sound, and the evening constellations had swept far to the west before I awoke.

In this essay, Muir describes taking a walk in the wilds of the Sierra Nevada mountains. What key events does he describe? Do you see any similarities between the way Muir describes his walk and the ideas expressed by Emerson and Thoreau in the readings of chapter 7?

After thawing and resting in the morning sunshine, I sauntered home,—that is, back to the Tuolumne camp,—bearing away toward a cluster of peaks that hold the fountain snows of one of the north tributaries of Rush Creek. Here I discovered a group of beautiful glacier lakes, nestled together in a grand amphitheater. Toward evening, I crossed the divide separating the Mono waters from those of the Tuolumne, and entered the glacier basin that now holds the fountain snows of the stream that forms the upper Tuolumne cascades. This stream I traced down through its many dells and gorges, meadows and bogs, reaching the brink of the main Tuolumne at dusk.

A loud whoop for the artists was answered again and again. Their camp-fire came in sight, and half an hour afterward I was with them. They seemed unreasonably glad to see me. I had been absent only three days; nevertheless, though the weather was fine, they had already been weighing chances as to whether I would ever return, and trying to decide whether they should wait longer or begin to seek their way back to the lowlands. Now their curious troubles were over. They packed their precious sketches, and next morning we set out homeward bound, and in two days entered the Yosemite Valley from the north by way of Indian Cañon.

NAME ______________________________

Worksheet 6 Nature in 19th and 20th Century America

1. In his essay, "The Fight For Conservation," written in 1910, Gifford Pinchot argues that Americans must preserve their natural resources. In three or four sentences, summarize his main argument for conservation.

2. In "Hetch Hetchy," John Muir argues against building a damn in the Hetch Hetchy valley in the High Sierras. In three or four sentences, summarize his main argument against building the damn.

3. What similarities and differences do you see in the values and principles appealed to by Gifford and Muir in the two essays.

4. In the reading, "Near View of the High Sierras," John Muir describes an expedition in which he guides some painters into the mountains. After the painters find a spot to paint, Muir continues on. What happens to him near the top of the mountain? What enables him to go on?

5. Contrast the experience of the painters and the experience of Muir on this expedition. Does Muir experience an aspect of wildness that the painters do not? Explain.

Pulling It Together — Nature in 19th and 20th Century America

QUESTIONS FOR THOUGHT AND DISCUSSION

1. What themes do you see that connect the two authors included in this chapter? What contrasts do you observe between them?
2. If you were to make an argument to your state or federal representative in favor of protecting wilderness and/or wildlife, what arguments would you give? What aspects of nature do you think should be protected and why? Which of the writers in this chapter uses arguments most similar to yours?

EXERCISES

1. Both Gifford Pinchot and John Muir were friends of Theodore Roosevelt and influenced his attitude toward nature. Nevertheless, the controversy over the Hetch Hetchy Valley dam was one that put Pinchot and Muir on different sides. Do some research and try to find out what arguments Pinchot gave for damming the valley. Write a two or three page essay in which you explain the controversy, outline the arguments that were given for and against the dam, and state what was finally decided. Then say what you believe should have been done and explain your view.
2. Rachel Carson's book, *Silent Spring*, was highly influential in motivating the beginnings of the modern environmentalist movement. Do some research on Rachel Carson's book and its impact on American policy. Write out your findings on this page.

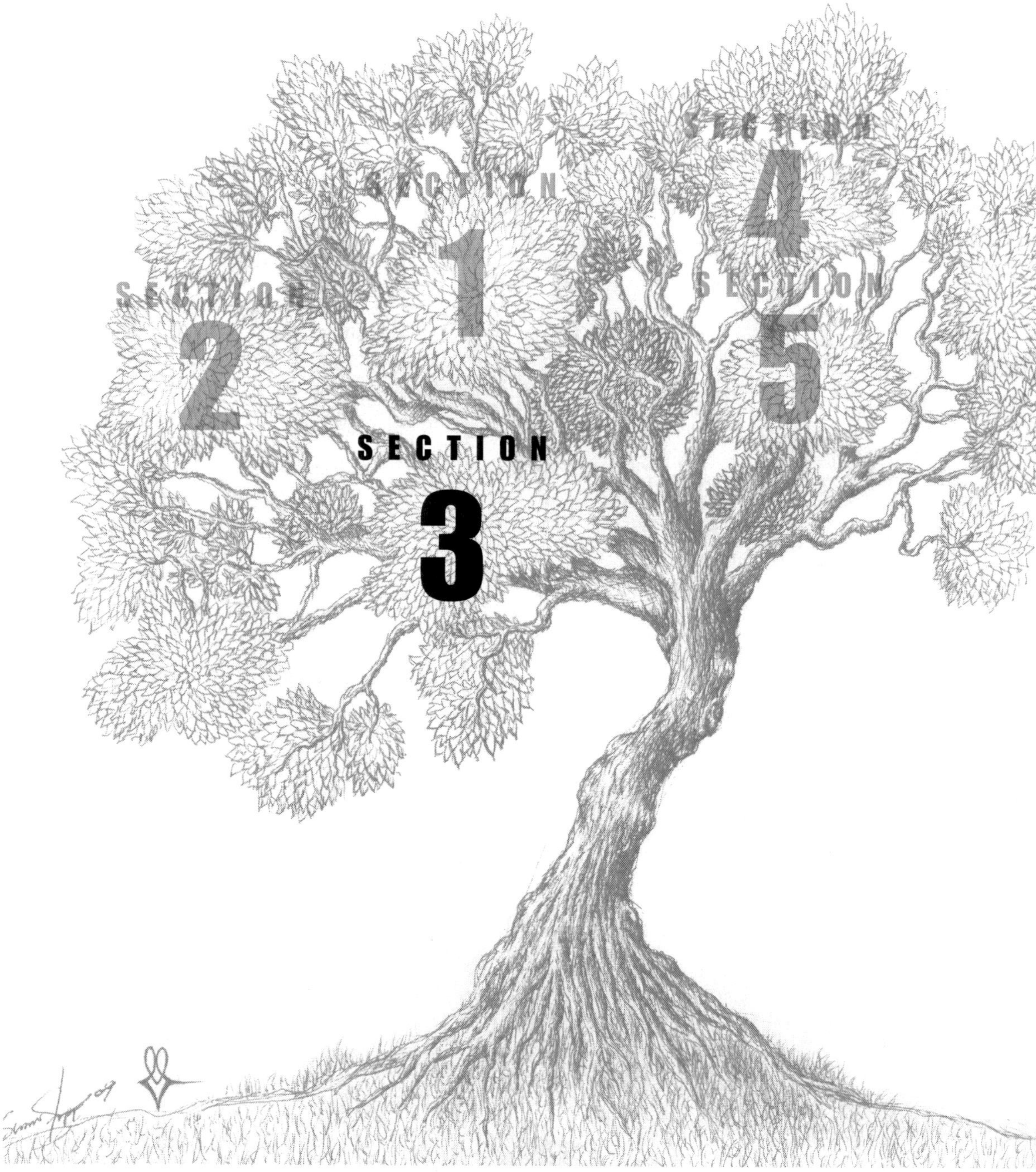

SECTION 3

Attitudes toward Nature in the World's Religions

Chapter

7

Creation Stewardship in the Abrahamic Religions

INTRODUCTION

In Section Two, we have been tracing attitudes and images of nature through the history of Western thought. Our investigation of medieval views of nature introduced the question of the relationship between Christianity and the human interaction with the environment. We now turn to more directly examine various religious perspectives on nature and the environment. Our exploration will include Christianity, but it will also include a number of other major religions of the world. As we explore various religious and cultural traditions, our primary focus will be not simply on the past, but on uncovering elements in each tradition that could shape attitudes and actions today. Around the world, among the leaders of virtually all living religious traditions, there is a deep concern that humankind learn to live in harmony with our nonhuman environment. In the readings presented here, scholars explore how their faith can contribute to this goal.

The religions of Judaism, Christianity, and Islam are closely related. Although they have significant differences, they also have much in common. They share the same basic monotheistic perspective, affirming the existence of one personal God, and only one God. Interestingly, all three see the ancient patriarch Abraham as the father of their tradition and as a paradigm of faith. The oldest of these traditions is Judaism. The roots of Judaism are told in the stories found in the Hebrew Bible, which Christians refer to as the Old Testament. Here we read of the call of Abraham and the formation of the Hebrews, later referred to as the Israelites, as God's chosen people. We read of the deliverance of the people from bondage in Egypt, the conquering of Canaan, and the establishment of a Kingdom under Saul, David, and Solomon. We read how the kingdom that had been united under thee three initial kings, broke up in political disputes and violence, and how the Jews were eventually defeated by enemy powers and sent into exile in foreign lands. The religion of the Jews, which had focused on the temple in Jerusalem, shifted its focus from the Temple to the Torah, God's Law. Synagogues were established, not for offering sacrifices as had been done at the Temple, but for studying God's Law. Although some Jews eventually returned to Jerusalem and rebuilt the temple, the focus of Judaism had shifted in important ways from being a Temple-centered religion to being a "religion of the book." The destruction of the temple in 70 CE by the Romans cemented this shift. Through the years, and in spite of persecution in many parts of the world, Jews have sought to be faithful to the the Law as it was handed down to Moses.

Christianity began as a sect within Judaism, founded by followers of a Jew named Jesus. Christians believe that Jesus was not just a prophet, but the Son of God, God incarnate, and "the Christ" (anointed one). In Jesus, God had invaded human history and established the Kingdom of God, not in the political sense that many Jews had expected, but as a spiritual kingdom to which all humans were invited. Christians continue to proclaim the "good news" that all are invited to enter into this kingdom through faith in Jesus. Through faith their sins will be forgiven, and they will be given a new life, not just after death, but in this world. Christians are to live out this new life in accord with God's kingdom principles. In the words of Jesus, his followers are to "seek first the kingdom of God and his righteousness," and God will provide for their needs. In the second reading in this chapter, Calvin DeWitt develops three principles that he believes are grounded in the Bible. Because his ideas are drawn in part from the Hebrew Bible (the Old Testament), Dewitt's ideas reflect some Jewish theories.

The third major religious tradition explored in this chapter is Islam. Islam was founded by Mohammed (520–632 CE), who Muslims believe was God's final prophet. Muslims accept the major prophets of both Judaism and Christianity as messengers from God, but they believe these traditions have been corrupted by errors that are corrected in the teachings of Mohammed. Islam shares both Judaism's and Christianity's teaching that there is only one God. It rejects, however, the Christian idea that God became a human being in Jesus and the related doctrine of the Trinity. The Islamic Scripture, the Quran, is believed by Muslims to have been revealed directly to Mohammed by God. Mohammed dictated God's word to scribes, who wrote it down to form the Quran. In the third reading in this chapter, Hyder Ihsan Mahasneh explains the foundations for environmental protection that he believes can be found in the Quran and Islamic law. As you read these three selections, look for similarities in the ways these authors, who represent two different religions, think about the human relationship to the natural environment. Although there are important differences in the details, there is significant agreement.

Ten Jewish Teachings on Judaism and the Environment

—Rabbi Lawrence Troster

Rabbi Lawrence Troster is Rabbinic Scholar in Residence for GreenFaith, an interfaith coalition for the environment that was founded in 1992. Greenfaith works with houses of worship, religious schools and people of all faiths to help them become better environmental stewards. GreenFaith's webpage, located at www.greenfaith.org, is a wonderful portal to excellent resources for exploring the relationship between religion and the environment and for information on how religious communities can translate care for the environment into action.

1. *God created the universe.*

This is the most fundamental concept of Judaism. Its implications are that only God has absolute ownership over Creation (Gen. 1-2, Psalm 24:1, I Chron. 29:10-16). Thus Judaism's worldview is theocentric not anthropocentric. The environmental implications are that humans must realize that they do not have unrestricted freedom to misuse Creation, as it does not belong to them. Everything we own, everything we use ultimately belongs to God. Even our own selves belong to God. As a prayer in the High Holiday liturgy proclaims, "The soul is Yours and the body is our handiwork." As we are "sojourners with You, mere transients like our ancestors; our days on earth are like a shadow. . . " (I Chronicles 29:15), we must always consider our use of Creation with a view to the larger good in both time (responsibility to future generations) and space (others on this world). We must also think beyond our own species to that of all Creation.

Everything we own, everything we use ultimately belongs to God. Even our own selves belong to God.

2. *God's Creation is good.*

In Genesis 1: 31 when God found all of Creation, "very good," this means several things. First of all it means that Creation is sufficient, structured and ordered (the rabbis called it *Seder Bereishit*, the Order of Creation). It is also harmonious. It exists to serve God(Psalm 148). This order reflects God's wisdom (Psalm 104:24), which is beyond human understanding (Psalm 92:6-7, Job 38-39). All of God's creations are consequently part of the Order of Creation and all are subject to its nature (Psalm 148). Humans are also part of the Order, which can be said to be a community of worshipers.

In its original sense, *tzelem Elohim*, means that humans were put on the earth to act as God's agents and to actualize God's presence in Creation.

3. *Human beings are created in the image of God.*

Human beings have a special place and role in the Order of Creation. Of all God's creations, only human beings have the power to disrupt Creation. This power, which gives them a kind of control over Creation, comes from special characteristics that no other creature possesses (Psalm 8). This idea is expressed in the concept that humans were created in the image of God (*tzelem Elohim*). In its original sense, tzelem Elohim, means that humans were put on the earth to act as God's agents and to actualize God's presence in Creation.

Human beings are supposed to exercise their power, consciousness and free will to be wise stewards of Creation.

Reprinted by permission of Rabbi Lawrence Troster. www.greenfaith.org

This also has ethical implications which stem from the fact that human beings have certain intrinsic dignities: infinite value, equality and uniqueness. It also means that human beings possess God-like capacities: power, consciousness, relationship, will, freedom and life. Human beings are supposed to exercise their power, consciousness and free will to be wise stewards of Creation. They should help to maintain the Order of Creation even while they are allowed to use it for their own benefit within certain limits established by God (Genesis 2:14). This balance applies to both human society as well to the natural world. Since the time of the expulsion from Garden of Eden, Creation has tended to be out of balance because of the human impulse towards inequality resulting from the misuse of its powers for selfish ends. The earth is morally sensitive to human misdeeds (Genesis 4, Leviticus 18:27-30).

4. *Humanity should view their place in Creation with love and awe.*

It may be said that there are two books of God's revelation to humanity: The Torah and Creation itself. The book of Creation can help us to perceive ourselves as "living breathing beings connected to the rhythms of the earth, the biogeochemical cycles, the grand and complex diversity of ecological systems." (Mitchell Thomashow, *Ecological Identity*) This knowledge is gained both through an understanding of Creation through scientific knowledge. In Judaism, this can be understood as the fulfillment of the commandments to love and to fear God (Deuteronomy 6:5,13). Rambam (Moses Maimonides, 1135-1204) interpreted these commandments in the following way:

It may be said that there are two books of God's revelation to humanity: The Torah and Creation itself.

> *"When a person observes God's works and God's great and marvelous creatures, and they see from them God wisdom that is without estimate or end, immediately they will love God, praise God and long with a great desire to know God's Great Name. . . And when a person thinks about these things they drawback and are afraid and realizes that they are small, lowly and obscure, endowed with slight and slender intelligence, standing in the presence of God who is perfect in knowledge." (Mishneh Torah, SepherMadah, Hikhot Yesodei Ha-Torah 2:1-2)*

Thus, when we study Creation with all the tools of modern science, we are filled with love and a sense of connection to a greater order of things. We feel a sense of wonder but also a sense of awe and humility as we perceive how small we are in the universe as well as within the history of evolution. Love and humility should then invoke in us a sense of reverence for Creation and modesty in our desire to use it. We should, according to Abraham Joshua Heschel see the world as God-centered, not human-centered. By putting God at the center of life, we see the sacred in everything and the natural world becomes a source of wonder and not only a resource for our use and abuse.

5. *The Sabbath and prayer help us to achieve this state of mind.*

The Sabbath is a way to begin to engender this sense of love and humility before Creation. It is also is a way to living a sustainable life. For one day out of seven, we limit our use of resources. We walk to attend synagogue and drive only when walking is not possible. We do not cook and we do not shop. We can use the day for relaxation, contemplation and to ask ourselves: what is the real purpose of human life? Are we here on earth only to get and to spend? As Rabbi Schorsch has written: "To rest is to acknowledge our limitations. Willful inactivity is a statement of subservience to a power greater than our own." (*To Till and to Tend*, page 20 Prayer also helps us to recognize that everything we are, everything we have and everything we use ultimately comes from God (*Babylonian Talmuid,* Brakhot 35a). When we say a blessing, we create a moment or holiness, a sacred pause.

The Sabbath is a way to begin to engender this sense of love and humility before Creation. It is also is a way to living a sustainable life.

Prayer also creates an awareness of the sacred by taking us out of ourselves and our artificial environments and allowing us to truly encounter natural phenomenon. Prayer creates a loss of control which allows us to " see the world in the mirror of the holy." (Heschel) We are then able to see the world as an object of divine concern and we can then place ourselves beyond self and more deeply within Creation.

6. *The Torah prohibits the wasteful consumption of anything.*

In Judaism, the halakhah (Jewish law) prohibits wasteful consumption. When we waste resources we are violating the mitzvah (commandment) of *Bal Tashhit* ("Do not destroy"). It is based on Deuteronomy 20:19-20:

> "When in your war against a city you have to besiege it a long time in order to capture it, you must not destroy its trees, wielding the ax against them. You may eat of them, but you must not cut them down. Are trees of the field human to withdraw before you into the besieged city? Only trees that you know do no yield food may be destroyed; you may cut them down for constructing siegeworks against the city that is waging war on you, until it has been reduced."

This law was expanded in later Jewish legal sources to include the prohibition of the wanton destruction of household goods, clothes, buildings, springs, food or the wasteful consumption of anything (see Rambam, *Mishneh Torah, Laws of Kings and Wars* 6:8, 10; Samson Raphael Hirsch, *Horeb*, 279-80). The underlying idea of this law is the recognition that everything we own belongs to God. When we consume in a wasteful manner, we damage Creation and violate our mandate to use Creation only for our legitimate benefit. Modesty in consumption is a value that Jews have held for centuries. For example one is not supposed to be excessive in eating and drinking or in the kind of clothes that one wears (Rambam, *Mishneh Torah, Laws of Discernment,* chapter 5). Jews are obligated to consider carefully our real needs whenever we purchase anything. We are obligated when we have a *simchah* (a celebration) to consider whether we need to have elaborate meals and wasteful decorations. We are obligated to consider our energy use and the sources from which it comes.

Modesty in consumption is a value that Jews have held for centuries.

7. *The Torah gives an obligation to save human life.*

The Jewish tradition mandates an obligation to save and preserve life (called in Jewish legal sources: *pikuach nefesh*) based on an interpretation of Leviticus 18:5, "You shall keep My laws and My rules, by the pursuit of which man shall live: I am the Lord (See *Babylonian Talmud* Sanhedrin 74a)." Jewish law forbids us from knowingly harming ourselves (Leviticus 19:28). There are also numerous sources mandated the proper disposal of waste is properly and that noxious products from industrial production must be kept far from human habitation (see for example, Deuteronomy 23:13-15, *Mishnah* Baba Batra 2:9) In the Jewish tradition, the public good overrides individual desires. While there are many useful and even lifesaving technologies that come from modern chemicals and materials, we have an obligation to be cautious in their use. *Pikuach nefesh* demands that we consider the impact of our use of chemicals and other materials, not only in the short term but also in the long term. For the Jewish tradition, the Precautionary Principle can be seen as a modern form of the warning not to tamper too much with the boundaries of Creation.

8. *The Torah prohibits the extinction of species and causing undo pain to non-human creatures.*

Our ancestors could not have anticipated the loss of biodiversity that the modern world has produced; from their perspective, there was no natural extinction rate of species. God, they believed, had created all species at one time and there could be no new creatures. Only humans could cause extinction and bring about the loss of one of the members of the Creation choir. In the Torah there is a law that says:

> "If along the road, you chance upon a bird's nest, in any tree or on the ground, with fledglings or eggs and the mother sitting over the fledglings or on the eggs, do not take the mother with her young. Let the mother go, and take only the young, in order that you may fare well and have along life." (Deuteronomy 22:6-7)

Ramban (Moses ben Nachman, Nachmanides, 1194-1270) in his commentary to the Torah wrote:

> "This also is an explanatory commandment of the prohibition *you shall not kill it [the mother] and its young both in one day* (Leviticus 22:28). The reason for both [commandments] is that we should not have a cruel heart and not be compassionate, or it may be that Scripture does not permit us to destroy a species altogether, although it permits slaughter [for food] within that group. Now the person who kills the mother and the young in one day or takes them when they are free to fly,[it is regarded] as though they have destroyed that species."

It is evident from the first chapter of Genesis and other Biblical texts (Psalm 104,148, and Job 38-41) that God takes care of, and takes pleasure in, the variety of life that makes up Creation. And although we might regard a species as unimportant or bothersome to human beings, God does not regard them so. The rabbis understood that we do not know God's purpose for every creature and that we should not regard any of them as superfluous. "Our Rabbis said: Even those things that you may regard as completely superfluous to Creation – such as fleas, gnats and flies—even they were included in Creation; and God's purpose is carried through everything—even through a snake, a scorpion, a gnat, a frog." (Breishit Rabbah 10:7) In environmental terms, every species has an inherent value beyond its instrumental or useful value to human beings. Related to this idea is the concept of *Tzar Baalei Chayyim*, the prohibition of hurting animals without good purpose (based on Deut. 22:6, 22:10, 25:4, Numbers 22:32, Exodus 20:8-10, Lev. 22:27-8). These concepts bring to our relationships with the non-human world limits and controls over our power and greed.

9. *Environmental Justice is a Jewish value.*

The Torah has numerous laws which attempt to redress the power and economic imbalances in human society and Creation. Examples are the Sabbatical year (Exodus 23:11, Leviticus 25:2-5, Deuteronomy 15:1-4) and the Jubilee (Leviticus 25:8-24) There is a whole program in the Torah for creating a balanced distribution of resources across society (Exodus 22:24-26, Leviticus 25:36-37, Deuteronomy 23:20-1, 24:6,10-13,17).This is an expression of the concept of *Tzedek*, which means righteousness, justice and equity. It is the value, which tries to correct the imbalances, which humans create in society and in the natural world. In the modern world globalization has strived to achieve the free movement of people, information, money, goods and services but it can also5create major disruptions in local cultures and environments. While globalization has created great wealth for millions of people, many millions more have been bypassed by its benefits and has had in some cases a negative impact upon the environment and human rights. The Jewish concept of Tzedek demands that we create a worldwide economy that is sustainable and that is equitable in the distribution of wealth and resources.

The Jewish concept of Tzedek demands that we create a worldwide economy that is sustainable and that is equitable in the distribution of wealth and resources.

10. *Tikkun Olam: The perfection/fixing of the world is in our hands.*

There is a midrash (Rabbinic commentary on the Bible) which Jewish environmentalists are fond of quoting:

> "When God created the first human beings, God led them around the garden of Eden and said: "Look at my works! See how beautiful they are—how excellent! For your sake I created them all. See to it that you do not spoil and destroy My world; for if you do, there will be no one else to repair it." (*Midrash Kohelet Rabbah*, 1 on Ecclesiastes 7:13)

In the Jewish liturgy there is a prayer called *Aleinu* in which we ask that the world be soon perfected under the sovereignty of God (*le-takein 'olam be-malkhut Shaddai*).*Tikkun 'olam*, the perfecting or the repairing of the world, has become a major theme in modern Jewish social justice theology. It is usually expressed as an activity, which must be done by humans in partnership with God. It is an important concept in light of the task ahead in environmentalism. In our ignorance and our greed, we have damaged the world and silenced many of the voices of the choir of Creation. Now we must fix it. There is no one else to repair it but us.

Take Good Care, It's God's Earth

—Calvin B. DeWitt

Calvin DeWitt is a Professor of Environmental Studies at the University of Wisconsin–Madison. DeWitt is an active Christian environmentalist. He has been a long-time activist and educator on environmental issues, especially among evangelical Christians. He and others founded the Evangelical Environmental Network. He has served as the Director of the Au Sable Institute for Environmental Studies in Michigan. He is the author of numerous books and articles on Christian environmentalism. His ethical belief that we should care for the creation God has given us is informed by his scientific understanding. Putting his ethics and his scientific knowledge together, he has also been an activist, helping shape his small village in Wisconsin into a model of environmentally sustainable small-town living.

For me it was a startling experience. On my way north to the remote farming community of Neerlandia on the 54th parallel in Alberta, Canada, a timber wolf crossed in front of my car. Several minutes later, after coming to the home of a third-generation hog farmer, I was walking a wooded path to his hog barn. "Do you have timber wolves around here?" I asked. His answer to my cautious question startled me about as much as the wolf. Wheeling, with face aglow, he replied excitedly, "Yes!" He then explained that his privilege and responsibility before God was to take care of all God's creatures on his land, including the wolves!

As we conversed in his modern low-slung hog barn I found he used no antibiotics in the hog food, giving medication only to sick animals in order to protect the consumers of pork; he let the runts live as long as the others "because they are God's creatures": and he rested his sows on beds of sweet hay "so they would live better lives." As for protecting his hogs from wolves, he explained it was his duty to keep the animals in the fold—by which he meant the neat outdoor pens with chain-linked fencing around and above. Yet, slung above a door casing was a rifle, readily accessible, "just in case."

This farmer leaves his barley fields fallow every two years since "the Bible requires a Sabbath for the land." He does so every two years rather than seven because "That's what my land needs." He raises only half the hogs he could to allow more time with is family, his books and his Bible, and to provide opportunity for another farmer to raise hogs too. After tea with his family he drove me past a 40-acre wetland he keeps for moose and dropped me at his neighbors' dairy farm where I found a family engaged in a similar stewardship of the land, and, of coyotes! Both families keep their land and its animals as creatures under their care; both are living out biblical teachings for keeping creation.

Probably none of us lives in a setting like Neerlandia. Yet each of us, whether in the inner city, countryside, suburbs or wilderness, has the opportunity to think and act in godly ways toward God's creation. I think the best and most important way to become friends of God's creation is to steep ourselves in the Scriptures. There we will discover the rich and compelling call to care for the earth. There is no escaping it, really, and the more we allow the lordship of Christ to operate in our minds as we read the Bible, the more we will find ourselves being converted to faithful stewardship.

It is true that failure to live by God's Word results in several degradations of creation. It is such sinful failure that brings us today's widespread species extinctions, pollution of land, air and water, global toxification of food, deforestation, and ultimately, human and cultural degradation. We are all familiar with some of the specifics: Iowa has lost one third of its topsoil and will become a pebble field in less than one hundred years; one third of all current species will be extinct 50 years from now at the current rate of rainforest destruction (for every 10 acres of forest cut, one is replaced); more than 40% of the world's people do not have adequate water supply; the rich consume at least seven times more energy than the poor; one in ten persons is severely malnourished, on in five (more than a billion) people live in absolute poverty. We cannot separate our destructive patterns toward the earth from the destruction experienced by people. All of creation is tied together in an intricate web of dependency, and when we disobey God's call to care for creation, we also disobey God's commandment to care for our brothers and sisters.

All of creation is tied together in an intricate web of dependency, and when we disobey God's call to care for creation, we also disobey God's commandment to care for our brothers and sisters.

THREE PRINCIPLES FOR CREATION STEWARDSHIP

The Bible is our rule for faith and practice, and clearly in our times our faith and practice must apply to God's creation, including us, other creatures, and the environments and habitats we share. Creation of course is much more inclusive than what we call "the environment," since it includes all creatures – all the works of God, including ourselves. Thus being an "environmentalist" is not quite what a steward under God aspires to be: it is a bit too confining. While incorporating "the environment" and perhaps joining with "environmentalists" in many ways, the steward under God deals with the whole creation: land and children, forests and stars, fruitful trees and aged people. The creation includes ourselves; we are not independent of it. While made in the image of God, and imaging God in our care for creation, we too are creatures – part and parcel of creation. While being God's creatures, and dependent upon earth's bounty, we also are stewards made in God's image. We, through our lives and actions, image God's sustaining and reconciling love for the world. And wonderfully, the Bible gives us compelling direction in the exercise of this love.

I believe if we are careful in our commitment to understand the Scripture's call to creation, we will implant three biblical principles for environmental stewardship into our hearts. We will learn that all Christians **must** keep creation, **must** give creation its Sabbaths, and **must** preserve creation's fruitfulness.

THE EARTHKEEPING PRINCIPLE

We must keep creation. As God keeps believing people, so should God's people keep Creation.

At the conclusion of many church services we conclude with the blessing of Aaron: "May the Lord bless you and keep you. . . " (Num 6:24). The word "keep" is the Hebrew word "shamar," meaning a loving, caring keeping. When we invoke this blessing we are asking God to keep us with all our relationships intact; the processes and structures of our own bodies, the interconnections with our family and loved ones, the relationships with our community and our connections with soil, water, and air, and our relationships to the sustaining processes of all creation. God's expectation of Adam and Eve was that they apply this kind of keeping to the garden (Gen 2:15). In addition to serving and dressing (*'abad*) the garden the were also to keep (*shamar*) the garden with all the dynamic relational fullness that this word means. Such keeping is not preservation as applied to the pickles in a jar; it is the kind of keeping we ask God to give us. When in accord with Genesis 2:15, we *keep* the creation, we make sure that the creatures under our care are maintained with all their proper connections—connection with members of the same species, with the many other species with which the interact, with the soil, air and water upon which they depend. As followers of the final Adam, Jesus Christ, through whom all things in creation hold together, we image God's love by being earthkeepers.

What does it mean to "keep" God's Creation?

we image God's love by being earthkeepers.

THE SABBATH PRINCIPLE

We must give creation its Sabbath rests. As human beings and animals are to be given their times of Sabbath rest (EX 20; Deut 5) so also the rest of creation (Ex 23; Lev 25-26).

We all know God's law for the Sabbath of the week during which we are expected to give ourselves and the people and animals under our care their Sabbath rests (Ex 20 and Deut 5). We know from the teachings of Jesus how the Sabbath is made to benefit those who experience it rather than simply to fulfill a legal requirement (Mark 2:27). Lesser known to most of us is another Sabbath law. God's Sabbath law for the land given in Exodus 23 and Leviticus 25 and 26. Gods' commandment is that people, land, and the creatures of creation must not be relentlessly pressed.

As we should observe the Sabbath of the week to help us get "off the treadmill" of continuous work to help us "get things together" again, so too must we give the land its Sabbath rests. In keeping with the teaching of Jesus Christ that the Sabbath is mad for the ones served by it – not the other way around—this mean giving creation necessary rests, intentional nurture, and active restoration. It may mean resting agricultural land every two years, rather than every seven (as in Neerlandia); it may mean leaving particularly sensitive land – such as steep lands that become fluid when wet (as in Malibu)—in continuous sabbath. The land and the creatures must be protected from relentless exploitation, and must be given what is needed for rejuvenation. The Sabbath for the land goes beyond agricultures to include all of Creation – including our use of water and air, discharges in to them, our exhausts, smoke, sewage, and throwing things "away."

How does Sabbath keeping reflect our treatment of the environment?

The Sabbath law for the land is reinforced with a severe warning. Failure to give land and creatures their needed rest will result in people no longer being supported by land and creatures. States Leviticus, ". . . if you will not listen to me and carry out all these commends. . . (Lev 26:14) Your land will be laid waste. . ." (Lev 26:33). And it goes on to say that "then the land will enjoy its Sabbath years all the time that it lies desolate. . . then the land will rest and enjoy its Sabbaths. . . the rest it did not have when you lived in it" (Lev 26:34-35). Followers of Jesus Christ, imagers of God's love, are Sabbath Observers.

THE FRUITFULNESS PRINCIPLE

We must preserve creation's fruitfulness. We may enjoy the fruit of God's Creation but must not destroy its fruitfulness (Ezek 34:18).

We all know the difference between fruit and the ability of things to bring forth fruit, like flowers and fish and people. The Scriptures allow us to make careful use of the fruit of the creation, but do not permit destruction of creation's ablity to be fruitful. We know the biblical blessing, "Be fruitful and multiply and replenish the earth." But most of us have not stopped to think carefully about to whom this blessing was given. True enough, it is given to people (Gen 1:28), but it is first given to other creatures. In this first chapter of the Bible–perhaps the most read chapter of the Bible—we often overlook or forget that the blessing of fruitfulness is first given to the fish of the sea and birds of the air in Genesis 1:22. Thus, it is God's will that the **whole of creation** be fruitful, not just people. And thus human fruitfulness man *not* be at the expense of God's will if our fruitfulness denies God's blessing to the other creatures.

What is the "fruitfulness" principle?

Noah, of course, understood fruitfulness very well for he was the first explicitly charged to preserve the fruitfulness of the creatures. Noah obediently brought a sufficient number of each kind of animal on the ark in order to preserve their lines—their genetic continuity. People of Noah's day, who themselves were destructive of people and the earth, were themselves destroyed by the flood—an early reflection of Revelation 11:18 where we learn that in the echelon God again will destroy those who destroy the earth.

As God's fruitful work brings fruit to Creation, giving to land and life what satisfies, so too should ours. As God provides for the creatures, so should we people who were created to reflect God and whose image we bear. Imaging God, we should provide for the creatures, not mess up creation, or degrade its ability to function ad God intends. "Is it not enough for you to feed on the good pasture? Must you also trample the rest of your pasture with your feet? Is it not enough for you to drink clear water? Must you also muddy the rest with your feet?" (Ezek 34:18). We must preserve the integrity of water, land and air; we must preserve the integrity of the biosphere. And with Noah, we should preserve the fruitfulness of earth's creatures, especially those threatened with extinction, species whose interactions with each other and with land and water, form the fabric of the biosphere. Followers of Jesus Christ, imagers of God's love, are Preservers of Creation's Fruitfulness.

KNOWLEDGE OF CREATION

Knowledge of God's requirements for Creation stewardship is not enough. In order to apply this knowledge effectively and productively we must also understand how God's creation works. Creation is a marvelous teacher, and we must be open to its teaching. First, creation tells of God's divinity and everlasting power, leaving us without excuse (Rom 1:20; Ps19), but it also tells us about its operations. We learn from it the nature dynamics, and functions of river flood plains, volcanoes and slippery slopes, earthquake fault zones, swamps and bogs; we learn the habitat requirements for creatures and of suitable and unsuitable lands for human habitation. We also learn from creation what troubles its operations and functions, what brings about the degradations of its blessed fruitfulness.

In developing such knowledge, we must also exercise discernment. Unfortunately, the call to care for Creation has been muffled by voices motivated by agendas contrary to the Scripture. Strangely enough, too often it is a *Christian* leader calling us to abandon creation! We must learn and remember the difference between the primary, **refereed literature** which is subjected to the kind of discipline that raises the probability of its truth, the **gray literature** which may be authoritative but may be influenced by the policies of the agencies that produce it, and the **popular literature** whose primary goal may be to maximize sales by overstatement, understatement, denial, and polarization. We also must seek to develop our abilities and find reliable sources among us to discern hearsay from disciplined reports, talk-show chats from refereed scientific papers; we must distinguish between broad-brush generalizations and eyewitness reports of actual situations.

Couldn't we be content with much less?

There is much to be gained today (for some) by denying the environmental problems that are everywhere evident, but there is much more to be lost by such denial. The steward of today must be a prophet, and the prophet of today need only to honestly describe the present. We must forthrightly report what we have done and are doing in and to creation; we must not be devious.

PUTTING KNOWLEDGE AND BELIEF INTO PRACTICE

Knowing God's requirements of us, knowing how the world works, and knowing creation's degradations, we must proceed to put our knowledge and belief into practice. If we do not do so, we fail and God's stewards of creation. Jesus asks, "Why do you call me Lord, Lord, and do on do what I say?" Ezekiel points to those who hear the message of the Lord but do not put it into practice. . . they are nothing more than those who sing love songs with a beautiful voice and play an instrument well (Ezek 33:31-32). And James also deals with the need to act richly on our knowledge and beliefs (James 1:22-25).

What of course the Scriptures are telling us is that we should not confine our publication of what we believe to song and writings, but must be certain that its major publication is in our lives and in the landscape. Our churches and places of vocation should clearly publish themselves by appearance, reflecting what an old Plain Farmer poem states in its title, "I'd Rather See a Sermon." We are not alone in this publication, of course, for the heavens declare the glory of God with words that go throughout the earth; all nature bears convincing
itness to God's divinity and eternal power the tress clap their hands. Our lives must not diminish nor mute
erful testimony. Instead, we must live in integrity, dissolving the smoggy veil that shrouds
incing witness reconciling what has been degraded, and breaking forth with light and truth to
and all creatures more fully and more beautifully praise God from whom all blessings flow.

An Islamic Faith Statement on Conservation

Hyder Ihsan Mahasneh is a biologist and Islamic scholar and was the first African head of the Kenya National Parks Service. He was appointed by the Muslim World League to compile this paper.

HUMANS AND THE ENVIRONMENT

Humanity's most primordial concepts of religion relate to the environment. Human history on planet Earth is, on a geological scale, very short indeed. Planet Earth itself is a mere 3,800 million years old; human beings only appeared one million or maybe two million years ago.

Most of the physical patterns of planet Earth were probably in place, broadly speaking, by the time humans evolved. Apart from what they first saw, they also probably witnessed some spectacular changes themselves. They must, at the very least, have gone through one Ice Age and seen some graphic volcanic eruptions—assuming they were able to avoid the consequences. The environment, therefore, very probably induced the first thoughts of a Super-Being—a God, if you like—whose manifestations lay in human beings' immediate surroundings.

The environment also provided another dimension in humanity's relationship with nature. To survive in a given environment, humans have to adjust what they take from that environment to what can give them sustainable yields on (at the very least) an annual basis. In effect this meant that early humans had to learn to conserve at an early age. Being largely dependent on what was available rather than on what they could cultivate, they entered into a partnership with the environment.

To take more than the regenerative capacity of the environment could lead to serious subsequent exhaustion—quite rightly seen as harsh retribution from an angry God. The converse situation of exploitation with moderation led to sustained yields, which were (again, quite rightly) taken as having pleased God.

THE INDUSTRIAL REVOLUTION

This relationship between conservation and religion is thus not only a natural one but also probably as old as the proverbial hills. But when we quickly open most of the pages of human history on planet Earth and come to the past 300 years or so, we find the advent of the Industrial Revolution. It made possible the production of large quantities of goods in a very short time. That meant that raw materials in ever-increasing quantities had to be found to feed the hungry mills ready to convert them into finished or semi-finished goods.

> **The paradox of "progress" today is the easily perceived correlation between complex consumer societies and the degeneration of the human being.**

The consequences were many—economic, social, and environmental. The material achievements of the human race in the past 100 or so years have overshadowed the contributions made by all past civilizations.

The Industrial Revolution that took place in Europe in the eighteenth and nineteenth centuries exacted a high social and environmental cost. Now these

---

costs are even higher and more universal, being manifestly so in the great urban centers of the world. The paradox of "progress" today is the easily perceived correlation between complex consumer societies and the degeneration of the human being. Or as John Seymour puts it:

We see men now wherever we look, so blinded by arrogance and the worship of man as God that they are doing things no one but the insane would do. . . men maddened by the belief that they are both omniscient and omnipotent, that they are indeed God.

THE PURSUIT OF MONEY

The Industrial Revolution also proclaimed a new revival of another God: Mammon. Mammon regrettably has no respect for environmental integrity—nor do his followers. The last 250 years have seen a growing decimation of ever more pristine areas of nature to feed the insatiable industrial cuckoo and its resultant consumerism. Forests—particularly tropical forests—have been systematically hewn down, the seas ransacked, the lands made totally dependent on a host of inorganic fertilizers and pesticides for food production. Wastes galore have filled the seas, the rivers, and the lakes, not to mention the landfills.

Mammon regrettably has no respect for environmental integrity—nor do his followers.

We must also take note that the "unmatched material progress" of this century that we are often fond of talking about has only been possible for the few: that is, the population of the northern hemisphere and a small minority among the peoples of the South. This is usually translated as less than 25 percent of the world's population consuming over 75 percent of the world's resources.

This rate of consumption by a minority of the human species has caused unparalleled climatic change, ecosystem disintegration, and species extinction. As a report by the World Wide Fund for Nature observes:

Loss of biodiversity worldwide, and the combination of global warming with other human pressures will present the greatest challenge in conservation for decades to come.

This would lead us to conclude that there is a profound and inherent contradiction in the efforts made by the "North" to keep ahead of the rest as consumers, and the push by the remaining 75 percent of the world's population to catch up. Given this scenario, if just Eastern Europe or Russia or India or China managed to raise its standard of living by just a few percentage points, then the consequences of putting this extra load on the earth's ecosystem, which is already under severe strain, would be catastrophic.

This is the background against which the followers of the relatively ancient, environmentally conscious (indeed environmentally concerned) God have gathered to reexamine and to restate their own commitment to environmental integrity from their own individual religion's standpoint. We for our part will look at the underpinnings of conservation in Islam.

ISLAM AND CONSERVATION

There are several Islamic principles that, when taken individually, seem to have little bearing on conservation. Together, however, they add up to a clear concept of the Islamic view on conservation.

TAWHEED

The first Islamic principle that relates to conservation is that of the Oneness of Allah, or Tawheed. This principle is absolutely fundamental to Islam. Every Muslim must believe in this Oneness of Allah. It is said by some Ulamaa that some two-thirds of Prophet Muhammad's (SAW) early preaching—and indeed of the Qur'an itself—were and are dedicated purely to endorsing this very Oneness of Allah. One indivisible God means to a Muslim that there is no separate deity for each of the many attributes that to Muslims belong to the One Universal God who is also God of the Universe.

Muslims belong to the One Universal God who is also God of the Universe.

Tawheed is the monotheistic principle of Islam and it begins by declaring that "there is no God but God" (the second half of this declaration asserts that "Muhammad is His Messenger"). We are for the present concerned with the first part, which affirms that there is nothing other than the Absolute, the Eternal, All Powerful Creator. This is the bedrock statement of the Oneness of the Creator from which stems everything else.

It is the primordial testimony of the unity of all creation and the interlocking grid of the natural order of which man is intrinsically a part.

God says in the Qur'an:

Say: He is Allah the One and Only; Allah the Eternal Absolute; He begetteth not nor is He begotten; And there is none like unto Him. (112.001-4)

God is Real, not an abstract idea or concept; He is One, the Everlasting Refuge for all creation.

MAN'S RELATION TO GOD

The emphasis on Tawheed is significant in itself but it is even more relevant to the present discussion by virtue of defining a Muslim's relationship to Allah. The Omniscience and Omnipotence of Allah means, by definition, that a Muslim's relationship to Allah is total. To Him—and to Him only—should humans refer for all their needs: physical, mental, and spiritual. Indeed, Allah would not have it any other way. As He says in the Qur'an:

Allah forgiveth not that partners should be set up with him; but He forgiveth anything else to whom He pleaseth; to set up partners with Allah is to devise a sin most heinous in deed. 004.048.

But Allah is not only the One Indivisible God. He is also the Universal God as well as the Lord of the Universe:

Praise be to Allah, Lord of the Worlds. 001.002.

And again:

Say: "Allah's guidance is the [only] guidance and we have been directed to submit ourselves to the Lord of the worlds.

To establish regular prayers and to fear Allah; for it is to Him that we shall be gathered together.

It is He Who created the heavens and the earth in true [proportions]: the day He saith "Be" Behold! it is. His Word is the truth. His will be the dominion the day the trumpet will be blown. He knoweth the Unseen as well as that which is open. For He is the Wise well acquainted [with all things]. 006.071-3.

TO ALLAH BELONGS THE EARTH AND THE HEAVENS

Yet another principle that underpins Islamic commitment to the conservation of nature and natural resources is the principle of divine ownership of all that exists on earth and in the heavens—animate and inanimate. There are countless verses in the Holy Qur'an that state this. A few are given below.

In the celebrated Ayatul Kursiyy:

Allah! There is no Allah but He the living the Self subsisting Eternal. No slumber can seize him nor sleep. His are all things in the heavens and on earth. Who is there can intercede in His presence except as He permitteth? He knoweth what [appeareth to his creatures as] before or after or behind them. Nor shall they compass aught of his knowledge except as He willeth. His throne doth extend over the heavens and the earth and He feeleth no fatigue in guarding and preserving them. For He is the Most High the Supreme [in glory]. 002.255.

And again:

To Him belong all things in the heavens and on earth. And enough is Allah as a Disposer of affairs. 004.171

To Him belongeth all that dwelleth [or lurketh] in the night and the day. For He is the One Who heareth and knoweth all things. 006.013.

To Him belongs what is in the heavens and on earth and all between them and all beneath the soil. 020.006.

To Him belongs what is in the heavens and on earth and all between them and all beneath the soil. 020.006.

To Him belong all [creatures] in the heavens and on earth: even those who are in His [very] Presence are not too proud to serve Him nor are they [ever] weary [of His service)]. 021.019.

But we are reminded that all things animate and inanimate, in their own ways, submit themselves to the Glory of Allah. There are many verses in the Qur'an about this:

To Him belongs every being that is in the heavens and on earth: all are devoutly obedient to Him. 030.026.

And again:

*Whatever is in the heavens and on earth doth declare the Praises and Glory of Allah the Sovereign the Holy One the Exalted in Might the Wise.*062.001.

Thus Allah, the One Indivisible God, the Universal God and the Lord of the Universe is the Owner also of all that is in the universe, including man. After all, we are reminded to say constantly:

Be sure We shall test you with something of fear and hunger some loss in goods or lives or the fruits [of your toil] but give glad tidings to those who patiently persevere.
Who say when afflicted with calamity: "To Allah we belong and to Him is our return." 002.155-6

The above set of principles—all taken from Islam's ultimate authority, the Holy Qu'ran—define the perspectives of the relationship of humanity to God and of God to the environment in its totality. A second set of principles that the Holy Qur'an enunciates prescribe man's relationship to the environment after, of course, humanity has accepted the preceding principles.

MAN AND THE KHALIFA

The appointment of people as Khalifa, or guardians, is the sacred duty God has given to the human race.

The most important of this second set of principles is that which defines the human role and responsibilities in the natural order that Allah provided. The appointment of people as Khalifa, or guardians, is the sacred duty God has given to the human race. The appointment of humanity to this elevated position gives rise to the one occasion when the Angels actually questioned Allah's decision as seen in the following verses:

Behold thy Lord said to the angels: "I will create a vice-regent on earth." They said "Wilt thou place therein one who will make mischief therein and shed blood? Whilst we do celebrate Thy praises and glorify Thy holy [name]?" He said: "I know what ye know not."

And He taught Adam the nature of all things; then He placed them before the angels and said: "Tell Me the nature of these if ye are right."

They said: "Glory to Thee of knowledge we have none save that Thou hast taught us: in truth it is Thou who art perfect in knowledge and wisdom."

He said: "O Adam I tell them their natures." When he had told them Allah said: "Did I not tell you that I know the secrets of heaven and earth and I know what ye reveal and what ye conceal?"

*And behold We said to the angels: "Bow down to Adam"; and they bowed down, not so Iblis he refused and was haughty he was of those who reject Faith.*002.030-34

Clearly Allah preferred unprogrammed free will of humanity to the pre-programmed goodness of Angels!

And again:

It is He who hath made you [His] agents inheritors of the earth: He hath raised you in ranks some above others: that he may try you in the gifts He hath given you: for thy Lord is quick in punishment: yet He is indeed Oft-Forgiving Most Merciful. 006.165.

The exercise of the vice regency is defined in the Qur'an by another set of principles in which man's privileges as well as his responsibilities are clearly defined.

MIZAAN

One of the most important attributes conferred on human beings is the faculty of reasoning. This, above all, might well be the deciding fact in their appointment as God's vice regents on earth. Here are the relevant verses:

[Allah] Most Gracious!
It is He Who has taught the Qur'an.
He has created man:
He has taught him speech [and Intelligence]
The sun and the moon follow courses [exactly] computed;
And the herbs and the trees—both [alike] bow in adoration.
And the firmament has He raised high and He has set up the balance [of Justice]
In order that ye may not transgress [due] balance.
So establish weight with justice and fall not short in the balance.
It is He Who has spread out the earth for [His] creatures:
Therein is fruit and date-palms producing spathes [enclosing dates]:
Also corn with [its] leaves and stalk for fodder and sweet-smelling plants.
Then which of the favors of your Lord will ye deny? (055.001-013)

Humans were not created to function exclusively on instinct. The "explanation" was taught to us because we had the capacity to reason and understand.

There is order and purpose in the whole pattern of creation. The Sun and Moon following stable orbits make life possible. The whole universe is in submission to the Creator—the stars that enable us to steer courses and the trees that give us sustenance, shelter and other uses. The world functions only because creation follows a preordained pattern. Man then has a responsibility by virtue of being able to reason, to behave justly, "to transgress not in the balance." We owe this to ourselves as much as for the rest of creation.

Man then has a responsibility by virtue of being able to reason, to behave justly, "to transgress not in the balance." We owe this to ourselves as much as for the rest of creation.

JUSTICE

The capacity to reason and to balance intellectual judgment would in itself be insufficient without the additional moral commitment to Justice. And this is what the Qur'an prescribes for Muslims:

O ye who believe! Stand out firmly for justice as witnesses to Allah even as against yourselves or your parents or your kin and whether it be [against] rich or poor: for Allah can best protect both. Follow not the lusts [of your hearts] lest ye swerve and if ye distort [justice] or decline to do justice verily Allah is well acquainted with all that ye do. 004.135.

And again:

Whoever recommends and helps a good cause becomes a partner therein: and whoever recommends and helps an evil cause shares in its burden: and Allah hath power over all things. 004.085.

Allah doth command you to render back your trusts to those to whom they are due; and when ye judge between man and man that ye judge with justice: verily how excellent is the teaching which He giveth you! for Allah is He who heareth and seeth all things. 004.058

And again:

O ye who believe! stand out firmly for Allah as witnesses to fair dealing and let not the hatred of others to you make you swerve to wrong and depart from justice. Be just: that is next to Piety: and fear Allah for Allah is well-acquainted with all that ye do.005.009.

*[They are fond of] listening to falsehood of devouring anything forbidden. If they do come to thee either judge between them or decline to interfere. If thou decline they cannot hurt thee in the least. If thou judge judge in equity between them; for Allah loveth those who judge in equity.*005.045.

USE BUT DO NOT ABUSE

Several times in the Qur'an, man is invited to make use of the nourishing goods that Allah has placed on earth for him, but abuse—particularly through extravagance and excess—is strictly forbidden. Sometimes these principles are stated in one breath, so to speak. Sometimes they are stated separately. But the message is the same, as the following verse indicates:

O children of Adam!. . . eat and drink: but waste not by excess for Allah loveth not the wasters. 007.031.

There are as many invitations to partake of nature as provided for man and for other creatures of the earth as there are for the avoidance of wasteful extravagance. Time and again, Allah reminds us that He loveth not wasters.

It is He who produceth gardens with trellises and without and dates and tilth with produce of all kinds and olives and pomegranates similar [in kind] and different [in variety]: eat of their fruit in their season but render the dues that are proper on the day that the harvest is gathered. But waste not by excess: for Allah loveth not the wasters. 006.141.

FITRA

Fitra can be taken as perhaps the most direct injunction by Allah to man to conserve the environment and not to change the balance of His creation. This is specifically contained in the verse below:

So set thou thy face steadily and truly to the Faith: [Establish] Allah's handiwork according to the pattern on which He has made mankind: no change [let there be] in the work [wrought] by Allah: that is the standard Religion: but most among mankind understand not. 030.030.

Thus, Islam teaches that humanity is an integral part of the environment; it is part of the creation of Almighty God. We remain deeply locked into the natural domain despite the fact that there is talk of bringing the environment to the people as though we were independent of it.

Islam teaches that humanity is an integral part of the environment; it is part of the creation of Almighty God.

The power given to man by God is seen in Islam to be limited by the responsibilities he bears, not only toward God and other men and women, but also toward the rest of creation.

Seyyed Hossein Nasr says: *"The Divine Law (al shariah) is explicit in extending the religious duties of man to the natural order and the environment."*

CONCLUSION

As we indicated at the beginning, there are several Qur'anic principles that, taken separately, do not have an obvious connection with conservation. But taken in their totality, they state in clear terms that Allah, the One True God is the Universal God and the Creator of the Universe and indeed, the Owner of the Universe. To Him belong all the animate and inanimate objects, all of whom should or do submit themselves to Him.

Allah, in His Wisdom, appointed humans, the creatures that He has conferred with the faculty of reason and with free will, to be His vice regents on earth. And while Allah has invited people to partake of the fruits of the earth for their rightful nourishment and enjoyment, He has also directed them not to waste that which Allah has provided for him—for He loveth not wasters.

Furthermore, Allah has also ordered humans to administer his responsibilities with Justice. Above all, people should conserve the balance of Allah's creation on Earth. By virtue of their intelligence, humanity (when it believes in the One Universal Allah, the Creator of the Universe) is the only creation of Allah to be entrusted with the overall responsibility of maintaining planet Earth in the overall balanced ecology that man found.

If biologists believe that humans are the greatest agents of ecological change on the surface of the earth, is it not humans who, drawn from the brink, will—for their own good—abandon Mammon and listen to the prescriptions of God on the conservation of their environment and the environment of all the creatures on earth? The Islamic answer to this question is decisively in the affirmative.

This was printed, along with Statements from ten other faiths, in <u>*Faith in Conservation*</u> *by Martin Palmer with Victoria Finlay, published by the World Bank in 2003.*

NAME ____________________

Worksheet 7 Creation stewardship in the Abrahamic Religions

1. In Reader 7.1, Rabbi Troster, summarizes 10 traditional Jewish teachers which hear on the human relationship to our environment. List 3 of these teachers that you believe give us a reason to care about the natural world.

2. DeWitt lists four principles of stewardship that he derives from the Bible. Write out each one (not just the name, but the principle) and paraphrase in your own works what each one implies concerning our responsibilities.

   a.

   b.

   c.

   d.

3. Summarize as much as you can of DeWitt's article by explaining in three of four sentences how he would define "stewardship."

4. **The Islamic Faith statement (7.3) argues that Islam supports human care for the environment. How does this reading describe the proper human-environment relationship?**

5. **List three examples of teachings from the Quran that Reading 7.3 says support environmental protection.**

6. **What common themes do you see in Troster's Jewish perspective, DeWitt's Christian perspective, and the Islamic Faith statement perspective?**

Pulling It Together

Creation stewardship in the Abrahamic Religions

QUESTIONS FOR THOUGHT AND DISCUSSION

1. The articles in this chapter cover three monotheistic religions. What similarities do you see between the Jewish, Christian and Muslim views expressed in these articles? What difference do you see?
2. Some have criticized religions, such as Judaism, Christianity, and Islam, which suggest that humans are "higher" than animal and plants, arguing that this teachings encourages behavior that is harmful to the environment. Do the articles in this chapter support this criticism? Why or why not?

EXERCISES

1. Writing Assignment: Chapter 2 includes several passages from the Bible which speak to the theme of God's relationship to nature. Most of these passages are from the Hebrew Bible, which is the Christian Old Testament. In other words, they are in both the Jewish and the Christian Bibles. In the present chapter, the articles by Troster and DeWitt represent Jewish and Christian perspectives. Explain how Troester and DeWitt interpret the key biblical passages, especially Genesis 1-3. In what ways do their interpretations agree? What differences of interpretation to you see between Troster and DeWitt?
2. Research Activity: Find out what religious people believe about the relationship between the environment and their faith. First, ask you friends who are Christians, Jews and/or Muslims what they believe their religions teach about our relationship to nature? Second, make appointments with two or more religious leaders from local churches, synagogues and/or mosques (perhaps a leader from your religious group) and ask them to tell you what they believe is humankind's relationship to nature and responsibility to nature. Write up you findings to turn in and discuss in class.

Chapter

8

Nature in Eastern Religions

INTRODUCTION

Judaism, Christianity and Islam arose in the part of the world we call the "Middle East." These traditions in multiple forms have spread west to Africa, Europe, and the Americans. In the East, however, we find several other strong religious traditions with ancient roots. In this chapter we will consider Eastern religions. They can be grouped into two families.

The First family of religions we will explore has its roots in ancient India. The nation of India is large and diverse. Hinduism, a term that refers simply to India, and is an invention of the British, is used to cover the multitude of religious traditions of India. At heart, Hinduism is a combining of many indigenous religious beliefs, traditions, myths, and practices that began to develop as far back as around 6500 BCE. Unlike the Western mind, which seems to want to separate one religion and way of thinking distinctly from others, the Indian mind has no problem with accepting a wide diversity of beliefs. Even the concept of God, central to the monotheistic religions, varies within Hinduism from the belief that all reality is God (pantheism) to the belief that there are many gods (polytheism), to the belief that there is no God (atheism). Central to modern Hinduism is the concept of Dharma, or "law," and Hinduism is well known for its concepts of karma and reincarnation.

Buddhism, although currently more prominent in other nations than in India, is a member of the same family as Hinduism. Buddhism was founded by an Indian teacher named Siddhartha Gautama, who lived in India around 400 BCE. Like Taoism and Confucianism, Buddhism began, it seems, as more of a philosophical approach to life than as a religion. Siddhartha, according to tradition, was a wealthy prince who led a relatively isolated life. But one day he ventured out of the confines of his home and encountered poverty, sickness, and suffering. He was so struck by what he saw that he left his life (and his wife!) behind and went in search of an answer to human suffering. After failing to find the answer in several places, including traditional religion, Siddhartha eventually found enlightenment while meditating under a tree. He then began to help others find a similar enlightenment. Consistent with its origins, Buddhism is a philosophy of life that seeks to enable humans to respond to suffering, focused not on some divine reality, but on everyday human life. Reflecting Buddhism's roots in traditional Indian thought, a look at the Buddhist perspective on reality reveals important elements shared with Hinduism, including the notion of reincarnation and related concepts. In working through this chapter, you will explore Hindu and Buddhist ideas through a research project designed to help you discover the attitudes and practices of these religions toward nature.

The second family consists of two traditional philosophies of China: Daoism and Confucianism. The founding of Daoism (sometimes spelled "Taoism"), is traditionally attributed to an ancient sage name Laozi, who is said to be the author of Dao De Ching. Chinese tradition places Laozi in the sixth century BCE. Confucianism is said to have arisen in the fifth century BCE with the teaching of another ancient Chinese philosopher named Kong Fuzi, whose name was westernized to Confucius by Europeans. In their early years, both Daoism and Confucianism were more like philosophical systems than religions. Later, however, religion characterized by various rituals and practices developed around the ideas of each of the two philosophers. Both philosophies employed the concept of Dao, sometimes translated as "Way." Some have even translated it as "nature." In Daoism, the focus is on the eternal flow of the Dao, which encompasses all things. The key to life is to not strive against the Dao, but to harmonize one's life with it. Confucianism on the other hand, finds the focus of the Dao to be a set of social and cultural values and relationships. The Dao for each person, according to Confucianism, is determined by his or her position in life, a position that comes with responsibilities toward one's elders, one's family, and one's nation. Reading 8.3 focuses on Daoism.

All of the philosophical and religious traditions explored in this chapter have perspectives that differ in important ways from the ideas of the monotheistic religions we explored in Chapter 7. As you read and study Eastern religions, look for specific areas of contrast and similarities between these religions and the Western religions. Whatever your religious beliefs, or lack of beliefs, ask what you can learn from these traditions.

Hindu Environmentalism: Traditional and Contemporary Resources

—Christopher Key Chapple

Christopher Key Chapple is the Navin and Pratima Doshi Professor of Indic and Comparative Theology at Loyola Marymount University. He has published several books, including Karma and Creativity (1986), a co-translation of the Yoga Sutras of Patanjali (1991) and Nonviolence to Animals, Earth, and Self in Asian Traditions (1993), Hinduism and Ecology (2000), a co-edited volume, Jainism and Ecology: Nonviolence in the Web of Life (2002) and Reconciling Yogas (2003). He is the Associate Editor of Encyclopedia and Nature (2005).

The Hindu religion, perhaps more than any other, is closely linked with a sense of place. It was born on the river banks of the Indus and Ganges rivers, and in recent memory, if a Hindu left the subcontinent to travel across the sea, his or her religious status was compromised. Hinduism holds India's mountains, rivers, and trees sacred, regarding them to be infused with individual spirits (jîva) and suffused with an all-pervasive universal consciousness (brahman). Its religious metaphors speak of continuity and reciprocity; its cyclic concept of time does not allow for a fixed moment of creation, nor does it presage a final cataclysmic moment of destruction.

Hinduism holds India's mountains, rivers, and trees sacred, regarding them to be infused with individual spirits (jîva) and suffused with an all-pervasive universal consciousness (brahman).

India remains a largely agricultural society. Its products have traditionally posed no great ecological threat. Folk wisdom recognizes the integral nature of waste. The people work to integrate garbage into the cycle of life processes. The lowly is seen as the support for the lofty; both exist in mutual interdependence. People of low status gather the dung of India's omnipresent cow, shaping it into patties that they then sell for cooking fuel. Other people gather scraps of garbage to be reused. Colonies of scavengers surround dump sites, turning offal into treasures. Viewed idyllically, this practice of recycling waste provides an ideal model for ecological living.

A friend of mine once bolted from a bus after a long bumpy ride in India, and blurted to the closest stranger, "Where is the closest latrine?" The stranger, waving his arms expansively, pronounced, "The whole world is a latrine," and sent my friend scurrying to some nearby bushes. Waste in India cannot be separated from landscape; India's mother earth receives and reprocesses all things.

India, with one of the world's fastest growing industrial economies, presents special challenges and resources for the environmentalist. On the one hand, it embraces the comforts of modernity with the passion of 1950s America. On the other hand, it offers a traditional worldview that, if properly applied, can help counter some of the ill effects of the technological trance that has robbed so much from the psyche and physical environment of industrialized nations. To make correlations between traditional worldviews and modern realities is risky business at best, but in this essay I would like to propose that for India and even for the West it might be helpful to tap the conceptual resources of Hinduism for the development of an environmental ethic.

TRADITIONAL RESOURCES FOR HINDU ENVIRONMENTALISM

From the earliest strata of Indian history we find evidence of a city culture attuned to and respectful of natural rhythms; the seals of the Indus valley cities of Mohenjodaro and Harappa depict a meditating figure, thought by some scholars to be the first representation of the god Siva, peacefully surrounded with lush vegetation and peaceful animals. When the Aryan tribes advanced into India from the northwest, they composed songs extolling the powers and wonders of the earth. The Rig Veda lavishes high praise upon various natural forces, regarding them to be divinities worthy of worship. The rivers *(Gangā, Yamunā, Sarasvatī, Sindhu)* and the earth *(Prthitī)* are regarded to be goddesses, while the winds *(Maruts)* and fire *(Agni)* are invoked as male deities. From these earliest hymns an image of the human person arises that sees a continuity between the individual and nature in the creation of the cosmos. The *Puruṣa Sūkta* states:

> ***The moon was born from his mind;***
> ***His eyes gave birth to the sun;***
> ***Indra and Agni came from his mouth;***
> ***And Vayu [the wind] from his breath was born.***
> ***From his navel the midair rose;***
> ***The sky arose from his head;***
> ***From feet, the earth; from ears, the directions.***
> ***Thus they formed the worlds.***[1]

In this passage, an identity or correlation is proclaimed between the external world and the individual human person. This perspective speaks of an intimacy between human beings and their environment. The unknown author likens our thoughts to the moon, our eyes to the sun, and the powers of victory and fire to the human mouth. Our body itself finds its roots in the earth, stretching upward into the sky. In the Judeo-Christian story, God made men and women in his own image; in this primordial tale, which includes no creator god, the forces that compose the universe are also found within the human body.

This perspective speaks of an intimacy between human beings and their environment.

This sense of continuity with nature is also found in the *Brhadāraṇyaka Upaniṣad,* which draws a simile between a tree and a person:

> ***As a tree of the forest,***
> ***Just so, surely, is man.***
> ***His hairs are leaves,***
> ***His skin the outer bark.***
> ***From his skin blood,***
> ***Sap from the bark flows forth.***
> ***From him when pierced comes forth A stream, as from the tree when struck.***
> ***His pieces of flesh are under-layers of wood.***
> ***The fiber is muscle-like, strong.***
> ***The bones are wood within.***
> ***The marrow is made resembling pith.***[2]

Both these images of the human body establish a strong kinship between the self and the world of nature, indicating a worldview that holds inherent respect for the nonhuman realms of existence. The Atharva Veda, the source of India's traditional medicine known as Ayurveda, includes passages in praise of the earth, asking

---

[1]Rig Veda 10:190, 13-14, trans. Antonio T. de Nicolás, *Meditations through the Rg Veda: Four Dimensional Man* (New York: Nicolas Hays, 1976).
[2]*Brhadāraṇyaka Upaniṣad* 3:9, 28, trans. Robert Ernest Hume, in *The Thirteen Principal Upanishads* (London: Oxford University Press, 1921).

for her beneficence and pledging protection in return. The text states: "The earth is the mother, and I the son of the earth!" (12:12).[3] The author attributes all wealth to the earth and appeals for her to be generous:

> The earth that holds treasures manifold in secret places, wealth, jewels, and gold shall she give to me; she that bestows wealth liberally, the kindly goddess, wealth shall she bestow upon us! (12:44)

Despite this yearning to benefit from her largesse, the author nonetheless harbors a desire not to hurt her, stating, "What, O earth, I dig out of thee, quickly shall that grow again: may I not, O pure one, pierce thy vital spot, (and) not thy heart!" (12:35). This plea indicates a sense of respect and care for the earth, expressing concern that the earth be made aware that the speaker will not hurt her. In yet another verse, the speaker states:

> Thy snowy mountain heights, and thy forests, O earth, shall be kind to us! The brown, the black, the red, the multi-coloured, the firm earth, that is protected by Indra, I have settled upon, not suppressed, not slain, not wounded. (12:11)

In recognition of mother earth's abundance, the Atharva Veda offers both praise of her power and assurances that she will not be harmed by human intervention.

The Hindu tradition, which finds its primary authority in the Vedic literature composed by the Indo-European invaders, offers diverse conceptual resources that can be used to promote ecological sensitivity. In the Vedic hymns, we have seen an intimate relationship between nature and the human order as through various personifications of the earth, water, thunderstorms, and so forth. The Vedic rituals, many of which are still performed today, invoke elemental forces. The *Sāṃkya* tradition reveres the five great elements *(maā bhūta)* of earth, water, fire, air, and space as the essential building blocks of physical reality. From the Upanishads and later Vedantic formulations, all things with form *(saguṇa)* are seen to be essentially not different from the universal consciousness, or ultimate reality; any thing with form can be an occasion to remember that Brahman which is beyond form *(nirguṇa)*. In the monistic Hindu model, the human order is seen as an extension of and utterly reliant upon the natural order. In the language of Vedanta, the Brahman is inseparable from its individual manifestations. As stated in the Bhagavad Gita, the person of knowledge "sees no difference between a learned Brahmin, a cow, an elephant, a dog, or an outcast."[4] There is no fundamental difference between ourselves and others; both are undergirded by the common substrate known as Brahman. To violate another creature is to violate Brahman itself. This ethos signals deep concern for harmony among life forms and can be used to advocate a minimal consumption of natural resources.

To violate another creature is to violate Brahman itself. This ethos signals deep concern for harmony among life forms and can be used to advocate a minimal consumption of natural resources.

Throughout the Hindu religious tradition, rituals and ceremonies celebrate myriad manifestations of nature. Pilgrimages abound wherein pious folk trek to sacred places such as mountain tops and the confluence of rivers. Daily observances include the veneration of the four elements and the recitation of ancient hymns that extol their power. Families often do not partake of a meal until food has been shared with birds, who fly through the open windows to receive their offerings. Special trees are placed "under worship," such as the Bodhi Tree under which the Buddha achieved his enlightenment, and the Tulsi tree, a species beloved by and closely associated with the worship of Krishna. In these and countless other examples, a continuity is assumed between the human world and that of nature.

GANDHIAN ECONOMICS

Mahatma Gandhi, who drew upon the existing tradition as learned primarily from Jainas (with input from Tolstoy and British vegetarians) to develop a campaign for national independence, included in some of his writings the basis for what could be interpreted as an environmentally friendly system of economics.

---

[3] *Hymns of the Atharva Veda,* trans. Maurice Bloomfield (1897; reprint, New York: Greenwood Press, 1969), 200; the verses quoted in my following text appear on pp.204, 203, and 200, respectively, in this edition.

[4] Bhagavad Gita 5:18, trans. B. Srinivasa Murthy (Long Beach, Calif.: Long Beach Publications, 1985).

He proposed the revitalization of village economies, based on the principles of nonviolence *(ahimsā)* and nonpossession *(aparigraha)*. The purpose of Gandhi's campaign was to make villages self-sufficient, able to cooperate through mutual trade without the importation of foreign produced goods. His means to achieve this was through spinning and weaving cloth and revitalizing other crafts within each village, requiring that schools include these skills as part of the curriculum. Although the aim of this program was to subvert the colonial economic dependence thrust upon India by the British, and despite the fact that Gandhi himself did not object to industrialization, the conservation of energy inherent in the system could be effectively used to counter some of the environmental peril posed in the postmodern world.

Gandhi warned that the urban model associated with industrialization would be detrimental to India's well-being:

> Industrialization on a mass scale will necessarily lead to passive or active exploitation of the villages as the problems of competition and marketing come in. Therefore, we have to concentrate on the village being self-contained.[5]

Gandhi envisioned a network of largely self-sufficient villages, where goods would be both locally produced and consumed:

> My idea of village *swaraj* is that it is a complete republic, independent of its neighbors for its own vital wants, and yet inter-dependent for many others in which dependence is a necessity. Thus, every village's first concern will be to grow its own food crops and cotton for its clothes. . . . My economic creed is a complete taboo in respect to all foreign commodities, whose importation is likely to prove harmful to our indigenous interests. This means that we may not in any circumstances import a commodity that can be adequately supplied from our country. (*VR*, 30)

Though trade policies have recently been changed in India to allow greater foreign investment, for the most part, particularly in regard to food and clothing, contemporary India has abided by Gandhi's precept and imports very few basic commodities.

Using an ironically prophetic voice, Gandhi urges that the people of India remember the sacredness of their land and resist the ills of urban life:

> This land of ours was once, we are told, the abode of the Gods. It is not possible to conceive Gods inhabiting a land which is made hideous by the smoke and the din of mill chimneys and factories, and whose roadways are traversed by rushing engines, dragging numerous cars crowded with men who know not for the most part what they are after, who are often absent-minded and whose tempers do not improve by being uncomfortably packed like sardines in boxes and finding themselves in the midst of utter strangers who would oust them if they could and whom they would, in their turn, oust similarly. I refer to these things because they are held to be symbolic of material progress. But they add not an atom to our happiness. (VR, 34)

By minimizing one's needs and the means used to produce those needs, life, according to Gandhi, holds more potential for happiness.

By minimizing one's needs and the means used to produce those needs, life, according to Gandhi, holds more potential for happiness. In contemporary times, such a lifestyle is an integral part of environmentalism.

CONTEMPORARY HINDU ENVIRONMENTALISM

During a trip to the Indian state of Kerala in 1981, I was stunned by the rampant pollution caused by the chemical manufacturers. We had spent a couple of hours traveling by bus to the birthplace of Sankaracharya, one of the great medieval Hindu philosophers, when, all of a sudden, the lush, verdant landscape gave way to brown stubble and the sweet tropical air became foul with chemical stench, revealing the worst of

[5]M. K. Gandhi, *The Village Reconstruction* (Bombay: Bharatiya Vidya Bhavan, 1966), 43; hereafter, *VR*, with page references cited in the text.

what one encounters in New Jersey or Niagara Falls. We approached the plant—ironically a producer of fertilizer—and were "welcomed" by a dour crowd of employees, who seemed none too thrilled with their factory-dominated existence. As we pulled away from the plant, I commented to one of our hosts that the pollution seemed extreme. He blithely replied that fortunately no one lived nearby. However, as we rounded a bend, a construction project was underway, well within the umbrella of foul air: new housing for factory workers. As we left the general locale and returned to the world of rice paddies, rickshaws, and coconuts, I could not help but fear for the long-term effects of apparently unregulated industry. A couple of years later, sadly, my sense of foreboding was confirmed: the Union Carbide disaster in Bhopal in 1984 took the lives of over 3800 persons in central India.

During this trip, I was struck with the violence to the environment incurred in India by its burgeoning industrialization. The level of awareness in this regard was close to nonexistent, much to my consternation and distress. However, since that time, a noticeable change has taken place: virtually every newspaper routinely includes a story of environmental interest nearly every day. Since the Bhopal Union Carbide disaster, two major centers have been established to serve as clearinghouses for environmental issues. In New Delhi, the Centre for Science and Environment, in addition to other activities, provides a news service that supplies India's many newspapers with stories of environmental and ecological interest. In Ahmedabad, the Centre for Environmental Education, located at the Nehru Foundation for Development, established in 1984, conducts an array of programs designed to enhance the general public's awareness of environmental issues. It offers workshops and produces materials that reach over ten thousand teachers per year. It operates a "News and Features Service" similar to that of the Centre for Science and Environment. It has initiated a rural education program to help stem the destruction of India's remaining forests. It conducts various urban programs, including the promotion of smokeless *chulha,* or woodburning stove, for cooking.[6] In 1986 it launched the Ganga Pollution Awareness Programme and in cooperation with the School of Forestry of the State University of New York, located in Syracuse, it produced a series of environmental films for children.[7]

Many movements in India have taken direct action in an effort to bring attention to environmental concerns. The Chipko movement in Uttar Pradesh involves local women saving trees by embracing them, staving off bulldozers.[8] In traditional Indian folk culture, the tree is sacred and intimately tied to survival of the land and its people. Vandana Shiva, a defender and spokesperson for the Chipko movement, tells the following story in this regard:

> Chipko started in the Himalayas . . . the source of the Ganges River . . . . The Ganges was a mother goddess, and here were prayers for her to be brought to earth . . . . She couldn't just come because her power was so strong that if she landed on earth, she would just destroy it. It's really symbolic of the way we get our monsoon rain. It comes so strong, so powerful, that if we don't have forest cover, we get landslides and floods. So, the god Shiva had to be requested to help in getting the Ganges down to earth. And Shiva laid out his hair, which was very matted, to break the force of the descent of the Ganga. Shiva's hair is basically seen by a lot of us in India as a metaphor for the vegetation and forests of the Himalayas. . . .
>
> That's the sort of concept people in India have constantly. And so when they see forests cut, they see the god's tresses being violated. When they see the Ganges being dammed, they see their sacred river being violated. When Indian movements live for a long time, they are very heavily based on these sorts of concepts.[9]

The Chipko movement, whose name means "to embrace," traces its origins to 1913, when there was a major movement to protect forest lands. It became revitalized in 1977 with a group of women in the Himalayan region who tied sacred threads and formed chains around trees in order to prevent their harvest. Women who dwell in forested areas have for millennia lived in relationship with trees; the forest

---

[6]Kartikeya V. Sarabhai, "Strategy for Environmental Education: An Approach for India," 12, a paper presented at the Annual Conference of the North American Association for Environmental Education, Washington, D.C., 1985.

[7]"Centre for Environment Education Annual Report, 1987–88," Nehru Foundation for Development, Ahmedabad.

[8]Information on this movement is included in the periodical publication *Worldwide Women in the Environment,* P.O. Box 40885, Washington, D.C. 20016.

[9]Ann Spanel, "Interview with Vandana Shiva," *Woman of Power* 9 (1988): 27; hereafter, "Interview," with page references cited in the text.

provides fodder, fertilizer, food, water, and fuel. The cutting of forests for purposes of monoculture interrupts this ecological balance and has caused great devastation in various parts of India. Although men are involved with Chipko, trees and their protection have long been identified with women in India. Vandana Shiva notes that in Hinduism:

> All the nature deities are always female, by and large, because all of them are considered *prakrti,* the female principle in Hindu cosmology. Also, all of them are nurturing mothers—the trees feed you, the streams feed you, the land feeds you, and everything that nurtures you is a mother. ("Interview," 27)

She tells the story of one woman who battled the development of a quarry which was proposed at the expense of a vast expanse of forest. The contractor had hired two hundred men to harass the demonstrators, who refused to leave, despite being beaten and pelted with stones. When asked, "What is it in you that gives you all this *shakti* [strength]?", the friend replied:

> "Can you see all this grass growing? We come to cut this grass and every year it grows back. And the power in that grass is the power in me. Do you see these trees growing? They are two hundred years old. Every year we lop these trees to feed our cattle and to keep our children alive, so that the children have milk, and still the trees keep growing and still keep nurturing us, and that *shakti* is in me. See this stream? Every year the rain comes, and it could just run off every year, but these trees stay live long after the rain goes, and they keep feeding us. Clear sparkling water better than anything you get in the cities, and I call it living water. Your water you get in the cities is dead; it comes from a tap. This living water gives me life. And that's my *shakti.*" ("Interview," 31)

This anecdote shows the simple beauty of a naturally ecological lifestyle, one that is assaulted by increasing urbanization and Western-style "development."

Vandana Shiva explicitly attacks the premises of third world development projects in her book *Staying Alive,* an eloquent appeal to reverse the drive for world homogenization based on the Western model. In addition to a more standard feminist critique, she also develops a theory of nature rooted in *prakrti* and *shakti,* thus utilizing conceptual resources indigenous to Indian tradition. She points out that development policies often entail a shift from holistic, ecologically sound subsistence farming largely conducted by women to cash-crop farming of one product, often enhanced by technology, that is dominated by men. She refers to this practice as "maldevelopment," stating that "it ruptures the co-operative unity of masculine and feminine, and places man, shorn of the feminine principle, above nature and women, and separated from both . . . Nature and women are turned into passive objects, to be used and exploited for the uncontrolled and uncontrollable desires of alienated man."[10] She contrasts the instantiated immediacy of *prakrti,* which views all things as part of a living continuum, with the deadness of things other than human as perceived in the Cartesian-scientific-technological model, wherein they are regarded only for their potential to be transformed into consumable goods. She criticizes the manipulation of seed technology and the development of inorganic fertilizers as potentially harmful to India's ecosystem.

In many ways the current lifestyle of India contains elements that support an environmental perspective. Most persons live within a short scooter ride or walking distance of work. Foodstuffs consumed by Indians are comprised of grains purchased in bulk from the market and cooked with vegetables procured from traveling greengrocers who push their carts through virtually every neighborhood all day long. Waste is collected and used for fertilizer.

Yet things are now beginning to change and modernize in India, for better and for worse. Inoculations and increased knowledge of sanitation have helped decrease the amount of infectious disease in India. But with modernity comes ironic absurdity. One of the great ecological debates for child-rearing Americans has been whether to use cloth or disposable diapers. The laundering of cloth diapers uses a great deal of water and potentially releases cleaning chemicals into the water system. Disposable diapers use tree and petroleum resources and overwhelm landfills. In my early trips to India, this problem simply did

---

[10]Vandana Shiva, *Staying Alive: Women, Ecology, and Development* (London: Zed Books, 1988), 6.

not exist. The hard tile floors of Indian homes allowed for easy cleanup of any toddler indiscretions, and children learned to use the toilet at a young age. However, during recent visits to India, the burgeoning advertising industry has struck gold with one new account. In both New Delhi and Bangalore, one can see towering, crudely painted billboards, strapped to bamboo, announcing "Nude is no longer in style," depicting a plump happy tot sporting a disposable diaper. Other signs of potential and real environmental degradation are beginning to appear. In New Delhi, yogurt is served in a disposable plastic container. Private automobiles have begun to proliferate throughout India. The advent of a consumer economy seems to be eroding the possibility for an ecologically sound form of development. Although industrialization and technologization of the subcontinent are modest by American standards, the sheer numbers of people entering into the middle class make it difficult for the same mistakes of Western development to be avoided. One small example is the automobile: India now produces its own small cars, and increasingly they are owned by individuals. By some accounts (and by personal experience) the Delhi area has perhaps the most polluted air in the world. Yet these newly produced vehicles have no emission controls, nor does there seem to be an interest in lobbying for them.

And yet as the general awareness of environmental ravage increases, the best solution seems to be for each individual to make a change in his or her lifestyle. In America, this requires doing with less. In India, it would require not taking on more. Americans now are educating themselves on how not to use the fabulous technology available to us, from household chemicals to nuclear weapons. One Indian housewife commented to me that although she now owns a refrigerator, she does not really use it, because she still buys her food fresh each day, and that her family would not consider buying a car because public transport is relatively inexpensive and convenient.

In the Hindu tradition, people have not been divorced from the earth: to think of themselves as separate from the ongoing and all-pervasive cycle of life and death would be inconceivable.

In the Hindu tradition, people have not been divorced from the earth: to think of themselves as separate from the ongoing and all-pervasive cycle of life and death would be inconceivable. India now faces the challenges of modernity, technology, consumerism, and technological ravage—in short, buying into a trance state where the world and one's relationship to it becomes estranged and objectified. If India can learn from the lessons of Bhopal and resist the enticements of unbridled industrialization, and follow the advice of those who advocate minimal consumption, then an authentically indigenous environmental ethic can emerge.

A Buddhist Faith Statement on Conservation

This statement was prepared by Kevin Fossey, Buddhist educator and representative of Engaged Buddhism in Europe; Somdech Preah Maha Ghosananda, Patriarch of Cambodian Buddhism; His Excellency Sri Kushok Bakula, 20th Reincarnation of the Buddha's Disciple Bakula, head of Ladakhi Buddhism, and initial rebuilder of Mongolian Buddhism; and Venerable Nhem Kim Teng, Patriarch of Vietnamese Buddhism.

DHARMA

All Buddhist teachings and practice come under the heading of Dharma, which means Truth and the path to Truth. The word Dharma also means "phenomena," and in this way we can consider everything to be within the sphere of the teachings. All outer and inner phenomena, the mind and its surrounding environment, are understood to be inseparable and interdependent. In his own lifetime the Buddha came to understand that the notion that one exists as an isolated entity is an illusion. All things are interrelated; we are interconnected and do not have autonomous existence. Buddha said, "This is because that is; this is not because that is not; this is born because that is born; this dies because that dies." The health of the whole is inseparably linked with the health of the parts, and the health of the parts is inseparably linked with the whole. Everything in life arises through causes and conditions.

The health of the whole is inseparably linked with the health of the parts, and the health of the parts is inseparably linked with the whole.

ECOLOGY AND BUDDHISM

Many Buddhist monks such as His Holiness the Dalai Lama, Venerable Thich Nhat Hanh, Venerable Kim Teng, and Venerable Phra Phrachak emphasize the natural relationship between deep ecology and Buddhism. According to the Vietnamese monk Venerable Thich Nhat Hanh:

> *"Buddhists believe that the reality of the interconnectedness of human beings, society and Nature will reveal itself more and more to us as we gradually recover—as we gradually cease to be possessed by anxiety, fear, and the dispersion of the mind. Among the three—human beings, society, and Nature—it is us who begin to effect change. But in order to effect change we must recover ourselves, one must be whole. Since this requires the kind of environment favorable to one's healing, one must seek the kind of lifestyle that is free from the destruction of one's humanness. Efforts to change the environment*

In order to protect the environment we must protect ourselves. We protect ourselves by opposing selfishness with generosity, ignorance with wisdom, and hatred with loving kindness.

and to change oneself are both necessary. But we know how difficult it is to change the environment if individuals themselves are not in a state of equilibrium."

In order to protect the environment we must protect ourselves. We protect ourselves by opposing selfishness with generosity, ignorance with wisdom, and hatred with loving kindness. Selflessness, mindfulness, compassion, and wisdom are the essence of Buddhism. We train in Buddhist meditation which enables us to be aware of the effects of our actions, including those destructive to our environment. Mindfulness and clear comprehension are at the heart of Buddhist meditation. Peace is realized when we are mindful of each and every step.

In the words of Maha Ghosananda:

"When we respect the environment, then nature will be good to us. When our hearts are good, then the sky will be good to us. The trees are like our mother and father, they feed us, nourish us, and provide us with everything; the fruit, leaves, the branches, the trunk. They give us food and satisfy many of our needs. So we spread the Dharma (truth) of protecting ourselves and protecting our environment, which is the Dharma of the Buddha. When we accept that we are part of a great human family—that every being has the nature of Buddha—then we will sit, talk, make peace. I pray that this realization will spread throughout our troubled world and bring humankind and the earth to its fullest flowering. I pray that all of us will realize peace in this lifetime and save all beings from suffering.

"The suffering of the world has been deep. From this suffering comes great compassion. Great compassion makes a peaceful heart. A peaceful heart makes a peaceful person. A peaceful person makes a peaceful family. A peaceful family makes a peaceful community. A peaceful community makes a peaceful nation. A peaceful nation makes a peaceful world. May all beings live in happiness and peace."

BUDDHISM AS AN ECOLOGICAL RELIGION OR A RELIGIOUS ECOLOGY

The relationship between Buddhist ideals and the natural world can be explored within three contexts:

1. Nature as teacher
2. Nature as a spiritual force
3. Nature as a way of life.

NATURE AS TEACHER

"Like the Buddha, we too should look around us and be observant, because everything in the world is ready to teach us. With even a little intuitive wisdom we will be able to see clearly through the ways of the world. We will come to understand that everything in the world is a teacher. Trees and vines, for example, can all reveal the true nature of reality. With wisdom there is no need to question anyone, no need to study. We can learn from Nature enough to be enlightened, because everything follows the way of Truth. It does not diverge from Truth". (Ajahn Chah, Forest Sangha Newsletter)

Buddha taught that respect for life and the natural world is essential. By living simply one can be in harmony with other creatures and learn to appreciate the interconnectedness of all that lives. This simplicity of life involves developing openness to our environment and relating to the world with awareness and responsive perception. It enables us to enjoy without possessing, and mutually benefit each other without manipulation.

By living simply one can be in harmony with other creatures and learn to appreciate the interconnectedness of all that lives.

However, the Buddha was no romantic idealist. He also saw and realized that every living thing is suffering. He saw creatures struggling for survival in a precarious world. He saw death and fear, the strong preying on the weak and the devastation of thousands of beings as one lonely figure plowed the earth to reap the harvest. He also saw impermanence. As Ajahn Chah has written:

> ***"Take trees for example. . . first they come into being, then they grow and mature, constantly changing, until they finally die as every tree must. In the same way, people and animals are born, grow and change during their lifetimes until they eventually die. The multitudinous changes which occur during this transition from birth to death show the Way of Dharma. That is to say, all things are impermanent, having decay and dissolution as their natural condition". (Buddha-Nature)***

Nature is not independent and unchanging and neither are we. Change is the very essence of nature. In the words of Stephen Batchelor:

> ***"We each believe we are a solid and lasting self rather than a short-term bundle of thoughts, feelings and impulses". (The Sands of the Ganges)***

We do not exist independently, separate from everything else—all things in the universe come into existence, "arise" as a result of particular conditions. It is surely a mistake to see fulfillment in terms of external or personal development alone.

Buddha taught us to live simply, to cherish tranquility, to appreciate the natural cycle of life. In this universe of energies, everything affects everything else. Nature is an ecosystem in which trees affect climate, the soil, and the animals, just as the climate affects the trees, the soil, the animals and so on. The ocean, the sky, the air are all interrelated, and interdependent—water is life and air is life.

A result of Buddhist practice is that one does not feel that one's existence is so much more important than anyone else's. The notions of ego clinging, the importance of the individual and emphasis on self is, in the West, a dominant outlook which is moving to the East as "development" and consumerism spread. Instead of looking at things as a seamless undivided whole we tend to categorize and compartmentalize. Instead of seeing nature as our great teacher we waste and do not replenish and forget that Buddha learned his "wisdom from nature."

Once we treat nature as our friend, to cherish it, then we can see the need to change from the attitude of dominating nature to an attitude of working with nature—we are an intrinsic part of all existence rather than seeing ourselves as in control of it.

Once we treat nature as our friend, to cherish it, then we can see the need to change from the attitude of dominating nature to an attitude of working with nature—we are an intrinsic part of all existence rather than seeing ourselves as in control of it.

NATURE AS A SPIRITUAL FORCE

For Shantiveda in eighth-century India, dwelling in nature was obviously preferable to living in a monastery or town:

> ***"When shall I come to dwell in forests***
> ***Amongst the deer, the birds and the trees,***
> ***That say nothing unpleasant***
> ***And are delightful to associate with."***
> (A Guide to the Bodhisattva's Way of Life)

Patrul Rinpoche, one of the greatest Tibetan Buddhist teachers of the nineteenth century, writes:

> ***"Base your mind on the Dharma,***
> ***Base your Dharma on a humble life,***
> ***Base your humble life on the thought of death,***
> ***Base your death on a lonely cave."***
> (The Words of My Perfect Teacher)

The Buddha taught that the balance of nature is achieved by the functions of the forest. Survival of the forest is vital to the survival of natural harmony, balance, morality, and environment.

Buddhist teachers and masters have constantly reminded us of the importance of living in tune with nature, to respect all life, to make time for meditation practice, to live simply and use nature as a spiritual force. Buddha stressed the four boundless qualities: loving-kindness, compassion, sympathetic joy (delight in the well-being of others), and equanimity (impartiality).

RETREATS

Venerable Asabho has spoken of the value of living in retreat in Hammer Woods, Chithurst in England. The forest has its own rhythms and after a few days the metabolism and sleeping patterns adjust and the senses begin to sharpen to this new and unfamiliar setting. Ear and nose play a more important role when not having any comforts of life—gas, electricity, artificial light and the like.

Living in the fast and furious pace of the twentieth century our true nature is often dulled by the massive sensory impact unavoidable in modern urbanized living. Living close to nature is a very healing experience—to have few activities, few distractions. Learning to trust yourself and being more of a friend than a judge one develops a lightness of being, a light confidence. One realizes the truth of the notion of impermanence—the sound of animals, the texture of trees, the subtle changes in the forest and land, the subtle changes in your own mind. A retreat, or simply living in the forest with nature,

> *"helps people get back to earth, to calm you down—just living with the unhurried rhythms of nature. With nature, everything—birth, growth, degeneration and decay is just as it is, and in that holistic sense everything is all right. Touching lightly is the right touch, the natural touch in which blame, praise, crises, retreats, progress, delays are just as it is and so all right". **(Talks given at Chithurst Buddhist Monastery)***

Living in this way we can appreciate the fragility of all we love, the fickleness of security. Retreat and solitude complement our religious practice and give the opportunity of deepening, refining, and strengthening the mind. By being mindful about the daily routine one pays attention to the flow of life—to see nature as a positive, joyful, spiritual force.

NATURE AS A WAY OF LIFE

The Buddha commended frugality as a virtue in its own right. Skillful living avoids waste and we should try to recycle as much as we can. Buddhism advocates a simple, gentle, nonaggressive attitude toward nature—reverence for all forms of nature must be cultivated.

Buddha used examples from nature to teach. In his stories the plant and animal worlds are treated as part of our inheritance, even as part of ourselves. As Krishnamurti said, "We are the world, the world is us." By starting to look at ourselves and the lives we are living we may come to appreciate that the real solution to the environmental crisis begins with us. Craving and greed only bring unhappiness—simplicity, moderation, and the middle way bring liberation and hence equanimity and happiness. Our demands for material possessions can never be satisfied—we will always need to acquire more, there is not enough in the universe to truly satisfy us and give us complete satisfaction and contentment, and no government can fulfil all our desires for security.

By starting to look at ourselves and the lives we are living we may come to appreciate that the real solution to the environmental crisis begins with us.

Buddhism, however, takes us away from the ethos of the individual and its bondage to materialism and consumerism. When we try to conquer greed and desire we can start to have inner peace and be at peace with those around us. The teaching of the Buddha, the reflections on Dharma, relate to life as it actually is. Namely to be mindful—receptive, open, sensitive, and not fixed to any one thing, but able to fix on things according to what is needed in that time and at that place.

THE RIGHT ACTIONS

By developing the right actions of not killing, stealing, or committing misconduct in sexual desires perhaps we can begin to live with nature, without breaking it or injuring the rhythm of life. In our livelihoods, by seeking work that does not harm other beings and refraining from trading in weapons, breathing things, meat, alcohol, and poisons, we can feel more at one with nature.

Our minds can be so full, so hyperactive, we never allow ourselves a chance to slow down to be aware of our thoughts, feelings, and emotions, to live fully in the present moment. We need to live as the Buddha taught us to live, in peace and harmony with nature, but this must start with ourselves. If we are going to save this planet we need to seek a new ecological order, to look at the life we lead and then work together for the benefit of all; unless we work together no solution can be found. By moving away from self-centeredness, sharing wealth more, being more responsible for ourselves, and agreeing to live more simply, we can help decrease much of the suffering in the world. As the Indian philosopher Nagarjuna said, "Things derive their being and nature by mutual dependence and are nothing in themselves."

By moving away from self-centeredness, sharing wealth more, being more responsible for ourselves, and agreeing to live more simply, we can help decrease much of the suffering in the world.

A Daoist Faith Statement on Conservation

The China Daoist Association, based at White Cloud Temple in Beijing, is the leading body representing all Daoists in mainland China. This piece is an authoritative statement by the Association.

Daoism emerged on the basis of what are known as the One Hundred Schools of Thought during the period 770–221 B.C. Starting with the formal setting up of Daoist organizations in the East Han period (A.D. 25–220), the faith has a history of nearly 2,000 years. Daoism has been one of the main components of Chinese traditional culture, and it has exerted great influence on the Chinese people's way of thinking, working, and acting. It is no exaggeration to say that in every Chinese person's consciousness and subconscious, the factors of Daoism exist to a greater or lesser degree.

Because of its deep cultural roots and its great social impact, Daoism is now one of the five recognized religions in China (the others are Buddhism, Catholicism, Islam and Protestantism). Even more, the influence of Daoism has already transcended the Chinese-speaking world and has attracted international attention.

According to our statistics, more than 1,000 Daoist temples have now opened to the public (this number does not include those in Taiwan, Hong Kong, and Macao), and about 10,000 Daoists live in such communities. There are about 100 Daoist associations all over China, affiliated to the China Daoist Association. Several colleges have also been established to train Daoists, and many books and periodicals on the study and teaching of Daoism have been published. All Daoists work hard in order that Daoism should develop and flourish. They take an active part in mobilizing the masses, carrying forward the best in Daoist tradition, and working for the benefit of human society.

Like every major world religion, Daoism has its own outlook on the universe, human life, ideals of virtue, and ultimate purpose. Due to its distinctive cultural and historical background, it has its own striking characteristics. It can be briefly summarized in the following two precepts:

1. **Give respect to the Dao above everything else.**

Dao simply means "the way." Daoism considers that Dao is the origin of everything, and Dao is the ultimate aim of all Daoists. This is the most fundamental tenet of Daoism. Dao is the way of Heaven, Earth, and Humanity. The Dao took form in the being of the Grandmother Goddess. She came to Earth to enlighten humanity. She taught the people to let everything grow according to its own course without any interference. This is called the way of no action, no selfishness (wu-wei), and this principle is an important rule for Daoists. It teaches them to be very plain and modest, and not to struggle with others for personal gain in their material life. This kind of virtue is the ideal spiritual kingdom for which the followers of Daoism long.

2. **Give great value to life.**

Daoism pursues immortality. It regards life as the most valuable thing. Master Zhang said that life is another expression of Dao, and the study of Dao includes the study of how to extend one's life. With this principle in mind, many Daoists have undertaken considerable exploration in this regard. They believe that life is not

---

controlled by Heaven, but by human beings themselves. People can prolong life through meditation and exercise. The exercises include both the moral and the physical sides. People should train their will, discard selfishness and the pursuit of fame, do good deeds, and seek to become a model of virtue (De).

Daoism considers that the enhancement of virtue is the precondition and the first aim of practicing the Dao. The achievement of immortality is a reward from the gods for practicing worthy acts. With a high moral sense and with systematic exercise in accordance with the Daoist method and philosophy of life, people can keep sufficient life essence and energy in their bodies all their lives. The Daoist exercise of achieving immortality has proved very effective in practice. It can keep people younger and in good health. But there is one point that cannot be neglected: a peaceful and harmonious natural environment is a very important external condition.

DAOIST IDEAS ABOUT NATURE

With the deepening world environmental crisis, more and more people have come to realize that the problem of the environment not only is brought about by modern industry and technology, but also has a deep connection with people's world outlook, with their sense of value, and with the way they structure knowledge. Some people's ways of thinking have, in certain ways, unbalanced the harmonious relationship between human beings and nature, and overstressed the power and influence of the human will. People think that nature can be rapaciously exploited.

This philosophy is the ideological root of the current serious environmental and ecological crisis. On the one hand, it brings about high productivity; on the other hand, it brings about an exaggerated sense of one's own importance. Confronted with the destruction of the Earth, we have to conduct a thorough self-examination on this way of thinking.

We believe that Daoism has teachings that can be used to counteract the shortcomings of currently prevailing values. Daoism looks upon humanity as the most intelligent and creative entity in the universe (which is seen as encompassing humanity, Heaven, and Earth within the Dao).

THE FOUR MAIN PRINCIPLES

There are four main principles that should guide the relationship between humanity and nature:

1. In the Dao De Jing (Tao Te Ching), the basic classic of Daoism, there is this verse: "Humanity follows the Earth, the Earth follows Heaven, Heaven follows the Dao, and the Dao follows what is natural." This means that the whole of humanity should attach great importance to the Earth and should obey its rule of movement. The Earth has to respect the changes of Heaven, and Heaven must abide by the Dao. And the Dao follows the natural course of development of everything. So we can see that what human beings can do with nature is to help everything grow according to its own way. We should cultivate in people's minds the way of no action in relation to nature, and let nature be itself.

the whole of humanity should attach great importance to the Earth and should obey its rule of movement.

2. In Daoism, everything is composed of two opposite forces known as Yin and Yang. Yin represents the female, the cold, the soft and so forth; Yang represents the male, the hot, the hard and so on. The two forces are in constant struggle within everything. When they reach harmony, the energy of life is created. From this we can see how important harmony is to nature. Someone who understands this point will see and act intelligently. Otherwise, people will probably violate the law of nature and destroy the harmony of nature.

It is obvious that in the long run, the excessive use of nature will bring about disaster, even the extinction of humanity.

There are generally two kinds of attitude toward the treatment of nature, as is said in another classic of Daoism, Bao Pu Zi (written in the fourth century A.D.).

One attitude is to make full use of nature, the other is to observe and follow nature's way. Those who have only a superficial understanding of the relationship between humanity and nature will recklessly exploit nature. Those who have a deep understanding of the relationship will treat nature well and learn from it. For example, some Daoists have studied the way of the crane and the turtle, and have imitated their methods of exercise to build up their own constitutions. It is obvious that in the long run, the excessive use of nature will bring about disaster, even the extinction of humanity.

3. People should take into full consideration the limits of nature's sustaining power, so that when they pursue their own development, they have a correct standard of success. If anything runs counter to the harmony and balance of nature, even if it is of great immediate interest and profit, people should restrain themselves from doing it, so as to prevent nature's punishment. Furthermore, insatiable human desire will lead to the overexploitation of natural resources. So people should remember that to be too successful is to be on the path to defeat.

4. Daoism has a unique sense of value in that it judges affluence by the number of different species. If all things in the universe grow well, then a society is a community of affluence. If not, this kingdom is on the decline. This view encourages both government and people to take good care of nature. This thought is a very special contribution by Daoism to the conservation of nature.

CONCLUSION

To sum up, many Daoist ideas still have positive significance for the present world. We sincerely hope that the thoughts of all religions that are conducive to the human being will be promoted, and will be used to help humanity build harmonious relationships between people and nature. In this way eternal peace and development can be maintained in the world.

This was printed, along with Statements from ten other faiths, in Faith in Conservation by Martin Palmer with Victoria Finlay, published by the World Bank in 2003.

NAME ____________________

Worksheet 8

Nature in Eastern Religions

1. In "Hindu Environmentalism: Traditional and Contemporary Resources," Christopher Chapple argues that traditional Hinduism teaches that there is a strong continuity or connection between humans and nature. Summarize one example he gives of how Hindu tradition supports this idea.

2. According to Chapple, what is one example of contemporary Hindu environmentalism. Describe it in two sentences.

3. According the "A Buddhist Faith Statement on Conservation," what do Buddhists believe are interconnected? Which of these can begin to effect change?

4. Briefly summarize what the Buddhist statement says about a) nature as teacher, b) nature as spiritual force, and c) nature as a way of life.

   a.

   b.

   c.

5. **According to "A Daoist Faith Statement on Conservation," what is the Dao?**

6. **What is the "way of no-action"? According to it, how should people live?**

7. **This article explains that there are "two kinds of attitude toward the treatment of nature." What are they? According to Daoism, which kind of attitude is best for us to adopt? What will happen if we do not?**

8. **The Daoist statement says that people should restrain themselves from doing anything that "runs counter the harmony and balance of nature." What are some ways this might place a limit on human activity? What does this imply about how we respond to our human desires?**

9. **What does the reading say should be the goal of both government and people?**

10. **What similarities do you see between Hinduism, Buddhism, and Daoism concerning the human relationship to nature?**

Pulling It Together — Nature in Eastern Religions

QUESTIONS FOR THOUGHT AND DISCUSSION:

1. Hinduism, Buddhism, and Daoism are religions that are most prominent in Asia. What similar themes do you see in these religions, especially with regard to their views of nature?
2. What differences do you see in the teachings of these religions about nature?
3. What implications do each of these belief systems have for how humans should treat the natural environment?

EXERCISE

Write an essay on the following topic: Compare and contrast ideas of nature and the relationship of humans to nature in Western thought (Hebrew/Jewish, Christian, Greek, etc.) with those in Eastern thought (Daoism, Hinduism, Buddhism). Illustrate your points with specific examples from the different religions whenever possible. Do you think these different views imply different perspectives on how humans should treat the natural world? Do they each support care for the world? How?

Chapter 9

Nature in the Worldviews of Native Traditions

INTRODUCTION

We have been exploring the ideas and practices of the major religions of the world with a particular interest in their beliefs and attitudes toward nature, considering in particular how each religion provides a basis for care for the environment. In addition to these major religious traditions, there are many smaller, indigenous religions around the world. In every place, it seems, people have ways of seeing and living in the world with religious dimensions. Many of these people live in much closer communion with the natural world, especially in cultures that have been less influenced by mass communication and trade. Among people who live close to nature, one finds rich traditions, with much to say about how humans should relate to the nonhuman world.

Our reading in this chapter is by J. Baird Callicott, a leading thinker in environmental ethics today. He compares traditional American Indian attitudes toward nature with those commonly held in Western European culture, including the dominant views in America. One way to understand a different perspective is to compare and contrast it with a perspective with which one is more familiar. Callicott helps us do just that by summarizing the typical "Western" perspective we explored in Section 2, and comparing that view with the views of various Native American peoples. There is a danger in speaking of the Native American view as if all of America's native tribal groups held the same beliefs. Native Americans tribal groups have differing traditions and practices. Yet there is also benefit in looking for common threads among native traditions. Callicott does a good job of citing beliefs of different tribes, recognizing differences, yet finding common themes, helping us to compare those themes with attitudes that have dominated Western European and American thought.

Traditional American Indian and Western European Attitudes Toward Nature: An Overview

—*J. Baird Callicott*

J. Baird Callicott is at the University of North Texas. For over forty years, Callicott has been a pioneer in the development of environmental ethics as a focus of research and reflection within the field of philosophy. He is the co-Editor-in-Chief of the Encyclopedia of Environmental Ethics and Philosophy. Callicott is an interpreter and defender of "the land ethic," first proposed by Aldo Leopold. In addition to writing numerous articles and books on Leopold's work, he has also published on ideas of and attitudes toward nature in various non-Western cultures. His books include Earth's Insights: A Multicultural Survey of Ecological Ethics from the Mediterranean Basin to the Australian Outback (1997) and American Indian Environmental Ethics: An Ojibwa Case Study(2003). In this article, Callicott compares the traditional attitudes of Native Americans toward nature with those that have dominated the development of European American culture.

A generalized traditional Western world view is compared with a generalized traditional American Indian world view in respect to the practical relations implied by either to nature. The Western tradition pictures nature as material, mechanical, and devoid of spirit (reserving that exclusively for humans), while the American Indian tradition pictures nature throughout as an extended family or society of living, ensouled beings. The former picture invites unrestrained exploitation of nonhuman nature, while the latter provides the foundations for ethical restraint in relation to nonhuman nature. This conclusion is defended against disclaimers by Calvin Martin and Tom Regan.

As you read make a list of elements of American Indian attitudes that Callicott argues set their view apart from traditional Western views.

I

In this paper I sketch (in broadest outline) the picture of nature endemic to two very different intellectual traditions: the familiar, globally dominant Western European civilization, on the one hand, and the presently beleaguered tribal cultures of the American Indians, on the other. I argue that the world view typical of American Indian peoples has included and supported an environmental ethic, while that of Europeans has encouraged human alienation from the natural environment and an exploitative practical relationship with it. I thus represent a romantic point of view; I argue that the North American "savages" were indeed more noble than "civilized" Europeans, at least in their outlook toward nature.

I do not enter into this discussion unaware of the difficulties and limitations which present themselves at the very outset. In the first place, there is no *one* thing that can be called *the* American Indian belief system. The aboriginal peoples of the North American continent lived in environments quite different from one another and had culturally adapted to these environments in quite different ways. For each tribe there were a cycle of myths and a set of ceremonies, and from these materials one might abstract *for each* a particular view of nature. However, recognition of the diversity and variety of American Indian cultures should not obscure a complementary unity to be found among them. Despite great internal differences there were common characteristics which culturally united American Indian peoples. Joseph Epes Brown claims that

> ***this common binding thread is found in beliefs and attitudes held by the people in the quality of their relationships to the natural environment. All American Indian peoples possessed what***

---

has been called a metaphysic of nature; all manifest a reverence for the myriad forms and forces of the natural world specific to their immediate environment; and for all, their rich complexes of rites and ceremonies are expressed in terms which have reference to or utilize the forms of the natural world.[1]

Writing from a self-declared antiromantic perspective, Calvin Martin has more recently confirmed Brown's conjecture:

> *What we are dealing with are two issues: the ideology of Indian land-use and the practical results of that ideology. Actually, there was a great diversity of ideologies, reflecting distinct cultural and ecological contexts. It is thus more than a little artificial to identify a single, monolithic ideology, as though all Native Americans were traditionally inspired by a universal ethos. Still, there were certain elements which many if not all these ideologies seemed to share, the most outstanding being a genuine respect for the welfare of other life-forms.*[2]

A second obvious difficulty bedeviling any discussion of American Indian views of nature is our limited ability accurately to reconstruct the abstract culture of New World peoples prior to their contact with (and influence from) Europeans. Documentary records of pre-contact Indian thought simply do not exist. American Indian metaphysics existed embedded in oral traditions. Left alone an oral culture may be very tenacious and persistent. If radically stressed, it may prove to be very fragile and liable to total extinction. Hence, *contemporary* accounts by contemporary American Indians of *traditional* American Indian philosophy are vulnerable to the charge of inauthenticity, since for several generations American Indian cultures, cultures preserved in the living memory of their members, have been both ubiquitously and violently disturbed by transplanted European civilization.

We ought, therefore, perhaps to rely where possible upon the earliest written observations of Europeans concerning American Indian belief. The accounts of the North American "savages" by sixteenth, seventeenth, and eighteenth-century Europeans, however, are invariably distorted by ethnocentrism, which to the cosmopolitan twentieth-century student appears so hopelessly abject as to be more entertaining than illuminating. The written observations of Europeans who first encountered American Indian cultures provide rather an instructive record of the implicit European metaphysic. Since Indians were not loyal to the Christian religion, it was assumed that they had to be mindfully servants of Satan, and that the spirits about which they talked and the powers which their shamans attempted to direct had to be so many demons from Hell. Concerning the Feast of the Dead among the Huron, Brebeuf wrote in 1636 that "nothing has ever better pictured for me the confusion among the damned"[3] His account, incidentally, is very informative and detailed concerning the physical requirements and artifacts of this ceremony, but the rigidity of his own system of belief makes it impossible for him to enter sympathetically that of the Huron.

Reconstructing the traditional Indian attitude toward nature is, therefore, to some extent a speculative matter. On the other hand, we must not abandon the inquiry as utterly hopeless. Post-contact American Indians do tell of their traditions and their conceptual heritage. Among the best of these nostalgic memoirs is Neihardt's classic, *Black Elk Speaks,* one of the most important and authentic resources available for the reconstruction of an American Indian attitude toward nature. The explorers', missionaries', and fur traders' accounts of woodland Indian attitudes are also useful, despite their ethnocentrism, since we may correct for the distortion of their biases and prejudices. Using these two sorts of sources, first contact European records and transcribed personal recollections of tribal beliefs by spiritually favored Indians, plus disciplined and methodical modern ethnographic reports, we may achieve a fairly reliable reconstruction of traditional Indian attitudes toward nature.

---

[1]Joseph E. Brown, "Modes of Contemplation through Action: North American Indians," *Main Currents in Modern Thought* 30 (1973–74): 60.

[2]Calvin Martin, *Keepers of the Game: Indian Animal Relationship and the Fur Trade* (Berkeley and Los Angeles: University of California Press, 1978), p. 186.

[3]W. Vernon Kinietz, *Indians of the Western Great Lakes, 1615–1760* (Ann Arbor: University of Michigan Press, 1965), p. 115.

II

On the European side of the ledger, at first glance, an analogous pluralism appears to confound any generalizations. How different are the lands, the languages, the life styles of, say, the Swedes and the Spanish, or the Slavic and Gaelic peoples. If anything, Europeans appear to be a more ethnically diverse, motley collection of folks than Indians. Europeans, however, have enjoyed a collective intellectual *history* which the Indians, for better or worse, have not. Moreover, for many centuries a common learned language was shared by scholars of every civilized European country. This alone constitutes an enormously unifying force. The intellectual history of Europe has been, to be sure, dialectical and disputatious, but the pendulum of opinion has swung between well-defined limits and certain universally accepted assumptions have prevailed.

As the European style of thought was set by the Greeks of classical antiquity, I begin with them. I treat modern science, i.e., modern European natural philosophy, as a continuation and extrapolation of certain concepts originating with the fifth and fourth-century Greeks. Greek ideas about nature were remarkably rich and varied, but only some of these ideas, for historical reasons which cannot be explored in this discussion, inspired and informed modern natural philosophy. They became institutionalized in the modern Western world view. It is upon them, accordingly, that I especially focus.

Mythopoeic Greek cosmology had curious affinities with some of the central cosmological concepts of the American Indians. Sky and Earth (Uranus and Gaia) are represented by Hesiod in the *Theogony* as male and female parents (Father and Mother) of the first generation of gods and either directly or indirectly of all natural beings. Some Ionian Greeks in the city of Miletus apparently became disenchanted with traditional Greek mythology and embarked upon speculations of their own. Everything, they said, is water or air. Things change because of the struggle of the Hot with the Cold and the Wet with the Dry. The implicit question—what is the nature of that out of which all things come and into which all things are resolved?—proved to be both fascinating and fruitful. After about one hundred and fifty years of uninterrupted controversy, Leucippus and Democritus, with characteristic Ionian simplicity and force, brought this line of thought to a brilliant culmination in the atomic theory of matter. The atom was conceived by them to be an indestructible and internally changeless particle, "so small as to escape sensation." There are infinitely many of these. They have substance, that is they are solid or "full," and possess shape and relative size. All other qualities of things normally disclosed by perception exist, according to Democritus, only by "convention," not by "nature." In the terms of later philosophical jargon, characteristics of things such as flavor, odor, color, and sound were regarded as *secondary* qualities, the subjective effects of the primary qualities on the sensory patient. Complementary to the concept of the atom is the concept of the *void*—free, homogeneous, isotropic space. The atoms move haphazardly about in this space. Macroscopic objects are assemblages of atoms; they are wholes exactly equal to the sum of their parts. These undergo generation and destruction, which were conceived as the association and dissociation of the atomic parts. The atomists claimed to reduce all the phenomena of nature to this simple dichotomy: the "full" and the "empty," "thing" and "no-thing," the atom and space.

Thomas Kuhn succinctly comments that "early in the seventeenth century atomism experienced an immense revival . . . . atomism was firmly merged with Copernicanism as a fundamental tenet of the 'new philosophy' which directed the scientific imagination."[4] The consolidated Newtonian world view included as one of its cornerstones the atomists' concept of free space, thinly occupied by moving particles or "corpuscles," as the early moderns called them. It was one of Newton's greatest achievements to supply a quantitative model of the regular motion of the putative material particles. These famous "laws of motion" made it possible to represent phenomena not only materially, but also mechanically.

That the order of nature can be successfully disclosed only by means of a quantitative description, a rational account in the most literal sense of that word, is itself, of course, an idea which originated in

---

[4]Thomas S. Kuhn, *The Copernican Revolution: Planetary Astronomy and the Development of Western Thought* (Cambridge: Harvard University Press, 1957), p. 237.

sixthcentury Greece with Pythagoras. Pythagoras' insight had such tremendous scientific potential that it led Plato to eulogize it as Promethean, a veritable theft from the gods of the key to the secrets of the cosmos. It was cultivated and developed by the subsequent Pythagorean school and by Plato himself in the *Timaeus*, a work which enjoyed enormous popularity during the Renaissance. Modern philosophy of nature might be oversimply, but, nonetheless, not incorrectly portrayed as a merger of the Pythagorean intuition that the structure of the world order is determined according to ratio, to quantitative proportions, and the Democritean ontology of void space (so very amenable to geometrical analysis) and material particles. The intellectual elegance and predictive power of the Newtonian natural philosophy resulted, as Kuhn suggests, in it becoming virtually institutionalized in the nascent European scientific community. Its actual and potential application to practical matters, to problems of engineering and tinkering, also made it a popular, working picture of nature, gladly and roundly embraced by all Europeans participating in enlightenment.

That the order of nature can be successfully disclosed only by means of a quantitative description, a rational account in the most literal sense of that word, is itself, of course, an idea which originated in sixth-century Greece with Pythagoras.

Paul Santmire characterized the modern European attitude toward nature as it took root in the American soil in the nineteenth century as follows:

> ***Nature is analogous to a machine; or in the more popular version nature is a machine. Nature is composed of hard, irreducible particles which have neither color nor smell nor taste. . . . beauty and value in nature are in the eye of the beholder. Nature is the dead res extensa, perceived by the mind, which observes nature from a position of objective detachment. Nature in itself is basically a self-sufficient, self-enclosed complex of merely physical forces acting on colorless, tasteless, and odorless particles of hard, dead matter. That is the mechanical view of nature as it was popularly accepted in the circles of the educated [white Americans] in the nineteenth century.***[5]

Santmire's comments bring to our attention a complementary feature of the European world view of particular interest to our overall discussion. If no qualms were felt about picturing rivers and mountains, trees, and (among the legions of Cartesians) even animals as inert, material, mechanical "objects," the line was drawn at the human mind. Democritus and later Hobbes had attempted a thoroughgoing and self-consistent materialism, but this intrusion of matter into the very soul of man did not catch on—everything else, maybe, but not the human ego.

The conception of the soul as not only separate and distinct from the body, but as essentially alien to it (i.e., of an entirely different, antagonistic nature) was also first introduced into Western thought by Pythagoras. Pythagoras conceived the soul to be a fallen divinity, incarcerated in the physical world as retribution for some unspecified sin. The goal in life for the Pythagoreans was to earn the release of the soul from the physical world upon death and to reunite the soul with its proper (divine) companions. The Pythagoreans accomplished this by several methods: asceticism, ritual purification, and intellectual exercise, particularly in mathematics. This led Plato, who was more than passingly influenced by Pythagoras, to (half) joke in the *Phaedo* that philosophy is the study of death, an exercise in the disentanglement of the soul from the body. The Pythagoreans and Plato indeed inverted the concepts of life and death. In the *Cratylus*, for example, Plato alleges that the word *body (sôma)* was derived from *tomb (sêma)*. The body is thus the tomb of the soul as well as its place of imprisonment.

The Pythagorean/Platonic concept of the soul as immortal and other-worldly, essentially foreign to the hostile physical world, has profoundly influenced the European attitude toward nature. It was not only revived in a particularly extreme form by Descartes in the seventeenth century, but it became popularized much earlier in Pauline Christianity. The essential self, the part of a person by means of which he or she perceives and thinks, and in which resides virtue or vice, is not of this world and has more in common with god(s) than with nature. If the natural world is the place of trial and temptation for the soul, if the

---

[5]H. Paul Santmire, "Historical Dimensions of the American Crisis," reprinted from *Dialog* (Summer 1970) in *Western Man and Environmental Ethics*, ed. Ian G. Barbour (Menlo Park: Addison-Wesley Publishing Co., 1973), pp. 70–71.

body is the prison and the tomb of the soul, then nature must be despised as the source of all misery and corruption, a place of fear and loathing: "a joyless place where murder and vengeance dwell, and swarms of other fates—wasting diseases, putrefactions and fluxes—roam in darkness over the meadow of doom."[6]

If the natural world is the place of trial and temptation for the soul, if the body is the prison and the tomb of the soul, then nature must be despised as the source of all misery and corruption, a place of fear and loathing.

So what attitude overall to nature does modern classical European natural philosophy convey? In sum, nature is an inert, material, and mechanical continuum exhaustively described by means of the arid formulae of pure mathematics. In relation to nature the human person is a lonely exile sojourning in a strange and hostile world, alien not only to his physical environment, but to his own body, both of which he is encouraged to fear and attempt to conquer. Add to this Cartesian picture the Judaic themes forthrightly set out in *Genesis* (which Lynn White Jr., Ian McHarg, and others have so thoroughly criticized)—themes of dominion of man over nature, of its subjugation and domestication, of the *imago dei*, and so on—and we have a very volatile mixture of ingredients set to explode in an all-out war on nature, a war which in the twentieth century has very nearly been won. To the victors, of course, belong the spoils!

In relation to nature the human person is a lonely exile sojourning in a strange and hostile world, alien not only to his physical environment, but to his own body, both of which he is encouraged to fear and attempt to conquer.

This world view might have been ameliorated by Greek biological theory beginning with Plato and reaching fruition with Aristotle, but unfortunately it too proved to be atomistic and ecologically blind. Plato accounted for the existence of kinds of individuals, i.e., species, by means of his theory of ideas. Each individual or specimen "participated," according to Plato, in a certain essence or form and it derived its specific characteristics from the form in which it participates. The impression of the natural world conveyed by the theory of ideas is that the various species are determined by the static logical-mathematical order of the formal domain and then the individual organisms, each with its preordained essence, are loosed into the physical arena to interact clumsily, catch-as-catch-can. Nature is thus perceived, like a roomful of furniture, as a collection, a mere aggregate of individuals of various types, relating to one another in an accidental and altogether external fashion. This picture of the world is an "atomism" of a most subtle and insidious sort. It breaks a highly integrated functional system into separate, discrete, and functionally unrelated sets of particulars. Pragmatically, approaching the world through this model—which we might call "conceptual" in contradistinction to "material" atomism—it is possible radically to rearrange parts of the landscape without the least concern for upsetting its functional integrity and organic unity. Certain species may be replaced by others (e.g., grain for wildflowers in prairie biomes) or removed altogether (e.g., predator extermination) without consequence, theoretically, for the function of the whole.

Aristotle recoiled, of course, from the otherworldliness of Plato's philosophy, both from his theory of the soul and his theory of ideas. Aristotle, moreover, was a sensitive empirical biologist and did almost as much to advance biology as a science as Pythagoras did for mathematics and harmonics. Aristotle's subsequent influence on biological thought has been immense. Biology today still bears the personal stamp of his genius, especially in the system of classification (as modified and refined by Linnaeus) of organisms according to species, genus, family, order, class, phylum, and kingdom. This hierarchy of universals was not real or actual, according to Aristotle (to his everlasting credit); only individual organisms fully existed. However, Aristotle's taxonomical hierarchy (in isolation from evolutionary and ecological theory) resulted in a view of living nature which was, if that is possible, more ecologically blind than Plato's. Relations among things again are, in Aristotle's biological theory, accidental and inessential. A thing's essence is determined by its logical relations within the taxonomical schema rather than, as in ecological theory, by its working relations with other things in its environment—its trophic niche, its thermal and chemical requirements, and so on. As Aldo Leopold expresses this point with characteristic bluntness: "The species

[6]Empedocles, *Purifications,* DK 31 B 121, in *An Introduction to Early Greek Philosophy*, trans. John Mansley Robinson (New York: Houghton Mifflin, 1968), p. 152.

of a layer [in the biotic pyramid] are alike not in where they came from or in what they look like, but rather in what they eat."[7] Evolutionary and ecological theory suggest, metaphysically, that the essences of things, the specific characteristics of species, are a function of their relations with other things. To convey a very un-Aristotelian thought in an Aristotelian manner of speech, relations are "prior" to the things related, and systemic wholes are "prior" to their component parts. A taxonomical view of the biotic world untransformed by evolutionary and ecological theory has the same ecologically misrepresentative feature as Plato's theory of forms: nature is seen as an aggregate of individuals, divided into various types, which have no functional connection with one another. And the *practical* consequences are the same. The biotic mantle may be dealt with in a heavy-handed fashion, rearranged to suit one's fancy without danger of dysfunctions. If anything, Aristotle's taxonomical representation of nature has had a more insidious influence upon the Western mind than Plato's real universals, since the latter could be dismissed, as often they were, as abstracted Olympians in a charming and noble philosophical romance, while metaphysical taxonomy went unchallenged as "empirical" and "scientific."

We should also not forget another Aristotelian legacy, the natural *hierarchy*, according to which the world is arranged into "lower" and "higher" forms. Aristotle's teleology required that the lower forms exist for the sake of the higher forms. The practical t]endencies of this idea are too obvious to require further elaboration. Since human beings are placed at the top of the pyramid, everything else exists for the sake of them.

Since human beings are placed at the top of the pyramid, everything else exists for the sake of them.

III

The late John Fire Lame Deer, a reflective Sioux Indian, comments, straight to the point, in his biographical and philosophical narrative, *Lame Deer: Seeker of Visions*, that although the whites (i.e., members of the European cultural tradition) imagine earth, rocks, water, and wind to be dead, they nevertheless "are very much alive."[8] In the previous section I tried to explain in what sense nature, as the *res extensa*, is conceived as "dead" in the mainstream of European natural thought. To say that rocks and rivers are dead is perhaps misleading since what is now dead once was alive. Rather in the usual European view of things such objects are considered inert. But what does Lame Deer mean when he says that they are "very much alive"?

He doesn't explain this provocative assertion as discursively as one might wish, but he provides examples, dozens of examples, of what he calls the "power" in various natural entities. According to Lame Deer, "Every man needs a stone . . . . You ask stones for aid to find things which are lost or missing. Stones can give warning of an enemy, of approaching misfortune."[9] Butterflies, coyotes, grasshoppers, eagles, owls, deer, especially elk and bear all talk and possess and convey power. "You have to listen to all these creatures, listen with your mind. They have secrets to tell."[10]

It would seem that for Lame Deer the "aliveness" of natural entities (including stones which to most Europeans are merely "material objects" and epitomize lifelessness) means that they have a share in the same consciousness that we human beings enjoy. Granted, animals and plants (if not stones and rivers) are recognized to be "alive" by conventional European conceptualization, but they lack awareness in a mode and degree comparable to human awareness. Among the Cartesians, as I mentioned earlier, even animal behavior was regarded as altogether automatic, resembling in every way the behavior of a machine. A somewhat more liberal and enlightened view allows that animals have a dim sort of consciousness, but get around largely by "instinct," a concept altogether lacking a clear definition and one very nearly as obscure as the notorious occult qualities (the "soporific virtues," and so on) of the Schoolmen. Of course, plants are regarded as, although alive, totally lacking in sentience. In any case, we hear that only human

---

[7]Aldo Leopold, *A Sand County Almanac and Sketches Here and There* (Oxford: Oxford University Press, 1949), p. 215.

[8]Richard Erdoes, *Lame Deer: Seeker of Visions* (New York: Simon & Schuster, 1976), pp. 108–09.

[9]Ibid., p. 101.

[10]Ibid., p. 124.

beings possess *self*-consciousness, that is, are aware that they are aware and can thus distinguish between themselves and everything else!

Every sophomore student of philosophy has learned, or should have, that solipsism is a redoubtable philosophical position, and corollary to that, that every characterization of other minds—human as well as nonhuman—is a matter of conjecture. The Indian attitude, as represented by Lame Deer, apparently was based upon the consideration that since human beings have a physical body *and* an associated consciousness (conceptually hypostatized or reified as "spirit"), all other bodily things, animals, plants, and, yes, even stones, were also similar in this respect. Indeed, this strikes me as an eminently reasonable assumption. I can no more directly perceive another human being's consciousness than I can that of an animal or plant. I *assume* that another human being is conscious since he or she is perceptibly very like me (in other respects) and I am conscious. To anyone not hopelessly prejudiced by the metaphysical *apartheid* policy of Christianity and Western thought generally, human beings closely resemble in anatomy, physiology, and behavior other forms of life. The variety of organic forms themselves are clearly closely related and the organic world, in turn, is continuous with the whole of nature. Virtually all things might be supposed, without the least strain upon credence, like ourselves, to be "alive," i.e., conscious, aware, or possessed of spirit.

Virtually all things might be supposed, without the least strain upon credence, like ourselves, to be "alive," i.e., conscious, aware, or possessed of spirit.

Lame Deer offers a brief, but most revealing and suggestive metaphysical explanation:

> ***Nothing is so small and unimportant but it has a spirit given it by Wakan Tanka. Tunkan is what you might call a stone god, but he is also a part of the Great Spirit. The gods are separate beings, but they are all united in Wakan Tanka. It is hard to understand—something like the Holy Trinity. You can't explain it except by going back to the "circles within circles" idea, the spirit splitting itself up into stones, trees, tiny insects even, making them all wakan by his ever-presence. And in turn all these myriad of things which makes up the universe flowing back to their source, united in one Grandfather Spirit.***[11]

This Lakota pantheism presents a conception of the world which is, to be sure, dualistic, but it is important to emphasize that, unlike the Pythagorean-Platonic-Cartesian tradition, it is not an *antagonistic* dualism in which body and spirit are conceived in contrary terms and pitted against one another in a moral struggle. Further, and most importantly for my subsequent remarks, the pervasiveness of spirit in nature, a spirit *in everything* which is a splinter of the Great Spirit, facilitates a perception of the human and natural realms as unified and akin.

Not only does everything have a spirit, in the last analysis all things are related together as members of one universal family, born of one father, the sky, the Great Spirit, and one mother, the Earth herself.

Consider, complementary to this pan-psychism, the basics of Siouan cosmogony. Black Elk rhetorically asks, "Is not the sky a father and the earth a mother, and are not all living things with feet or wings or roots their children?"[12] Accordingly, Black Elk prays, "Give me the strength to walk the soft earth, a relative to all that is!"[13] He speaks of the great natural kingdom as, simply, "green things," "the wings of the air," "the four-leggeds," and "the two-legged."[14] Not only does everything have a spirit, in the last analysis all things are related together as members of one universal family, born of one father, the sky, the Great Spirit, and one mother, the Earth herself.

More is popularly known about the Sioux metaphysical vision than about those of most other American Indian peoples. The concept of the Great Spirit and of the Earth Mother and the family-like relatedness of all creatures seems, however, to have been very nearly a universal American Indian idea, and likewise the

---

[11]Ibid., pp. 102–03.

[12]John G. Neihardt, *Black Elk Speaks* (Lincoln: University of Nebraska Press, 1932), p. 3.

[13]Ibid., p. 6.

[14]Ibid., p. 7.

concept of a spiritual dimension or aspect to all natural things. N. Scott Momaday remarked, "'The earth is our mother. The sky is our father.' This concept of nature, which is at the center of the Native American world view, is familiar to us all. But it may well be that we do not understand entirely what the concept is in its ethical and philosophical implications."[15] And Ruth Underhill has written that "for the old time Indian, the world did not consist of inanimate materials . . . . It was alive, and everything in it could help or harm him."[16]

Concerning the Ojibwa Indians, who speak an Algonkian language and at the time of first contact maintained only hostile relations with the Sioux, Diamond Jenness reports:

> ***Thus, then, the Parry Island Ojibwa interprets his own being; and exactly the same interpretation he applies to everything around him. Not only men, but animals, trees, even rocks and water are tripartite, possessing bodies, souls, and shadows. They all have a life like the life in human beings, even if they have all been gifted with different powers and attributes. Consider the animals which most closely resemble human beings; they see and hear as we do, and clearly they reason about what they observe. The tree must have a life somewhat like our own, although it lacks the power of locomotion. . . . Water runs; it too must possess life, it too must have a soul and a shadow. Then observe how certain minerals cause the neighboring rocks to decompose and become loose and friable; evidently rocks too have power, and power means life, and life involves a soul and shadow. All things then have souls and shadows. And all things die. But their souls are reincarnated again, and what were dead return to life.***[17]

Irving Hallowell has noted an especially significant consequence of the pan-spiritualism among the Ojibwa: "Not only animate properties," he writes, "but even 'person' attributes may be projected upon objects which to us clearly belong to a physical inanimate category."[18] Central to the concept of a person is the possibility of entering into social relations. Nonhuman persons may be spoken with, may be honored or insulted, may become allies or adversaries, no less than human persons.

Nonhuman persons may be spoken with, may be honored or insulted, may become allies or adversaries, no less than human persons.

The French fur traders and missionaries of the seventeenth century in the Great Lakes region were singularly impressed by the devotion of the savages with whom they lived to dreams. In 1648, Ragueneau speaking of the Huron, according to Kinietz, first suggested that dreams were "the language of the souls."[19] This expression lacks precision, but I think it goes very much to the core of the phenomenon. Through dreams and most dramatically through visions, one came into direct contact with the spirits of both human and nonhuman persons, as it were, naked of bodily vestments. In words somewhat reminiscent of Ragueneau's, Hallowell comments, "it is in dreams that the individual comes into direct communication with the *atiso'kanak*, the powerful 'persons' of the other-than-human class."[20] Given the animistic or pan-spiritualistic world view of the Indians, acute sensitivity and pragmatic response to dreaming makes perfectly good sense.

Dreams and waking experiences are sharply discriminated, but the theater of action disclosed in dreams and visions is continuous with and often the same as the ordinary world. In contrast to the psychologized contemporary Western view in which dreams are images of sorts (like afterimages) existing only "in the mind," the American Indian while dreaming experiences reality, often the same reality as in waking experience, in another form of consciousness, as it were, by means of another sensory modality.

---

[15]N. Scott Momaday, "A First American Views His Land," *National Geographic* 149 (1976): 14.

[16]Ruth M. Underhill, *Red Man's Religion: Beliefs and Practices of the Indians North of Mexico* (Chicago: University of Chicago Press, 1965), p. 40.

[17]Diamond Jenness, *The Ojibwa Indians of Parry Island, Their Social and Religious Life,* Canadian Department of Mines Bulletin no. 78, Museum of Canada Anthropological Series, no. 17 (Ottawa, 1935), pp. 20–21. The (Parry Island) Ojibwa, Jenness earlier details, divided spirit into two parts—soul and shadow—though, as Jenness admits, the distinction between the soul and shadow was far from clear and frequently confused by the people themselves.

[18]A. Irving Hallowell, "Ojibwa Ontology, Behavior, and World View," *Culture in History: Essays in Honor of Paul Radin*, ed. S. Diamond (New York: Columbia University Press, 1960), p. 26.

[19]Kinietz, *Indians of the Western Great Lakes*, p. 126.

[20]Hallowell, "Ojibwa Ontology," p. 19.

As one lies asleep and experiences people and other animals, places and so on, it is natural to suppose that one's spirit becomes temporarily dissociated from the body and moves about encountering other spirits. Or, as Hallowell says, "when a human being is asleep and dreaming his *otcatcakwin* (vital part, soul), which is the core of the self, may become detached from the body (*miyo*). Viewed by another human being, a person's body may be easily located and observed in space. But his vital part may be somewhere else."[21] Dreaming indeed may be one element in the art of American Indian sorcery ("bear walking" among the Ojibwa). If the state of consciousness in dreams may be seized and controlled, and the phenomenal content of dreams volitionally directed, then the sorcerer may go where he wishes to spy upon his enemies or perhaps affect them in some malevolent way. It follows that dreams should have a higher degree of "truth" than ordinary waking experiences, since in the dream experience the person and everyone he meets is present in spirit, in essential self. This, notice, is precisely contrary to the European assumption that dreams are "false" or illusory and altogether private or subjective. E.g., in the second meditation Descartes, casting around for an example of the highest absurdity, says that it is, "as though I were to say 'I am awake now, and discern some truth; but I do not see it clearly enough; so I will set about going to sleep, so that my dreams may give me a truer and clearer picture of the fact.'" Yet this, in all seriousness, is precisely what the Indian does. The following episode from Hallowell's discussion may serve as illustration. A boy claimed that during a thunderstorm he saw a thunderbird. His elders were skeptical, since to see a thunderbird in such fashion, i.e., with the waking eye, was almost unheard of. He was believed, however, when a man who had dreamed of the thunderbird was consulted and the boy's description was *"verified!"*[22]

In one's practical dealings in such a world it is necessary to one's well-being and that of one's family and tribe to maintain good social relations not only with proximate human persons, one's immediate tribal neighbors, but also with the nonhuman persons abounding in the immediate environment. For example, Hallowell reports that among the Ojibwa "when bears were sought out in their dens in the spring they were addressed, asked to come out so that they could be killed, and an apology was offered to them."[23]

The Ojibwa, the Sioux, and if we may safely generalize, most American Indians, lived in a world which was peopled not only by human persons, but by persons and personalities associated with all natural phenomena. In characterizing the American Indian attitude toward nature with an eye to its eventual comparison to ecological attitudes and conservation values and precepts I have tried to limit the discussion to concepts so fundamental and pervasive as to be capable of generalization. In sum, I have claimed that the typical traditional American Indian attitude was to regard all features of the environment as enspirited. These entities possessed a consciousness, reason, and volition, no less intense and complete than a human being's. The Earth itself, the sky, the winds, rocks, streams, trees, insects, birds, and all other animals therefore had personalities and were thus as fully persons as other human beings. In dreams and visions the spirits of things were directly encountered and could become powerful allies to the dreamer or visionary. We may therefore say that the Indian's social circle, his community, included all the nonhuman natural entities in his locale as well as his fellow clansmen and tribesmen.

The Ojibwa, the Sioux, and if we may safely generalize, most American Indians, lived in a world which was peopled not only by human persons, but by persons and personalities associated with all natural phenomena.

Now a most significant conceptual connection obtains in all cultures between the concept of a person, on the one hand, and certain behavioral restraints, on the other. Toward persons it is necessary, whether for genuinely ethical or purely prudential reasons, to act in a careful and circumspect manner. Among the Ojibwa, for example, according to Hallowell, "a moral distinction is drawn between the kind of conduct demanded by the primary necessities of securing a livelihood, or defending oneself against agression, and unnecessary acts of cruelty. The moral values implied document the consistency of the principle of *mutual obligations* which is inherent in all interactions with 'persons' throughout the Ojibwa world."[24]

---

[21]Ibid., p. 41.

[22]Ibid., p. 32.

[23]Ibid., p. 35.

[24]Ibid., p. 47 (emphasis added).

The implicit overall metaphysic of American Indian cultures locates human beings in a larger *social*, as well as physical, environment. People belong not only to a human community, but to a community of all nature as well. Existence in this larger society, just as existence in a family and tribal context, place people in an environment in which reciprocal responsibilities and mutual obligations are taken for granted and assumed without question or reflection. Moreover, a person's basic cosmological representations in moments of meditation or cosmic reflection place him or her in a world all parts of which are united through ties of kinship. All creatures, be they elemental, green, finned, winged, or legged, are children of one father and one mother. One blood flows through all; one spirit has divided itself and enlivened all things with a consciousness that is essentially the same. The world around, though immense and overwhelmingly diversified and complex, is bound together through bonds of kinship, mutuality, and reciprocity. It is a world in which a person might feel at home, a relative to all that is, comfortable and secure, as one feels as a child in the midst of a large family. As Brown reports:

People belong not only to a human community, but to a community of all nature as well.

The world around, though immense and overwhelmingly diversified and complex, is bound together through bonds of kinship, mutuality, and reciprocity.

> ***But very early in life the child began to realize that wisdom was all about and everywhere and that there were many things to know. There was no such thing as emptiness in the world. Even in the sky there were no vacant places. Everywhere there was life, visible and invisible, and every object gave us great interest to life. Even without human companionship one was never alone. The world teemed with life and wisdom, there was no complete solitude for the Lakota (Luther Standing Bear).***[25]

IV

I turn now to the claim made at the beginning of this discussion, viz., that in its practical consequences the American Indian view of nature is on the whole more productive of a cooperative symbiosis of people with their environment than is the view of nature predominant in the Western European tradition.

Respecting the latter, Ian McHarg writes that "it requires little effort to mobilize a sweeping indictment of the physical environment which is [Western] man's creation [and] it takes little more to identify the source of the value system which is the culprit."[26] According to McHarg, the culprit is "the Judeo-Christian-Humanist view which is so unknowing of nature and of man, which has bred and sustained his simple-minded anthropocentricism."[27]

Popular ecologists and environmentalists (perhaps most notably, Rachel Carson and Barry Commoner, along with McHarg and Lynn White, Jr.) have with almost loving attention recited a litany of environmental ills, spoken of "chlorinated hydrocarbons," "phosphate detergents," "nuclear tinkering," and "the gratified bulldozer" in language once reserved for detailing the precincts of Hell and abominating its seductive Prince. Given the frequency with which we are reminded of the symptoms of strain in the global biosphere and the apocalyptic rhetoric in which they are usually cast I may be excused if I omit this particular step from the present argument. Let us stipulate that modern technological civilization (European in its origins) has been neither restrained nor especially delicate in manipulating the natural world.

With somewhat more humor than other advocates of environmental reform Aldo Leopold characterized the modern Western approach to nature thus: "By and large our present problem is one of attitudes and implements. We are remodeling the Alhambra with a steam shovel, and we are proud of our yardage. We shall hardly relinquish the shovel, which after all has many good points, but we are in need of gentler and more objective criteria for its successful use."[28] So far as the historical roots of the environmental crisis

---

[25]Brown, "Modes of Contemplation," p. 64.

[26]Ian McHarg, "Values, Process, Form," from *The Fitness of Man's Environment* (Washington, D.C.: Smithsonian Institution Press, 1968), reprinted in Robert Disch, ed., *The Ecological Conscience* (Englewood Cliffs: Prentice-Hall, 1970), p. 25.

[27]Ibid., p. 98.

[28]Leopold, *Sand County Almanac*, pp. 225–26.

are concerned, I have here suggested that the much maligned attitudes arising out of the Judeo-Christian tradition have not been so potent a force in the work of remodeling as the tradition of Western natural philosophy originating among the ancient Greeks and consolidated in modern scientific thought. At least the latter has been as formative of the cultural milieu, one artifact of which is the steam shovel itself, as the former; and together, mixed and blended, so to speak, they create a mentality in which unrestrained environmental exploitation and degradation could almost be predicted in advance.

It seems obvious (especially to philosophers and historians of ideas) that attitudes and values *do* directly "determine" behavior by setting goals (e.g., to subdue the Earth, to have dominion) and, through a conceptual representation of the world, providing means (e.g., mechanics and other applied sciences). Skepticism regarding this assumption, however, has been forthcoming. Yi-Fu Tuan says in "Discrepancies between Environmental Attitude and Behavior: Examples from Europe and China":

> ***We may believe that a world-view which puts nature in subservience to man will lead to the exploitation of nature by man; and one that regards man as simply a component in nature will entail a modest view of his rights and capabilities, and so lead to the establishment of a harmonious relationship between man and his natural environment. But is this correct?***[29]

Yi-Fu Tuan thinks not. The evidence from Chinese experience which he cites, however, is ambiguous. Concerning European experience, he marshalls examples and cases in point of large scale transformations imposed, with serious ecological consequences, upon the Mediterranean environment by the Greeks and Romans. They were, of course, nominally pagans. He concludes this part of his discussion with the remark that "against this background of the vast transformations of nature in the pagan world, the inroads made in the early centuries of the Christian era were relatively modest."[30] I believe, nevertheless, that my discussion in part two of this paper has explained the environmental impact of Greek and Roman civilization consistently with the general thesis that world view substantially affects behavior! Among the Chinese before Westernization, the facts which Yi-Fu Tuan presents, indicate as many congruencies as discrepancies between the traditional Taoist and Buddhist attitude toward nature and Chinese environmental behavior.

A simple deterministic model will not suffice with respect to the question, do cultural attitudes and values really affect the collective behavior of a culture? On the one hand, it seems incredible to think that *all* our conceptualizations, our representations of the nature of nature, are, as it were, mere entertainment, sort of epiphenomena of the mind, while our actions proceed in some blind way from instinctive or genetically programmed sources. After all, our picture of nature defines our theater of action. It defines both the possibilities and limitations which circumscribe human endeavor. On the other hand, the facts of history and everyday experience do not support any simple cause and effect relationship between a given conceptual and valuational set and what people do. My own view is that it is basic to human nature to both consume and modify the natural environment. Representations of the order of nature and the proper relationship of people to that order may have either a tempering, restraining effect on manipulative and exploitative tendencies or they may have an accelerating, exacerbating effect. They also give form and direction to these inherently human drives and thus provide different cultures with their distinctive styles of doing things. It appears to me, further, that in the case of the predominant European mentality, shaped both by the Judeo-Christian and Greco-Roman images of nature and man, the effect was to accelerate the inherent human disposition to consume and modify surroundings. A kind of "take-off" or (to mix metaphors) "quantum leap" occurred, and Western European civilization was propelled for better or worse into its industrial, technological stage with a proportional increase in ecological and environmental distress. The decisive ingredient, the *sine qua non*, may have been the particulars of the European world view.

If the predominant traditional Chinese view of nature and man is, as it has been characterized by Yi-Fu Tuan, "quiescent" and "adaptive," the American Indian view of the world has been characterized as in

---

[29]Yi-Fu Tuan, "Discrepancies between Environmental Attitude and Behavior," in *Ecology and Religion in History,* eds. Spring and Spring (New York: Harper and Row, 1974), p. 92.

[30]Ibid., p. 98.

[31]Steward Udall, "First Americans, First Ecologists," *Look to the Mountain Top* (San Jose: Gousha Publications, 1972), p. 2.

essence "ecological," for example, by Steward Udall, in the *Quiet Crisis*. In "First Americans, First Ecologists" Udall nostalgically invokes the memory of Thoreau and attributes to his ghost the opinion that "the Indians were, in truth, the pioneer ecologists of this country."[31] To assert without qualification that the American Indians were ecologists is, to say the least, overly bold. Ecology is a part of biology, just as organic chemistry is a part of chemistry. It is a methodic and *quantitative* study of organisms in a contextual, functional relationship to conditions of their several ranges and habitats. Udall, of course, disclaims that he means to suggest that Indians were scientists. One might prefer to say that American Indians intuitively acquired an essentially ecological "outlook," "perspective," or "habit of mind." That would be roughly to say that Indians viewed nature as a matrix of mutually dependent functional components integrated systemically into an organic whole. It would suggest a kind of global or holistic viewpoint; it would also imply an acute sensitivity to the complex factors influencing the life cycles of living things.

To attribute to American Indians, on the one hand, a highly abstract conceptual schematism and, on the other, disinterested, systematic, disciplined, and meticulous observation of minutiae is to press the romantic interpretation of American Indian thought much too far. Much of the material which I have already cited *does* indicate that both woodland and plains Indians were careful students of their natural surroundings. Knowledge of animals and their ways, particularly those of utilitarian value, and knowledge of plants, especially edible and medicinal ones, is a well-known and much respected dimension of traditional Indian cultures. The American Indian pharmacopoeia alone certainly testifies to Indian botanical acumen. My impression, nonetheless, is that the typically Indian representation of nature is more animistic and symbolic than mechanical and functional. The "rules" governing hunting and fishing seem more cast in the direction of achieving the correct etiquette toward game species than *consciously* achieving maximum sustained yield of protein "resources." Medicinal plants were sought as much for their magical, symbolic, and representational virtues as for their chemical effects. Of course, in the case of hunting and fishing, proper manners *are* behavioral restraints and, more often than not, the outcome of their being followed—of correct social forms in respect to bear, beaver, and so on, being observed—was to limit exploitation and, therefore, incidentally, to achieve sustained yield.

To suggest that the Indians were (intuitive or natural or pioneer or even primitive) ecologists, in other words, strikes me as being very much like saying that Indian healers, like Black Elk, were intuitive (etc.) physicians. Indian medicine was not at an earlier stage of development than European medicine, as if moving along the same path some distance behind. It followed a different path altogether. As Black Elk explains, "It is from understanding that power comes; and the power in the [curing] ceremony was in understanding what it meant; for nothing can live well except in a manner that is suited to the way the sacred Power of the World lives and moves." [32] The power that Black Elk employed was, *in his view*, ceremonial and symbolic—what it *meant*, not what it *did* to the patient. The cure, thus, was effected through symbolism, not biological mechanism.

Very old in the Native American world view is the conviction that the earth is vital, that there is a spiritual dimension to it, a dimension in which man rightly exists. It follows logically that there are ethical imperatives in this matter.[33]

The general American Indian world view (at least the one central part of it to which I have called attention) deflected the inertia of day-to-day, year-to-year, subsistence in a way that resulted, on the average, in conservation. Conservation of resources may have been, but probably was not, a *consciously* posited goal, neither a personal ideal nor a tribal policy. *Deliberate* conservation would indeed, ironically, appear to be inconsistent with the spiritual and personal attributes which the Indians regarded as belonging to nature and natural things, since these are represented by most conservationists in the predominant Pinchot tradition as only commodities, subject to scarcity, and therefore in need of prudent "development" and "management." The American Indian posture toward nature was, I suggest, neither ecological nor conservative in the modern scientific sense so much as it was moral or ethical. Animals, plants, and minerals were treated as persons, and conceived to be coequal members of a natural social order. "Very old

[32]Neihardt, *Black Elk Speaks*, p. 212.

in the Native American world view is the conviction that the earth is vital, that there is a spiritual dimension to it, a dimension in which man rightly exists. It follows logically that there are ethical imperatives in this matter."[33]

My cautious claim that American Indians were neither deliberate conservationists nor ecologists in the conventional sense of these terms, but manifested rather a distinctly ethical attitude toward nature and the myriad variety of natural entities is based upon the following basic points. The American Indians, on the whole, viewed the natural world as enspirited. Natural beings therefore felt, perceived, deliberated, and responded voluntarily as persons. Persons are members of a social order (i.e., part of the operational concept of a person is the capacity for social interaction). Social interaction is limited by (culturally variable) behavioral restraints, rules of conduct, which we call, in sum, good manners, morals, and ethics. The American Indians, therefore, in Aldo Leopold's turn of phrase lived in accordance with a "land ethic." This view is also maintained by Scott Momaday:

To point to examples of wastage—buffaloes rotting on the plains under high cliffs or beaver all but trapped-out during the fur trade—which are supposed to deliver the *coup de grace* to all romantic illusions of the American Indian's reverence for nature[34] is very much like pointing to examples of murder and war in European history and concluding therefrom that Europeans were altogether without a humanistic ethic of any sort. What is lacking is a useful understanding of the function of ethics in human affairs. Ethics bear, as philosophers point out, a normative relation to behavior; they do not describe how people actually behave, but rather set out how people ought to behave. Therefore, people are free either to act in accordance with a given ethic or not. The fact that on some occasions some do not scarcely proves that ethics are not on the whole influential and effective behavioral restraints. The familiar Christian ethic has exerted a decisive influence within European civilization; it has inspired noble and even heroic deeds both of individuals and whole societies. The documented influence of the Christian ethic is not in the least diminished by monstrous crimes on the part of individuals. Nor do shameful episodes of national depravity, like the Spanish Inquisition, and genocide, as in Nazi Germany, refute the assertion that a humanistic ethic has palpably affected behavior among members of the European civilization and substantially shaped the character of that civilization itself. By parity of reasoning, examples of occasional destruction of nature on the pre-Columbian American continent and even the extirpation of species, especially during periods of enormous cultural stress, as in the fur trade era, do not, by themselves, refute the assertion that the American Indian lived not only by a tribal ethic but by a land ethic as well, the *overall* and *usual* effect of which was to establish a greater harmony between Indians and their environment than enjoyed by their European successors.

V

This conclusion would not, perhaps, require further elaboration or defense had it not recently been specifically denied by two authors, Calvin Martin and Tom Regan. In this brief polemical epilogue, I therefore undertake to defend it against their criticisms. Martin writes:

> ***Land-use was therefore not so much a moral issue for the Indian as technique animated by spiritual-social obligations and understandings.... There is nothing here to suggest morality; certainly there is nothing to suggest the presumptuous, condescending extension of ethics from man-to-man to man-to-land, as the Leopoldian land ethic implies. When Indians referred to other animal species as "people"—just a different sort of person from man—they were not being quaint. Nature was a community of such "people"—"people" for whom man had a great deal of genuine regard and with whom he had a contractual relationship to protect one another's interests and fulfill***

---

[33]Momaday, "First American Views," p. 18.

[34]The most scurrilous example of this sort of argument with which I am acquainted is Daniel A. Guthre's "Primitive Man's Relationship to Nature," *BioScience* 21 (July 1971): 721–23. In addition to rotting buffalo, Guthrie cites alleged extirpation of pleistocene megafauna by Paleo-Indians, c. 10,000 b.p. (as if that were relevant), and his cheapest shot of all, "the litter of bottles and junked cars to be found on Indian reservations today."

mutual needs. Man and Nature, in short, were joined by compact—not by ethical ties—a compact predicated on mutual esteem. This was the essence of the traditional Indian-land relationship.[35]

As we see, Martin denies point-blank the major conclusion reached in the previous discussion: that the American Indian world view in its general and common characteristics incorporated an environmental ethos or fostered an ethical attitude toward land and the plants and animals belonging to the land community. But his statement is very puzzling when we analyze it closely and compare it further with some of the things he says immediately before it, for example, that one shared aspect of traditional American Indian world views was "*a genuine respect* for the welfare of other life forms . . . ; [that] aboriginal man felt a *genuine kinship* and often *affection* for wildlife and plant-life . . . ; [and that] wildlife were revered and propitiated not only out of fear that their favors might be withheld, although there was some sense of that, but also because they were felt to be *inherently deserving of such regard*."[36]

What more do we mean by *morality* or *ethics*, one wonders, than a sense of respect, kinship, affection, regard, and esteem? At the risk of sounding trite, I suggest that it may be merely a "semantical difference" that is involved here. I have called these American Indian attitudes toward wildlife and plant life, mountains and rivers, sky and earth, "moral" or "ethical" attitudes, while Martin apparently wishes to hold the terms *moral* and *ethical* in reserve (though for what he does not say). He does say, "Ethics were involved only when either party broke regulations, if even then"; but he does not explain this cryptic remark or what precisely he means here by *ethics*.[37] My view of what counts as an ethical attitude is more Humean, I suppose, while Martin's is, perhaps, more Kantian. Following Hume, I am willing to label behavior toward nature "ethical" or "moral" which is motivated by esteem, respect, regard, kinship, affection, and sympathy; Kant, on the other hand, regarded all behavior motivated by "mere inclination" (i.e., sentiment or feeling), however unselfish, as lacking genuine moral worth. For Kant, to be counted as ethical an action must be inspired solely by unsentimental duty toward some abstract precept, some categorical imperative, issued by pure reason unsullied by any empirical content. Perhaps this exalted Kantian standard for ethical behavior lies behind Martin's disclaimer. But only by some special and highly technical definition of *ethics*, such as Kant's, can Martin's discussion be rescued from the allegation that it is plainly incoherent, indeed, that it is blatantly self-contradictory.

That issue having been, if not settled, at least clarified, let us consider another of Martin's disavowals, viz., that, more specifically, the overall traditional American Indian attitude toward nature was not akin to Aldo Leopold's land ethic. In this case, the evidence which Martin himself has so masterfully assembled and presented earlier in his book plainly contradicts this negative claim and I can see no way to rescue his discussion from self-contradiction by semantical distinctions or by any other interpretive concessions.

For purposes of illustrating the similarity of his impression of Indian attitudes toward nature and the attitudes toward nature recommended by Leopold's land ethic, consider the following sample from Martin's very full account:

> *Nature, as conceived by the traditional Ojibwa, was a congeries of societies: every animal, fish and plant species functioned in a society that was parallel in all respects to mankinds. . . .*[38]
>
> *As we extend these ideas further, we come to realize that the key to understanding the Indian's role within Nature lies within the notion of mutual obligation: man and nature both had to adhere to a prescribed behavior toward one another. . . .*[39]
>
> *"According to Cree ideology. . . hunting rests on a kind of social relationship between men and animals. Throughout the cycle of hunting rites men emphasize their respect by means of symbolic expressions of their subordination to animals. . . ."*[40]

---

[35]Martin, *Keepers*, p. 187 (emphasis added).
[36]Ibid., p. 186 (emphasis added).
[37]Ibid., p. 187.
[38]Ibid., p. 71 (emphasis added).
[39]Ibid., p. 77 (emphasis added).
[40]Ibid., p. 116 (emphasis added). Martin is here quoting Adrian Tanner with approval.

To see just how similar the precepts of Leopold's land ethic are to American Indian attitudes, we need only take a closer look at the foundations of the Leopoldian land ethic. First, the primary feature of the land ethic is the representation of nature as a congeries of societies and human/nonhuman relationships as essentially social: "The land ethic simply enlarges the boundary of the *community* to include soils, waters, plants, and animals, or collectively: the land."[41] Second, it is social membership to which ethics and ethical attitudes are correlative: "All ethics so far evolved rest upon a single premise: that the individual is a member of a community of interdependent parts."[42]

Third, Leopold takes a more Humean than Kantian approach to the concept of ethics and morality: "It is inconceivable to me," he writes, "that an ethical relation to land can exist without *love, respect,* and *admiration* for land and a high *regard* for its value."[43] As we see, according to the reconstruction that Martin himself so ably and persuasively presents (though he later unaccountably denies it), the traditional American Indian attitude toward nature was in its shared and general assumptions so similar to the essential notions of Leopold's land ethic as to be basically identical with it.

Martin does not explain why he finds Leopold's land ethic "presumptuous" and "condescending," nor shall I speculate on his reason for these epithets. I find Leopold's land ethic to be, if anything, the opposite of presumptuous and condescending. For the sake of comparison, there is a very different sort of "environmental ethic," the so-called animal liberation or animal rights ethic, that is based upon an arbitrary condition, "sentience," for moral considerability which all human beings paradigmatically exhibit, with moral standing then extended to higher "lower animals" on the grounds that they too manifest the same quality.[44] This humane ethic does seem to me to be condescending and presumptuous. In Leopold's land ethic, the *summum bonum* resides in the "biotic community" and moral value or moral standing devolves upon plants, animals, people, and even soils and waters by virtue of their membership in this (vastly) larger-than-human society.[45] As Leopold rather bluntly puts it, "A land ethic changes the role of *Homo sapiens* from conqueror of the land-community to plain member and citizen of it."[46] The privileged position of human beings in the natural order is, thus, in the Leopoldian land ethic, done away with in a single bold stroke. How can this be either presumptuous or condescending?

There is another unjustifiably skeptical remark which Calvin Martin makes as his parting shot at the neoromantic environmentalist view of Indians. He says that

> ***even if we absolve him of his ambiguous culpability in certain episodes of despoliation, invoking instead his pristine sentiments toward Nature [once more, are these not the very soul of an ethical attitude?], the Indian still remains a misfit guru. . . . The Indian's was a profoundly different cosmic vision when it came to interpreting Nature—a vision Western man would never adjust to. There can therefore be no salvation in the Indian's traditional conception of Nature for the troubled environmentalist.***[47]

This statement, though brief and scarcely defended, has had wide influence. For example, one reviewer for a distinguished journal devoted to scholarship on American Indians, though thoroughly critical of Martin's ethno-historical methods and his controversial hypothesis of an Indian-animal war, assumes without question or criticism that "his epilogue disparages effectively contemporary views about the 'ecological' Indian."[48]

After having so sharply contrasted the traditional Western European attitude toward nature with the traditional American Indian, perhaps I should wholeheartedly agree with Martin on this particular, especially as the basis of his claim goes back to cultural roots or fundamentals. Indeed, it is precisely because

---

[41]Leopold, *Sand County Almanac,* p. 204 (emphasis added).
[42]Ibid., p. 203.
[43]Ibid., p. 227 (emphasis added).
[44]Cf. J. Baird Callicott, "Animal Liberation: A Triangular Affair," *Environmental Ethics* 2 (1980): 311–38.
[45]Cf. ibid., p, 324.
[46]Leopold, *Sand County Almanac,* p. 204.
[47]Martin, *Keepers,* p. 188 (emphasis added).
[48]Kenneth M. Morrison, *American Indian Culture and Research Journal* 3 (1979): 78.

Western culture is grounded, according to Martin, in the Judeo-Christian tradition that "even if he [the Indian] were capable of leading us we could not follow."[49]

A full discussion of this issue goes far beyond the scope of this section. However, in view of Martin's pessimistic conclusion I will hazard a more optimistic suggestion. It may prove to be true that in its own fashion Western science, particularly ecology and the life sciences, but also physics and cosmology, is contributing to the development of a *new* Western world view remarkably similar in *some* ways (but only in some) to that more or less common to American Indian cultures. Science in the twentieth century has retreated from its traditional mechanistic and materialistic biases; indeed, twentieth-century science has been, in just this respect, "revolutionary." Popular Western culture still lags behind. Europeans and Euro-Americans remain, for the most part, nominally Christian and unregenerately materialistic and mechanistic, but the new biocentric and organic world view (embedded in a holistic cosmology and coupled with a field-theory ontology) has already begun to emerge. And there is, further, every reason to expect that eventually it will fully flower in the form of a wholly new popular culture. Present interest in environmental pollution, endangered species, popularized ecology, *and* American Indian environmental attitudes and values are all harbingers of this emerging consciousness.

By the foregoing criticisms I certainly do not intend to belittle Calvin Martin's major achievement in the main body of *Keepers of the Game,* which is a monumental contribution to the recent effort to reconstruct the outlines of the traditional American Indian outlook upon the world. Tom Regan, in a discussion which is very dependent upon Martin's, takes a step toward articulating the very real philosophical problem which Martin may have felt, but was able neither coherently to state nor effectively to resolve. According to Regan, "There is always . . . the possibility that it was fear of the keepers [spiritual warden's of game animals in woodland Indian metaphysics], not appreciation of nature's inherent values, that directed these people's behavior."[50] Regan, therefore, believes that this consideration contributes "an ineradicable layer of ambiguity to the respectful behavior of Native Peoples [toward nature]."[51]

With Regan's observations before us we can now formulate what may have been the actual basis of Martin's reluctance to call American Indian attitudes toward nature "ethical" or "moral." Indian self-interest alone may have dictated deference toward nature, since if such restraint were not forthcoming, nature, everywhere spiritually enlivened, as they believed, would withhold its sustenance or, worse, actively retaliate and the Indians would woefully suffer. Now, any pattern of behavior motivated by mere selfishness on all accounts (including Hume's, to say nothing of Kant's) is certainly not properly described as moral or ethical. Hence, for all their touted spiritualizing of nature and reverence and restraint in taking game and gathering plants, perhaps "there is nothing here to suggest morality." to quote Martin again.

Regan quite fairly points out that "the ambiguity of Amerind behavior is the ambiguity of human behavior, the ancient puzzle over whether, as humans, we are capable of acting out of disinterested respect for what we believe has value in its own right or whether, beneath all manner of ceremony, ritual and verbal glorification of the objects of our attention, there resides, in Kant's memorable words, 'the dear self', the true, the universal sovereign of our wills."[52] As this observation clearly suggests, to describe *any* human behavior whatsoever as "ethical" or "moral" may be naive and incautious. Personally, I think Regan and perhaps Martin (if this is in fact the basis of his reservations) are being overly cynical about human, nature. I am more inclined to agree with Hume when he says, respecting human motives of behavior, that while

> ***we may justly esteem our selfishness the most considerable, I am sensible, that, generally speaking, the representations of this quality have been carried much too far; and that the descriptions, which certain philosophers delight so much to form of mankind in this particular, are as wide***

[49]Martin, *Keepers,* p. 188.

[50]Tom Regan, "Environmental Ethics and the Ambiguity of the Native American Relationship with Nature," in Tom Regan, *All That Dwell Therein: Animal Rights and Environmental Ethics* (Berkeley, Los Angeles, and London: University of California Press, 1982), p. 234. Martin does, of course, mention such fears (see n. 38), but does not pursue the line of thought this psychological element suggests to Regan.

[51]Ibid.

[52]Ibid., p. 235.

> ***of nature as any accounts of monsters, which we meet with in fables and romances. So far from thinking that men have no affection for anything beyond themselves, I am of opinion, that tho' it be rare to meet with one, who loves any single person better than himself; yet 'tis rare to meet with one, in whom all the kind affections taken together, do not over-balance all the selfish.***[53]

Human beings are not, I think, incapable of acting from motives of affection, sympathy, regard, respect, fellow feeling, reverence, and so on, as well as from purely selfish motives; thus, I think human beings *per se* are not incapable of ethical or moral behavior.

Most folk ethics (as distinct from formal, philosophical theories) take account of and play upon both our moral and selfish sentiments. Take the familiar Christian ethic as an example. We are urged to love God and to love our neighbor as ourselves. Those whose moral sentiments "over-balance" the selfish probably do genuinely behave respectfully, for the most part, toward other persons because they love God and at least sympathize with (if not love) one another. On the other hand, there is an appeal to the dear self. If you do not obey God's commandments and at least act *as if* you respected other persons, God will punish you.

In this respect it seems to me that Indian land ethics are precisely analogous to Western humanitarian ethics. Nowhere in this discussion have I claimed that traditional American Indians were morally better than Westerners in the sense that they were more altruistic and Europeans and Euro-Americans more selfish. Rather, I have claimed that these two broad *cultural traditions* provide very different views of nature and thus very differently excite or stimulate the moral sentiments of their members. In persons belonging to both cultures there is, we may be sure, a mixture of selfishness and altruism. The ratio does not vary so much from culture to culture as from individual to individual. Some individuals in any culture may be very nearly devoid of benevolent feelings, while others are so filled with them that they willingly sacrifice themselves for the sake of their fellows. Hence, Indian land ethics, like the humanitarian religious folk ethic of Western culture, is in this respect bilateral.

In traditional American Indian cultures the animals and plants were commonly portrayed as fellow members of a Great Family or Great Society. They were "persons" worthy of respect, even affection. But if there were individuals (and in every human group there always are) incapable of other-oriented feelings, people who were narrowly selfish, then an appeal to fear of punishment was included, as it were, as a backup motivation to the same forms of action that others were motivated to do because of *moral* sentiments. Noble and generous American Indians would have been mortified to slaughter game animals wantonly and needlessly for sport alone, since the animals were fellow members of an extended family or extended society and had themselves been so generous and cooperative. On the other hand, less noble Indians might have been induced to submit to similar restraints because of fear of retribution. This retributive factor does not suggest, to me at any rate, that American Indian *world views* did not, therefore, include a land ethic (Martin's claim), or even that there is, therefore, an "ineradicable layer of ambiguity" in the average, manifestly restrained, behavior of traditional Indians toward nature (Regan's claim), an ambiguity which makes it impossible for us to decide if their

[53]David Hume, *A Treatise of Human Nature* (Oxford: Clarendon Press, 1960), pp. 486–87.

NAME ______________________________

Worksheet 9

Nature in the Worldview of Native Traditions

1. As you read through J. Baird Callicott's article, "Traditional American Indian and Western European Attitides toward Nature: An Overview," make a chart listing key themes regarding nature in Western thought on the left, and key themes in Native thought on the right. After making your two lists, draw lines between themes that are similar or different, labeling lines to indicate similarities and difference: (______different_______ or _______similar_____).

Key themes in Western attitudes toward nature: (List at least 5 or 6)	Key themes in American Indian attitudes toward nature: (List at least 5 or 6)

2. Think about what you learned about Eastern thought in the previous chapter. Do you see any similarities or differences between American Indian attitudes described by Callicott and these Eastern views?

Pulling It Together

Nature in the Worldview of Native Traditions

QUESTIONS FOR THOUGHT AND DISCUSSION

1. What aspects of the American Indian views described by Callicott might contribute to a more helpful view of nature?

2. Are there aspects of these traditions that you do not believe you could accept?

EXERCISES

There are many sources available on the internet and many resources in the library for information about different religions and their attitudes toward the environment. The following exercises involve research to learn more about different religious tradition.

1. This chapter has covered ideas from Native American perspectives. But there are a many resources on attitudes toward nature in other indigenous traditions. Write an essay in which you compare and contrast beliefs and attitudes about nature among various indigenous peoples (Native Americans, Native Hawaiians, other aboriginal peoples, such as in Africa or Australia) with beliefs and attitudes toward nature in the monotheistic religions (Judaism, Christianity, and Islam).

2. Write an essay in which you compare and contrast beliefs and attitudes about nature among various indigenous peoples (Native Americans, Native Hawaiians, other aboriginal peoples, such as in Africa or Australia) with beliefs and attitudes toward nature typical in the history of Western thought (Greek and Hebrew thought, modern scientific thought, etc.).

3. Write an essay in which you compare and contrast beliefs and attitudes about nature among various indigenous peoples (Native Americans, Native Hawaiians, other aboriginal peoples, such as in Africa or Australia) with beliefs and attitudes toward nature typical in Eastern thought and religion (Taoism, Confucianism, Hinduism and Buddhism).

SECTION 4

The Ethics and Philosophy of Nature in Recent Philosophy

Chapter 10

Philosophical Ethics and Anthropocentrism

INTRODUCTION

In Section One, we explored the way in which the concept of "nature" has developed and been used in the history of Western Thought. In Section Two, we tasted the ideas from the religious and traditional teachings of many cultures of the world, focusing in particular on how these ideas can help us with our need for a more ecofriendly approach to the earth. Historians and scholars of religion have given us a lot to think about. Another group of scholars who have done some serious thinking about nature are professional philosophers. In this chapter we take a look at theories of the value of nature developed by philosophers who have thought rigorously about ethics and various elements of the nonhuman world. At the center of our study of these theories will be two questions: "How does ethics apply to the non-human world?" and "Why is nature valuable?"

In this chapter, we begin by considering traditional philosophical approaches to ethics and their application to the environment. In the first reading, I introduce you to some basic concepts and three major schools of thought in philosophical ethics. We will not be able to consider each school and the arguments for and against it in much detail here, but it is important to at least have a basic understanding of the common approaches to understand the readings in the rest of this section. The standard approaches to ethics that I will discuss have almost always focused on human beings, taking what is valuable to humans as of central importance in deciding what is right or wrong. In the second essay, I briefly discuss this human-centered, or anthropocentric, approach. Many of the readings in this book are highly critical of anthropocentric ways of viewing the environment, arguing that for humans to put themselves at the center, treating themselves as more important or more valuable is arrogant. Environmentally minded scholars have argued that anthropocentrism, which places human beings at the top of the scale of moral value, is exactly what has caused the environmental crisis we face today. If we are going to reverse the degradation of the Earth, they argue, we must begin to see other creatures which share our planet as not below us, but as our equals. We must learn to stop thinking about what is good for humankind, and begin to think of the welfare of the entire community of life of which we are but members. Suggestions like these have already been introduced by the historians and religious leaders and scholars we have already read. Later in this section, we will encounter philosophers who seek to rethink ethics in light of the animals and the environment.

But not all philosophers and scholars are ready to abandon the focus on humans. In an important article, William H. Murdy considers challenges to anthropocentrism, criticizes these nonanthropocentric perspectives, and argues that the most reasonable approach to caring for the environment is an enlightened anthropocentrism. "Our current ecological problems do not stem from an anthropocentric attitude per se, but from one too narrowly conceived."[1] Murdy argues that we have come to a new place in our cultural evolution such that we recognized that we must care for the environment for our own good, and to do so we must affirm, not deny, that humans have evolved to a higher place from which to solve the problems we face. Sometimes those critical of anthropocentrism are guilty of attacking a "straw person." It is easy to criticize the view that humans can do whatever they want with nature regardless of the consequences. But that's not what the enlightened anthropocentrist would advocate at all. In this chapter we will explore traditional ethical theories and some of the objections that have been raised against anthropocentrism. As a critical reader, ask yourself whether these objections give us grounds for rejecting all versions of anthropocentrism. Are there good "human-centered" reasons for caring for the environment? After careful thought, I think you will discover there are.

---

[1] W.H. Murdy, Anthropocentrism: A Modern View," *Science*, Volume 187, Number 4182 (March 28, 1975) : 1168–1172.

A Brief Introduction to Philosophical Ethics

In conversations about ethical issues, many people's thoughts go in one of two directions, if not both: religion and relativism. Either ethics is based on religion, they think, or it is just a matter of personal opinion or cultural norms. But among philosophers, neither one of these alternatives is the most popular approach to ethics. In this essay, I will argue briefly that neither religion nor relativism adequately explains ethics and introduce you to three approaches to philosophical ethics that many find more acceptable. In the history of philosophy, and even today, each of these three approaches is defended as the most satisfactory way of explaining what is morally right and wrong and what is morally good and evil. In this essay we will not be able to review all of the arguments for and against each approach. I will, however, introduce you to one theory in each of the three general categories and point out briefly some aspects of each representative theory that make it attractive, and some aspects that are more problematic. In the end, I will suggest that each of the approaches can provide insights into the nature of ethics and can be helpful in applying ethics to everyday issues.

ETHICS AND RELIGION

First let's think about religion and ethics. Most people who argue that ethics is based on religion seem to have in mind the idea that what is right or wrong is determined by God's commands. What God commands us to do is the morally right thing to do, and what God forbids is the morally wrong thing to do. If there is no God to command, then there is no right or wrong. Religion is the only basis for ethics.

As far back as 400 BCE, the philosopher Plato raised a problem with this approach. The problem is this: why does God command the things God does command, and why does God forbid the things God forbids? In many of the world's religion, for example, there is a commandment not to murder. So according to those religions in which this injunction is said to be based on a command from God, God commands humans not to murder. Why does God make this command? One answer is that God just arbitrarily commands certain things and forbids others. In Christianity and other theistic religions, God forbade murder, and so murder is wrong. According to the view we are considering, there is no reason, other than the fact that God forbids it, that murder is wrong. If this is true, then if God had commanded us to murder one another, instead of forbidding it, then murder would be morally right, and refraining from murder would be morally wrong. That seems outlandish, doesn't it? It just doesn't seem like God could command murdering one another, and it doesn't seem that such a command would make it morally right. I'll be honest. I would be more apt to say that the God who commanded such a thing is immoral than to conclude that murdering one another is moral because God commands it.

The other possible answer is that God forbids murder *for some good reason.* For example, perhaps God forbids murder because murder harms people. Murdering a person deprives that person of something that is valuable (his or her life) and harms others as well. But if this is the reason God forbids murder, murder isn't wrong because God's forbids it, God forbids it because it is wrong. In other words, the wrongness of murder is not based on God's commands but on something else. In this example, the wrongness of murder is a result of the fact that murder deprives people of good; it harms them. In short, it looks like either God's commands are arbitrary, or there is some other feature of reality that makes some actions right and others wrong. Some theists are willing to "bite the bullet" and say God's commands are arbitrary. But most philosophers try to either find a way out of the dilemma or conclude that there is some other feature of reality that makes some things right and other things wrong. The problem I have been discussing is called the "Euthyphro Dilemma." If you're interested, I would encourage you to do some research and think about it some more. In this essay, however, I'm going to move on, past religion, and ask whehter there is a different way to think about ethics that does not depend upon God's commands.[1]

Ethical Relativism

The second common thought people have about ethics is that it's relative, either to personal beliefs or to culture. The claim is that what is right or wrong is determined by what an individual believes is right or wrong or by his or her culture, or perhaps by the culture in which the action in question is performed. Many philosophers don't accept this any more than they accept the idea that ethics is based on religion. There are several reasons. First, relativism is often just a discussion stopper. Let's say that we are discussing whether gambling is morally wrong. Some people believe that gambling is wrong, whereas others believe that gambling is morally permissible. Someone says, "Well, people have different beliefs, and whatever a person believes is right is right for him or her." You can't argue with the fact that people have different beliefs about what is right and wrong, so if moral rightness is determined by what each person believes, then there really is no answer to the question about whether anything is right or wrong. There's nothing to discuss. End of discussion. But notice, we do discuss moral issues. There are relevant things to say in deciding whether something is right and wrong. It is relevant, for example whether anyone gets hurt. It is relevant whether people have a right to do things they want, and so forth.

According to personal relativism, if the person believes it is morally okay, these actions would be morally okay. But they are not. Rape, child abuse, and other types of abuse are simply wrong, no matter what the person who does then believes.

In the example of gambling, many may be willing to live with the fact that whatever an individual believes is right is morally right for that person. But there are other types of actions that don't seem so open to this interpretation. Consider the violent crimes of rape or child abuse. If a person is abused in one of these ways, that is wrong. Unfortunately, there are cases in which the person who performs the action, the abuser, believes he or she is not doing anything wrong. Does that make it morally okay? According to personal relativism, if the person believes it is morally okay, these actions would be morally okay. But they are not. Rape, child abuse, and other types of abuse are simply wrong, no matter what the person who does then believes. In other words, personal relativism is inadequate as an explanation of what makes things moral or immoral.

Another problem is that personal relativism provides little help for an individual who is trying to decide what is right and wrong. Imagine that Jen is trying to decide for herself whether gambling is morally permissible. If what is right or wrong for Jen is determined by what Jen believes is right or wrong, then Jen would have no way to decide for herself. Because she does not yet have a belief about it, she cannot appeal to her own beliefs. Ethics is supposed to help us think about issues we have never thought about before, to help us make decisions about what is right and wrong. Personal relativism either simply puts an end to discussion or leaves us high and dry with no way to think about the issue. That's not a very helpful approach to ethics.

So far, most of what I have said about relativism is aimed at the personal variety of relativism. Let's consider relativism in its cultural variety. Is morality determined by culture? There is no doubt about the fact that there are different norms in different cultures, and that a behavior that is perfectly polite in one culture may be offensive in another. A particular hand motion, for example, may be a common form of greeting in culture A, but communicate a very rude message in culture B. It seems clear, then, that performing that hand motion is morally permissible, and perhaps even morally praiseworthy, in the context

[1]Plato's discussion of the "Euthyphro Problem" can be found in the dialogue, Euthyphro, by Plato. There are many translations available. One translation can be found in Plato, *Five Dialogues*, translated by G. M. A. Grube (Indianapolis: Hackett Publishing Co. 1981).

of culture A, but morally insensitive or even downright wrong in the context of culture B. Does this make morality relative to culture? Well, kind of, but not in a very deep sense. The underlying principle that one should be respectful of others is what is at work in our analysis of what is right or wrong in each culture. The reason it is wrong to greet a person in the specified way in culture B is that one should show respect for others, and the greeting does not show respect for the person greeted. The reason it is morally permissible in culture A is that such a greeting is in fact a sign of respect. In other words, the single principle that one should show respect for others has different applications in different cultures. The principle itself, however, is assumed to hold across cultures. Because the question of ethics is whether there are any principles that apply across cultures, cultural relativism must defend the idea that all moral principles are fully determined by culture, not simply that culture can affect the way in which morality is to be applied.

So let's turn to that question. Is ethics fully determined by culture? Again, many philosophers do not think so. Why? First, philosophers frequently evaluate a claim by asking whether the implications of the claim make sense, and in this case many argue that they don't. If ethics is fully determined by culture, then there are no moral principles or standards which hold across cultures. But if there are no moral standards or principles that can be applied to different cultures, then there is no way for us to evaluate a culture and decide that it has improved morally or that it has declined morally. This is because, in essence, when a culture changes significantly, it, in effect, becomes a different culture with different moral standards. Any moral standards are either part of the earlier culture or part of the later culture. Thus no standard can be applied to both to determine whether the later culture is morally better or morally improved. If cultural relativism is true, then it doesn't even make sense to talk about a culture improving morally.

There do seem to be ways that cultures can become morally better.

But there do seem to be ways that cultures can become morally better. To take American culture, there was a time when slavery was culturally accepted by the majority of Americans. According to the norms of that time, slavery was morally acceptable. But through the years, slavery was abolished. Was that a moral improvement? I think it was. People were mistreated before in ways that they are not today (although there are plenty of more subtle ways in which people are still oppressed). We are not perfect, and there may be ways in which we have declined morally as a culture, but in this respect it seems clear to me that we have made moral improvements. To speak of moral improvement, however, implies that there is some standard according to which we have made progress. It assumes that there is some standard, some "ruler," by which I can measure our culture in 1800 and our culture today, and see whether there has been an improvement. This is just what cultural relativism denies. It says that at each point in history right and wrong is determined by what the culture accepts as right or wrong at that point. In the past, slavery was morally okay. Today it is not. We've clearly changed our views. Is it a change for the better? Cultural relativism says no. In the past we lived according to one standard regarding slavery. Today we live according to another regarding slavery. Neither is better or worse than the other. I think it is pretty clear that this is wrong. It is morally better not to have slavery. I have said earlier that a task of ethics is to help us think about issues and decide what is right or wrong. Another task of ethics is to help us look at ourselves and our culture and ask ourselves how we fall short and how we could improve. Cultural relativism simply isn't up to the task. It provides no standard to use in our evaluation of ourselves.

Perhaps rather than falling back to relativism, we should talk about what we believe is really right or wrong. Are there principles we can agree upon? Is the moral diversity in the world, at the level of basic principles, really as wide as relativism suggests? There is evidence that not all moral principles are culturally relative, that there are standards that people of many cultures accept.

My final point is a practical one. The more I think about the purpose of ethics, the more I come to believe that human ethics, whether it arose from the fact that we were created by a good God or through evolutionary social development, has the goal of helping people get along in such a way that we can live good lives. Given this, how should we respond to the undeniable disagreement about ethical issues in our world? Should we fall back on cultural relativism and just say, "Well, we disagree about what's right or wrong, and there is nothing we can do about it." At first, this sounds fine. It appears, in fact, to be a kind of tolerance. But this appearance is deceptive. When we get in serious disputes about issues we really care about and have no way to talk them through and apply common standards, the only solution is to fight it out. The one who has the greatest power wins. Without ethics, all we have are opinions, and the opinion backed by the most power prevails. This does not create a very healthy world for humans to live in. Perhaps rather than falling back to relativism, we should talk about what we believe is really right or wrong. Are there principles we can agree upon? Is the moral diversity in the world, at the level of basic principles, really as wide as relativism suggests? There is evidence that not all moral principles are culturally relative, that there are standards that people of many cultures accept. Something very similar to the Golden Rule, the command to do to others as you would have them do to you, is found in most of the major religions of the world. Commands against lying, murder, and adultery, although spelled out in different cultural frameworks, are found within the ethical frameworks of many cultures. Is it really true that ethics is culturally relative? Or is there something that makes some things right and other things wrong across cultures?

And what about the emerging global culture? We live in a world in which people from different cultures are interacting every day. To some extent, as cultures mix, a global culture is emerging. I hope this will never mean that the rich cultural diversity of the world is lost or that we will all melt into one homogenous culture. But I also hope that with the emerging global culture we can agree upon a global ethic that can help us make the world a safe and just place for all of us to live. This will not occur if we simply fall back to cultural relativism. Philosophical ethics can play a role in helping us find common ethical ground. It can provide ways to think about ethics that can help us move past our differences.

EVALUATING MORAL THEORIES

Before I move on to discuss philosophical approaches to ethics, it will be helpful to briefly review a method many philosophers use to evaluate theories. Everyone, it seems, has some beliefs about what is right and wrong. But ethical theories go beyond just evaluating particular actions and states of affairs. Ethical theories attempt to explain why certain things are right and wrong. They attempt to unify ethical convictions under one or a few concepts such that the result is a coherent set of moral beliefs including statements of specific moral judgments and more general principles. With this in mind, philosophers aim at what John Rawls has called "reflective equilibrium" among moral beliefs. When a person's beliefs are in reflective equilibrium, all of her moral beliefs fit together systematically.

One way to test a moral theory, then, is to ask whether the principles or concepts it proposes cohere with the specific moral judgments one is committed to. For example, imagine that you observe me lying to my mother. Your natural reaction might well be to judge my action as morally wrong. But because you are a person who is slow to judge, you ask me why I lied to my mother, hoping that I had some good reason for my behavior. But, in fact, I simply enjoy lying to my parents, and I tell you that. Although you are kind and do not tell me, you conclude that my action is just plain wrong. It is not right to lie to one's parents simply because one enjoys it. Your belief that my action is wrong, and the belief in general that it is wrong to lie to one's parents for fun, are examples of moral judgments. In the course of life, we find ourselves making many moral judgments. Some of these judgments we reject after further thought, realizing that we did not fully understand the situations. But other judgments become more deeply held convictions. Philosophers seek theories that fit with these more deeply held convictions. If a philosopher proposes a theory, for example, that implies that it is morally okay to lie to one's parents simply because one enjoys it, and I believe this is wrong, there is incoherence between the theory and my deep convictions. If I were to accept this theory and continue to hold my convictions about lying to parents, then my moral believes would be in disequilibrium; they would not be coherent with one another. There would be two ways to bring my beliefs back into equilibrium. I might reject my convictions about lying and continue to hold the theory, or I might reject the theory, keep my convictions, and look for another theory that will adequately make sense of my deep convictions.

This method of weighing moral convictions and theories against one another is the way philosophers evaluate theories. In my discussion of relativism, for example, I asked you to consider whether you believed that a society without slavery was better than one with slavery. If you do, I argued, this belief is inconsistent with relativism, because relativism implies that no society can be said to be better than any others. I concluded that we should reject relativism in favor of a more adequate theory. Relativism just does not make sense of our deep convictions. Similarly, if ethics is based only on God's commands, then anything God might command would be morally obligatory. For example, this theory would imply that if God commanded us to murder, then it would be morally required of us to murder. But I argued that many cannot accept this. A theory that implies that God's command could make murder morally obligatory is unacceptable because it does not cohere with a deep conviction that murder is wrong, whether God commands it or not. A theory based on God's commands does not make sense of this deep conviction. Although it would be possible to reject the conviction that murder would always be wrong, even if God commands it, and thus bring one's believes into equilibrium with the theory, many are more inclined to reject the theory and look for another that explains why murder is wrong. Thus, both in the case of our discussion of relativism and our discussion of ethics based on God's commands, we were applying a method to evaluate moral theories. In the following discussion of philosophical theories of ethics, I will be employing this method as I discuss some of the arguments for and against each theory.

THREE SCHOOLS OF THOUGHT IN PHILOSOPHICAL ETHICS

In contemporary ethical theory, there are three general "schools" of philosophical ethics. Philosophers disagree about which approach provides the best overall theory of ethics. I believe that an understanding of all three schools can contribute to discussions in applied ethics, so my primary goal in the remainder of this essay will be to introduce you to each approach. Along the way I will point quickly to some of the arguments for and against each approach. The three approaches are these: consequentialist ethics, deontological ethics, and "virtue theory". To introduce them, I will explain each general school and one particular theory within each school.

Consequentialist Ethics

According to the consequentialist school of thought, the rightness or the wrongness of an action is determined by its consequences. If a particular action brings about the best possible consequences, then it is morally right. If an action has bad consequences, or consequences that are not as beneficial as those of some other action that could have been performed instead, then that action is morally wrong. That is the basic concept of consequentialist ethics. The feature of the world that makes some things right and other things wrong is the fact that actions have consequences. Some actions make the world better off. Other actions make the world worse off. The moral action will result in the best consequences.

But this doesn't get us far, of course, because to apply this concept, we must know how to define "best consequences." We need to know, in short, what makes consequences good or bad. The best-known example of the consequentialist approach is a family of theories called "utilitarianism." As an example, we will consider the classical form of utilitarianism that was articulated by Jeremy Bentham (1748–1832) and John Stuart Mill (1806–1873). In the rest of this essay, I will refer to classical utilitarianism simply as "utilitarianism." You should be aware, however, that there are recent developments and elaborations of the general utilitarian approach that a serious student will want to consider.

According to the consequentialist school of thought, the rightness or the wrongness of an action is determined by its consequences.

Utilitarians argue that the greatest good in life is pleasure. Unlike some philosophers, the classical utilitarians use the terms "pleasure" and "happiness" interchangeably. In defense of the claim that happiness, or pleasure, is the greatest good, Mill analyzes the structure of our desires, pointing out that there are things we desire for themselves, and there are other things we desire for the sake of what they can help us acquire. We all desire money, for example, but we do not desire money for itself. If money became useless for buying goods and services, we would no longer value it. As it is, we value it because there are a variety of things we can get with it. We can secure a place to live, food to eat, transportation, medical care, and so forth. But why do we want these things? Some of them, food and medical care, for example, we value because they help us to stay healthy. Why do we want to be healthy? Because an unhealthy life is painful. In contrast, a healthy life is necessary for full human happiness. In short, we want some things because we desire to be happy and to avoid being unhappy. Take the example of transportation. We want a car so we can do things with other people. We want to do things with them because they are our friends. We value friendship because it makes us happy. Note, once again we have arrived at happiness as the goal. Why do we want to be happy? Utilitarians would argue that this question makes no sense. We do not want happiness for any reason external to happiness itself. It is good in itself.

If right actions are those actions that produce the greatest amount of good, and the least amount of evil, and good is defined as happiness or the lack of unhappiness, then right actions are those that maximize the amount of happiness in the world.

Once the utilitarians have established happiness as the greatest good, it is a short step to applying the consequentialist principle. If right actions are those actions that produce the greatest amount of good, and the least amount of evil, and good is defined as happiness or the lack of unhappiness, then right actions are those that maximize the amount of happiness in the world. Actions that do not maximize the amount of happiness in the world are morally deficient. John Stuart Mill states what he calls the Greatest Happiness Principle like this: "actions are right in proportion as they tend to promote happiness, wrong as they tend to produce the reverse of happiness."[2]

Utilitarianism is an attractive theory. Pleasure is something that we all experience. It doesn't take a religious experience or special insight to realize that happiness is a good thing. Utilitarianism also seems to do a fairly good job of explaining why actions are right and wrong. Restraining another person by holding them is generally wrong. It makes them quite unhappy! But if it is necessary to restrain them to prevent them from causing major pain to another person, then restraining the first person is the right thing to do. Utilitarianism also explains the reason why we should behave differently in different cultures. What produces happiness in my own country may well produce unhappiness in another, so I must learn the culture and adjust if I am going to do the right thing.

What makes an action right or wrong, then, according to utilitarianism, is the kind of consequences that the action produces, in particular the amount of pleasure and pain it produces and that the alternative actions would produce. This in turn seems to provide a way to decide what is right or wrong as well. If one is in a quandary, one should weigh the consequences, make a list of the good consequences and the bad consequences, not just for oneself but for everyone affected by the decision, compare this list with similar lists for other possible actions, and choose the action that will lead to the greatest overall happiness. Unfortunately, in some situations, no choice leads to happiness, and one must "choose the lesser of evils," performing that action that produces the least unhappiness in the world.

It is important to emphasize that utilitarianism is not egoism. Egoism is also a consequentialist theory, but it says that the right thing to do is the action that causes the greatest happiness for the person performing the action. As it is expressed popularly, "Look out for number one." In contrast, utilitarians argue that we should look out for everyone. Because everyone's happiness is equally important, when I consider what is the right thing to do, I must take everyone's pleasure and pain into consideration, not just my own. There will be times when the right thing for me to do will not make me happy. The happiness of others will outweigh my own happiness. Ethics calls me, in these times, to sacrifice my own happiness and accept unhappiness, to produce happiness in others. On the other hand, utilitarianism does not say that I should always sacrifice my own happiness for the sake of others. My happiness is just as important as that of others. In sum, then, to decide what is right and wrong, one must seek to discover what will produce the greatest happiness overall, for all concerned.

Utilitarianism is simple and attractive as a moral theory, but it also has its drawbacks. I will suggest three. First, critics often argue that utilitarianism could justify treating people very badly. For example, if I could

[2]John Stuart Mill. *Utilitarianism*, edited by George Sher (Indianapolis: Hackett Publishing Co. 1979). 7.

find a cure for cancer by doing medical experiments on unwilling homeless people, it is conceivable that the amount of happiness resulting from the cure would outweigh the unhappiness caused to the people used, involuntarily, as subjects. Suppose further, that there was no other way to do these experiments except on humans and there were no humans who were willing to volunteer. Should I perform the experiments? It would seem that utilitarianism would say "yes." I would cause pain and suffering to these people. But a cure for cancer would prevent massive amounts of pain that would surely outweigh the pain produced by my research. Critics of utilitarianism argue that even if the experiments increased the amount of overall happiness in the world in a way no alternative actions could, the experiments would surely be wrong. It is simply wrong to use people against their will for such purposes. Take another example. One can imagine a situation in which it would be possible for a surgeon to save three people who need organ transplants by harvesting the organs of one relatively healthy patient. Imagine that the one whose life is sacrificed has few friends, has no family, and leads a depressed life. By ending one relatively miserable life, the surgeon makes possible three happy ones. This is what utilitarianism seems to condone. But surely, the critic objects, such actions could not be morally right. The first objection to utilitarianism is that it implies that things might be morally right that are clearly morally wrong, at least according to the critic. Virtually all of us have deep convictions that certain things would be wrong, even if they produced the greatest happiness. When a theory does not fit our deep commitments like this, philosophers have three choices. We could give up some of our deep commitments, deny that utilitarianism has these implications, or look for a different account of ethics. There are different philosophers who respond in each of these ways.

A second objection to utilitarianism is that it doesn't seem to take into consideration all of the relevant elements of an action. Consider my most recent example. If a Dr. Rainier sacrificed the life of one person to save three, that seems wrong. Yet if Sergeant Casper throws himself on a grenade to save three persons who will otherwise be killed, that seems morally praiseworthy. In both cases, one person is sacrificed to save three. In both cases it seems the good consequences outweigh the evil. So what's the difference? In one case, "the sacrificial victim" acts voluntarily to intentionally save others. In the other, the "victim" does not act voluntarily. In one case, a person is sacrificing herself to save others; in the other, a person is sacrificing someone else. Utilitarianism, however considers only the consequences. The differences of motivation seem relevant to morally evaluating these two actions, but they are not reflected in the consequences. Utilitarianism, in other words, does not provide all the concepts needed to evaluate actions.

Finally, utilitarianism faces a practicality problem. Although it sounds very reasonable to weigh all the happiness and unhappiness that would result from alternative actions, in many circumstances this is a very difficult thing to do. How can one know in advance what all the consequences of possible actions would be? How does one determine the level of happiness or unhappiness in those involved, such that the agent can weigh them? This objection may not be as problematic as it seems. Even Mill recognizes that in many circumstances in life the best approach to decision-making is to simply follow the rules that cultural tradition has given us, since tradition tends to identify moral rules which bring about the best consequences. Yet in controversial cases, Mill would have us apply the Greatest Happiness Principle itself. This, I am saying, would often not be a very easy thing to do. In fact, these controversial cases would probably be the most complex and difficult for which to determine all the consequences. Thus, in situations in which we need the help of moral theory, utilitarianism seems to come up short.

I have presented three objections to utilitarianism. But it would be a misunderstanding to take this as a quick dismissal of the theory. That is not my intention. There are utilitarian ethicists working today who have taken up the challenges presented in these objections with considerable intellectual rigor. You are encouraged to read elsewhere to consider their arguments.

Deontological Ethics

The second school of thought in contemporary ethics is deontological ethics. The word "deontology" comes from the Greek word *deon,* which means "duty." Thus, deontological theories are theories that focus on principles of duty. It is these principles that determine what is right and wrong. A right action is an action that is performed in accord with duty and from duty. But again, this doesn't get us very far until we spell out what our duties are. The most commonly discussed form of deontological ethics is that proposed by Emmanuel Kant (1724–1804). In this essay, I will briefly introduce Kant's theory as an example of deontological ethics.

deontological theories are theories that focus on principles of duty. It is these principles that determine what is right and wrong.

As an Enlightenment philosopher, Kant believed that human rational capacities set human beings apart from other life. Thus reason plays a crucial role in his ethics. At the foundation of Kant's view is the conviction that, because we are rational, we human beings have intrinsic value, value in ourselves. Whereas other creatures have instincts and inclinations, humans have reason. Our reason is both the source of our moral worth and the source of our duties to one another. Because all humans are rational, each human being has intrinsic value, and we have a duty to treat him or her with respect. What does this involve? It is reason itself, according to Kant, that specifies our duties. Unlike many philosophers since his time, Kant believed that because we have the same rational capacity, and it is reason that dictates our duties, we all have the same duties. In other words, there are principles or duties that apply to us all equally. The most basic of these principles Kant calls the Categorical Imperative.

Kant speaks as if there is one Categorical Imperative. But when he turns to spell it out, he does so in different ways that don't always seem equivalent. Here we will simply consider the most well-known ways in which Kant formulates the Categorical Imperative. They can be paraphrased like this:

> ***Universal Law Formulation: Always follow maxims (rules of action) that you could at the same time rationally desire to become universal laws of nature.***
>
> ***Ends-in-themselves Formulation: Always treat every human being, including yourself, as an end-in-itself, and never treat yourself or another human being as merely a means.***

Although these two formulations seem quite different, they have in common the conviction that human beings should all be treated with respect in virtue of their worth as rational beings. To consider how this is the case, we will consider each formulation in turn.

In the first formulation, Kant is assuming that as rational beings, humans act according to principles (often translated "maxims," and which I have paraphrased "rules of action"). We don't act just by instinct or inclination; we choose how we will act. And when we choose, we apply some rule of action. To act rightly, we should act only on rules that can be applied equally to all rational beings. Thus, if I am considering doing something, I should ask what my rule of action is, or, in other words, I should ask why I am going to do the action I am considering. I should then ask whether my rule or reason for action would justify the same action for all rational beings. If I cannot accept its application to all, I shouldn't apply it to myself. Here's an example: Imagine that I am thinking about cheating on an essay assignment by copying a complete paper from the internet. Am I acting rationally and morally? To decide, my first step is to formulate a rule that I am following. It would presumably go something like this: "Whenever I don't want to or am not able to write an assigned essay myself, and want to get a good grade anyway, I will copy a paper from the internet." The next step is to formulate this rule as if it were a universal law (of human action): "Whenever someone wants a good grade on an essay assignment but does not want to or is unable to write the essay, he or she will copy one from the internet." Finally, I ask myself whether I would want the world to always operate according to this rule *at the same time as I am performing the action being considered.* My goal in cheating on the assignment is to get a good grade. But if everyone cheated in the way I am considering, professors would have good ways to detect such cheating and would know it goes on all the time. Thus I would not get a good grade in a world in which my rule was a universal law. Given my goals, I could not rationally want the world to be this way. So it is my rational duty not to follow that rule. Kant's main insight is this: If I have a really good reason to do something as a rational being, then respect for others, who are also rational, should lead me to grant that my reason is a good reason for everyone else to do the same thing. If I can't grant that, then it must not be a good reason for me, either.

The second formulation is more straight-forward. Kant believes that all humans are rational and thus intrinsically valuable. Their value lies in the fact that they are rational and because of this can act according to duty, not just instinct and inclination. Out of respect for this unique human capacity, humans should be allowed to follow duty as dictated by their own reason. In order to do this, they must be allowed to make free moral decisions. This implies that they should be treated as beings who set their own ends, not as simply means or tools to be used for other peoples' ends. They should be treated as ends-in-themselves. Consider a clear example of a person being used merely as a means: slavery. Slavery is wrong because in slavery human beings are using other human beings as means, as tools, not allowing them to act freely as ends-in-themselves. Slaves are not treated as persons, but as property. And property is seen as something that can be used for the owners' ends. Slavery is one way to treat a person as a mere means. But to the extent that any action treats a human being as a thing and not as a person, that action is wrong. Lying and cheating are also ways of using people. Consider the act of turning in the copied paper. My professor's goal is to help me learn certain things and to give me the grade that is proper for my work. The college's goals include the aim of maintaining the integrity of the educational system and providing an opportunity for others to learn. By turning in work that is not my own, I am manipulating both my professor and the institution to help me achieve my goals against their will. This is wrong. I am not treating other persons with respect.

Enlisting someone else to help you meet your own goals is not always wrong. The key is whether you are also treating them with respect. If I hire a plumber at a fair wage to replace my bathroom fixtures, I am using her as a means to one of my goals. But I am not treating the plumber as a *mere* means. Formally or informally, I have agreed with the plumber to pay for her services. In doing so, I have allowed her to make her own decision whether to enter into the arrangement and to pursue her own life goals by earning money. In contrast to slavery, I am not forcing or manipulating another person to serve as a means. I am helping another pursue her goals by helping me. Similarly, if I hire a tutor to help me learn the material for a class, I am not using either my tutor or my teacher. I am not deceiving them or manipulating them in any way. Lying and cheating to get a good grade, on the other hand, manipulates others to give me something I desire. These actions are wrong because they use others in manipulative ways, without their consent. The notion of free and open consent is an important one for Kantian ethics. Open consent implies both that the decision maker has all the relevant facts and that she makes her decision without being coerced. Treating a person as an end-in-herself or himself, in other words, requires both truthfulness and consent.

The Categorical Imperative, as illustrated in these two formulations suggested by Kant, tells us our rational duty. It is our rational duty that determines which actions are right or wrong. There is another dimension of ethical action that should be mentioned before we leave our discussion of Kant. From what we have said so far, we can conclude that a right action is one that conforms to rational duty. This is, in fact, part of Kant's answer. But for Kant, there is more to being a morally praiseworthy person than doing the right things. Morality calls for us to do the right things *for the right reasons.* In Kant's mind, the right reason is duty. To be morally praiseworthy in one's actions, one must do the right things and do them because it is one's duty. In discussing utilitarianism, we noted that it did not seem to take into account the intentions of the actor in performing an action, only the consequences of the action itself. In stipulating that a morally praiseworthy action must be performed for the sake of doing one's duty, Kant is weighing in on the moral relevance of intentions. A right action is one that conforms to the principles of rational duty. A morally good person is one who performs the right actions, not because she realizes it will make her life more pleasant, or because she fears being punished for acting wrongly, but because it is the right thing to do.

In order to give you a proper taste of Kant's form of deontology, it was necessary to go into a bit of detail. The most important points to take away from this discussion are these: Kant believed that human beings are set apart from other animals by being rational. In virtue of this rationality, humans have inherent value. Thus, it is our rational duty to treat ourselves and other humans with respect. To treat humans with respect implies that others should receive equal treatment as inherently valuable beings and allowed to act freely to pursue their own goals. On the other hand, because humans have inherent value, actions that use humans merely for the benefit of the person performing the action or for others, is morally wrong.[3]

---

[3]See Immanuel Kant, *Grounding for the Metaphysics of Morals*, translated by James W. Ellington (Indianapolis: Hackett Publishing Co. 1982).

There are several features of Kantianism that make it attractive. Above all, Kantianism captures some of our deep commitments about ethics. Take for example the desire to be tolerant of others that motivates many relativists. Kantianism provides a theoretical foundation that relativism lacks for the idea that humans should be allowed to act freely and purposively. Human beings should be treated with respect, allowed to make their own free decisions, and not forced to behave in certain ways. If what they do is wrong, one can enter into dialogue with them and, on the basis of reason, try to convince them that they should change. But to simply condemn others because they are different is surely wrong. Similarly, Kantianism captures many of our intuitions about using homeless people, against their consent, to undergo medical experiments in order to find a cure for cancer. In spite of the fact that the results could well produce happiness that outweighs the unhappiness of those on whom the experiments are performed, the experiments fail to treat people with respect and are thus wrong. Finally, Kantian ethics provides a basis for arguing that humans have basic rights. Unlike relativism, which seems to have no way to condemn actions within our own culture or others, Kantianism can argue that to treat persons with proper respect, as ends-in-themselves, implies that humans have rights to be treated in certain ways and not be treated in other ways. Kantianism thus captures many intuitions about what is right and wrong and grounds them in a general theory.

On the other hand, Kantianism has its own critics. Consequentialists argue that Kantianism does not take into account the importance of consequences. Kant believes his theory has strict implications. One of these is that one should never lie. A classical objection is to respond with a thought experiment: What if the Nazi's are at your door asking whether you are hiding Jews? What should you do? Many people would argue that in circumstances like these, you are justified in lying. Why would that be so? Presumably it is because in this circumstance lying avoids evil consequences and promotes good consequences. Kant argues, however, that it would be wrong to lie, even in these circumstances. In a similar way, Kant's theory implies that it is wrong to manipulate, detain, or harm innocent people for the benefit of others, especially without their consent. Yet it is not hard to imagine situations in which it would seem necessary, perhaps to save the lives of many, to do just these things.

Another aspect of Kantianism that seems problematic to some is the type of motivation that it calls for. As we have seen, Kant wrote that a moral person will perform the right actions out of duty. Although this does speak to relevance of intentions, to many it doesn't seem to get things right. Imagine that Joe has developed the habit of giving generously to various worthy causes. One can imagine that he gives because he has a sense of duty to help others through these causes. Yet, one can also imagine that Joe just enjoys helping others. Imagine that the motivation is not duty at all, but that Joe enjoys giving to these causes. If this is the case, if he is doing it because he likes to give, does this make his actions less praiseworthy? Does this motivation make Joe less worthy of a person? Many would say it does not. In fact, many would say that the fact that Joe enjoys doing the right thing is evidence of his goodness as a person. Although Kant recognizes the importance of intentions, these people would suggest he has the wrong understanding of what constitutes good intentions.

I have pointed to two difficulties that are sometimes raised against Kantianism. As with Utilitarianism, many have defined Kant's theory and other deontological theories. If you are interested, I would urge you to explore the writings of others who have both defended and criticized deontological ethics. For now, however, we must move on the third school of thought that has dominated ethical theory in recent years.

Virtue Ethics

A third approach to ethics commonly defended by philosophers today is Virtue Ethics. Although the label "Virtue Ethics" is used of a diverse range of ethical theories, I am going to attempt to introduce this general approach by considering ethical themes from ancient Greek philosophy, focusing in particular on Aristotle, from whom many virtue theorists get inspiration. At the heart of Aristotle's ethics is the notion that human beings are to live up to their unique natures. Like Plato, his teacher, and many others of the ancient Greeks, Aristotle believed that human beings were set apart from all other animals by their rationality. Humanness, in Aristotle's view, can be characterized by the fact that people are rational animals. Thus the human ability to reason is the key to living a good human life.

At the heart of Aristotle's ethics is a functionalist understanding of excellence or virtue. Aristotle assumes that one can speak of the function of things, including animals, plants, and things made by humans. Consider a knife. How do you determine whether a knife is a good knife? The answer is found in understanding the function of a knife, which is to cut. You might say that to have "knife-nature" is to be able to cut, and to be a good or virtuous knife is to be able to cut well. Because the function of a knife is to cut, a good knife is a knife that cuts well. Aristotle reasons in a similar way when he turns to humans. Human beings are set apart by their ability to reason. Humanity's unique nature is to reason. So the good or virtuous human life will be a life characterized by the proper use of reason. Socrates had suggested to his fellow Athenians that "the unexamined life is not worth living." In other words, one should use one's rational nature to examine one's life. Plato had argued that a good, or just human life will be a life in which the desires and passions are controlled by reason. Aristotle continues in the same vein. The key to a virtuous life is reason.

Virtue ethics begins, then, with a conception of what it means to lead a good life. In the case of Aristotle, the good life is a life lived in accord with reason. But unlike the later philosophy of Kant, Aristotle did not spell out "living in accord with reason" in terms of rules or principles; he instead understood the good life to be a life characterized by virtues. Think once again about a good knife. It cuts well. We could spell this out by saying that the good knife has the virtue of sharpness. Perhaps we could add other characteristics to our description of a good knife, such as being balanced and easily handled by the one who is cutting. Balance, sharpness, and ease of handling are virtues that enable the knife to cut well. In a similar way, there are characteristics (virtues) that contribute to living a good human life, and there are other character traits (vices) that detract from living a good human life.

What are the key virtues necessary for living a good human life? Aristotle answers by distinguishing between two categories of virtues, intellectual virtues and moral virtues. If a person can attain a life of rational contemplation, that is the absolutely best life one could live. This sort of life involves the intellectual virtues of knowledge and wisdom. But there are other ways reason can be built into one's life as well, through developing moral virtues. It is with regard to these moral virtues that Aristotle makes his most unique contribution. Aristotle suggests that moral virtues can be analyzed as midpoints between extremes. For example, Plato had included courage as a central virtue of a good human life. Aristotle argues that courage can be seen as a "mean" or midpoint between extremes. Aristotle says that in a situation of danger, a human being can fall short in two different directions. On the one hand, he can be foolhardy and rush into danger irrationally. Foolhardiness is a vice; it detracts from living a good life. Similarly, he might be cowardly and so much afraid of danger that he holds back even when he must take risks. Cowardice is also a vice. Courage, Aristotle argues, is a midpoint between irrational foolhardiness and irrational cowardice. The life characterized by reason will be a life lived courageously.

What are the key virtues necessary for living a good human life?

Consider a second example: generosity. What does it mean to be generous? Aristotle finds the answer in speaking of a mean between extremes. On the one hand, to be so stingy that one never gives aid to others is a vice. On the other hand, if a person gives her money away whenever anyone asks, one is not living rationally either. To be generous, then, is to have the habit of giving just the amount appropriate for the circumstances, neither too little nor too much. Finally, consider one third example: sense of humor. A good human life requires that a person have a sense of humor. If one finds nothing humorous, one is not living a rationally balanced life. To lack a sense of humor is a vice. On the other hand, if a person finds everything humorous, that person is a buffoon. This also is not conducive to the good life. To live a good life is to find humor in the right proportion for the particular circumstances in which one finds oneself. It is to have a proper sense of humor, a mean between two extremes. To live by the mean in this way is to live rationally, and it thus contributes to a good human life. In sum, then, Aristotle argues that a good person is a person in whom certain character traits are present that contribute to a rational life. These character traits are what we call virtues. A good person will have moral virtues built into his character, and she or he will not be characterized by moral vices, traits that detract from rational living.[4]

Aristotle's ethical theory illustrates several key aspects of Virtue Ethics. First, Virtue Ethics focuses on a different question from the two approaches we have considered above. Both consequentialism and deontological ethics focus on the question "What actions are right, and what actions are wrong?" Utilitarianism answers this question by appealing to consequences. Kantianism answers this question by applying principles of duty. Virtue Ethics seeks an answer to the question "What is a good person?" In doing so, it focuses on two key concepts: the good life and virtue. For Aristotle the question about what it means to be a good person begins with a conception of the function of human life, with a particular understanding of human nature. From his conception of human nature, Aristotle derives a general understanding of what it means to be a good person, a person who lives well. To spell this out, however, is to identify virtues that enable living well.

Both consequentialism and deontological ethics focus on the question "What actions are right, and what actions are wrong?" Utilitarianism answers this question by appealing to consequences. Kantianism answers this question by applying principles of duty. Virtue Ethics seeks an answer to the question "What is a good person?"

The elements I have outlined in Aristotle's ethics are reflected in virtue theories in general. First, a virtue theory rests on a particular concept of human nature, or humanness. Based on this, a virtue theory will spell out what it means to live into one's full human nature, a task that will require the development of character traits or virtues. Unlike the utilitarian, who focuses on doing actions that produce good consequences for oneself and others, or the Kantian, who focuses on following the rules of rational duty, the virtue theorist focuses on what kind of life is worth living, what kind of person one wants to become, and what character traits must be nurtured to make this possible.

Virtue Ethics has attractive features. First, Virtue Ethics provides a framework for thinking about moral education. Human beings develop as morally responsible persons not by memorizing rules, or even by learning ways to calculate consequences of actions, but by developing character. But to have character is to have developed habits or traits that enable one to live well. Virtue ethics is supportive of educational philosophies that aim at developing character in students. And since the Greeks, those taking the virtue approach have emphasized that character or virtue cannot be developed overnight by an adult, but must be the result of appropriate training and practice in living. Again, this approach supports the idea that teachers and parents should give children opportunities to develop various virtues. Character, if it has been nurtured properly in a person's life, becomes "second nature." For the Greeks, the most important virtue of all was wisdom, and a person of wisdom is able to respond to each situation in life in a proper way, not because he or she knows the rules, but because he or she is a good person. It is not rules that are fundamental, but good living. How should one live? The best way to learn the answer is to observe a good person, to learn from her way of life, and to develop virtues within oneself. One learns to be a good person, in other words, by observing good models and practicing being good. All of these features, it seems to me, fit with the most effective types of moral education. We learn not from memorizing rules of right and wrong, but by observing role models and developing proper character traits through a long process of nurture that begins in childhood.

This leads to a second feature that I find attractive in virtue theory. Earlier we mentioned that an adequate theory of ethics must account for the role of intentions. Although people's intuitions vary, many ethicists find Aristotle's approach to intentions more intuitive than Kant's. Earlier we said that Kant argues that a person's action is not morally praiseworthy unless the person performs the action out of duty. For Kant, acting out of inclination is the opposite of acting according to reason. Thus, if a person likes to give to others, and gives because she likes to, that person is not giving according to a morally worthy intention. She

---

[4]For Aristotle's own account, see Aristotle. *Nichomachean Ethics* (Indianapolis: Hackett Publishing Co. 1985).

is acting on inclination, not duty. The person who gives even though she doesn't want to is a better person than one who gives because she enjoys helping others. Virtue ethics turns this around. To be a good person is to so develop the virtue of generosity that one gives naturally, and even enjoys it, and finds happiness in living the virtuous life. Thus, on the Aristotelian account, a good person is a person who enjoys giving because it has become a part of who she is. Far from being forced, proper generosity is simply a part of her life. Some find this a more plausible account of moral psychology than Kant's. My discussion has merely introduced the issue, however. I encourage you to study both Kant and Aristotle yourself as you think through your own approach to ethics.

I have suggested features that make Virtue Ethics attractive, but there are important criticisms and questions faced by virtue theory as well. First, if my analysis above is correct, each virtue theory begins with a particular concept of human nature and of the good life. If so, Virtue Ethics seems to throw us back to relativism, because concepts of the good life and human nature vary from one culture to another, from one historical period to another. Not only that, in the postmodern period in which we live, the very question of whether there is such a thing as "human nature" is hotly debated. If human beings evolved over a period of time, it becomes less clear that they have a fixed "nature," "purpose," or "function" upon which a concept of the good human life can rest.

A second question that can be raised concerning Virtue Ethics is whether it has what it takes to provide a foundation for ethical theory. The issue is whether it is possible to identify a concept of the good life without appealing to either principles of right or wrong or to goods such as happiness. Allegedly at the foundation of ethics is a concept of the good life. Virtue theorists tend to say that the way to identify the good life is to observe a good person. But how does one distinguish a good person, and a good person's life? Can any life be considered a good life simply be being accepted within a culture? Or are there general features that make one life better than another? One is tempted to answer in one of a couple of ways. On the one hand, one might argue that a good person is a person who treats self and others in the right ways? But if this is the answer, then it appears there is something at stake more fundamental than virtues, because this answer appeals to something like the deontologists' principles of right action to determine whether a person treats self and others in the right way. On the other hand, one might suggest that a good life is a life that is happy in the overall sense. But again, this seems like an appeal to something similar to consequentialism. What distinguishes virtues from vices? If the virtue theorist answers by saying that virtues are those character traits that tend to make oneself and others happy, and vices are character traits that tend to make oneself and others unhappy, then one is appealing to something like utilitarianism to evaluate virtues. Either of these answers, then, seems to appeal to some other, more fundamental, ethical concepts to identify both good lives and virtues, not providing an alternative to the other theories at all.

CONCLUSION

In my discussion of consequentialism, deontological ethics, and virtue theory, I have suggested arguments for and against three representative theories. I do not pretend to have given conclusive considerations for or against any of the theories. Each human being should think through his or her ethical philosophy and ask what he or she thinks makes the most sense. I would encourage you to explore each theory in greater detail and ask yourself which you find the most acceptable. It should also be noted that some ethicists find it helpful, particularly in applying ethics to specific issues, to employ the concepts of all of the theories in bringing out various dimensions of moral issues. One might, for example, examine the consequences of the actions being considered, the rights and responsibilities of the people involved, and the virtues that could be nurtured in relation to a particular moral issue. It is my hope that this introduction has helped you to understand influential approaches to ethical theory in philosophy today, both as a basis for more extensive exploration of the theories and as a foundation for using concepts highlighted by various theories to apply ethics to specific moral issues.

The Anthropocentric Tendency in Western Ethics

Since the middle of the twentieth century, humankind has become increasingly aware of the growing threat we have caused ourselves through our actions toward the nonhuman world around us. As the environmental movement has grown, almost everyone has come to agree that we have an ethical obligation to care for our environment. But what is the basis of this obligation? Do we have an ethical obligation to care of our environment because human life depends upon it? Is it an obligation to ourselves and to our descendents? In other words, is our obligation to care for the environment really an obligation to human beings, both now and in the future? Anthropocentrism is the view that answers "yes" to these questions. According to anthropocentrism, ethics focuses primarily on humans. On this view, it is humans who are valuable, whereas other things are morally valuable only indirectly, because of their value to humans. In slightly more technical terms, this approach claims that humans alone, of earthly things at least, are intrinsically valuable. All other elements of the natural world, including other animals and plants, are extrinsically (or instrumentally) valuable in as much as they are of value to humans.

Among environmentalists, anthropocentrism is a common focus of blame for the environmental crises we face. According to them, it is our human-centeredness that has caused us to ignore the needs of other living beings. Who are we, they ask, to think that human life is important in itself, whereas other forms of life are not? Anti-anthropocentrists don't agree on what makes nature and various elements of the natural environment valuable, but they do agree that nature, or the environment, has value in itself. We have an ethical obligation to animals, to plants, to species, and to other aspects of the environment, not just for our own good, but because one or more of these aspects of nature have value in themselves.

Among environmentalists, anthropocentrism is a common focus of blame for the environmental crises we face. According to them, it is our human-centeredness that has caused us to ignore the needs of other living beings.

There is no doubt that Western ethics, both religious and philosophical, has tended to be anthropocentric. In this brief essay, I will illustrate this anthropocentric tendency by quoting some significant passages from three prominent philosophers in the Western tradition. Second, I will summarize points made by anthropocentism's critics. My intention, throughout the essay, is simply to introduce the historic roots and issues raised by the anthropocentric element in the Western tradition.

WESTERN ANTHROPOCENTRISM: THREE HISTORICAL SNAPSHOTS

A common thread that runs from Greek philosophy through medieval thought and into the modern worldview affirms that humankind stands above the rest of the natural world in such a way that nonhuman elements are valuable, if they are valuable, because of their value to humans. To illustrate this strong anthropocentric thread, I will cite passages from three centrally important Western philosophers, one ancient, one medieval, and one early modern. I turn first to a passage from the *Politics* of Aristotle. Aristotle writes

> ***It is similarly clear that we must suppose that plants exist on account of animals. . . and the other animals for the sake of man, the tame ones because of their usefulness and as food, and if not all***

the wild ones, then most, on account of food and other assistance (they provide, in the form of) clothing and other tools which come from them. If, then, nature does nothing without an end and nothing in vain it is necessary that nature made all these things on account of men.[1]

A common thread that runs from Greek philosophy through medieval thought and into the modern worldview affirms that humankind stands above the rest of the natural world in such a way that nonhuman elements are valuable, if they are valuable, because of their value to humans.

According to Aristotle, plants and animals derive their value from their usefulness to humans. In fact he goes so far as to say that nature "made" other living things "on account of man." In the natural order, as Aristotle sees it, everything else is for humankind. He does not stand alone among the Greeks in placing human beings at the center of value. Plato before him, and Greek philosophers after Aristotle, all believed that the distinctive feature of humankind was reason, and that this feature gave humans the top place in the hierarchy of value.

Both the belief that humans were set apart by human rationality and the belief that humans were the focus of value in the world were common in the medieval period as well. Although this is a long period of philosophical development and diverse views, the view that humans have a place above other living things on a hierarchy of value was accepted by most thinkers. In the following passage, from Thomas Aquinas, we get one example of this hierarchical perspective from the later medieval period.

The very condition of the rational creature, in that it has dominion over its actions, requires that the care of providence should be bestowed on it for its own sake: whereas the condition of other things that have not dominion over their actions shows that they are cared for, not for their own sake, but as being directed to other things. Because that which acts only when moved by another, is like an instrument; whereas that which acts by itself, is like a principle agent. . . . Accordingly intellectual creatures are ruled by God, as though he cared for them for their own sake, while other creatures are ruled as being directed to rational creatures.[2]

Like Aristotle, Aquinas suggests that although humans have value in themselves, other living things have value only for the sake of humans. And it is reason that is the key factor that distinguishes humans from others. Humans, in Aquinas's terms, are "intellectual beings." Only intellectual beings are valuable in themselves. And for Aquinas, the category of intellectual beings does not include nonhuman animals. Thus regarding the treatment of animals, he writes

Hereby is refuted the error of those who said it is sinful for a man to kill dumb animals; for by divine providence they are intended for man's use in the natural order. Hence it is not wrong for man to make use of them, either by killing or in any other way whatever. For this reason the Lord said to Noe (Gen. ix. 3): As the green herbs I have delivered all flesh to you.[3]

Aquinas goes on to explain that even passages in the Bible that admonish humans not to muzzle the ox that treads the grain, do not indicate that it is wrong in itself to mistreat nonhuman animals, but that mistreating animals is wrong because such actions nurture a disposition toward cruelty to human beings, or because it harms the animal victim's human owner. In other words, for Aquinas harm to nonhumans is wrong, if it is wrong at all, only because it harms humans.

These anthropocentric themes that humans are rational and thus above and more valuable than other living things continue into early modern philosophy as well. Immanuel Kant, perhaps the most influential philosopher of the Enlightenment, has this explanation of our duties to animals:

. . . So far as animals are concerned, we have no direct duties. Animals are not self-conscious and are there merely as a means to an end. That end is man. We can ask, "Why do animals exist?" But

---

[1]Aristotle, Politics, Translated by Benjamin Jewett, Book I, Chapter VIII.

[2]Thomas Aquinas, Summa Contra Gentiles, Translated by English Dominican Fathers (Benziger Brothers, 1928), Book 3, Part 2, Chapter 112.

[3]Thomas Aquinas, Summa Contra Gentiles, Translated by English Dominican Fathers (Benziger Brothers, 1928), Book 3, Part 2, Chapter 112.

to ask, "Why does man exit?" is a meaningless question. Our duties toward animals are merely indirect duties toward humanity.

In this passage, Kant's reasoning is very similar to that of Aquinas. Kant notes that humans are self-conscious, a characteristic that for Kant is virtually synonymous with reason. Kant goes on to argue that because actions toward animals are analogous to actions toward humans, it follows that in cultivating kindness or cruelty to animals, we are cultivating kindness or cruelty to humans. Kant writes: "If he is not to stifle human feelings, he must practice kindness toward animals, for he who is cruel to animals becomes hard also in his dealing with men."[4]

In other words, humans should be kind to animals, but not because it is wrong in itself to mistreat animals themselves. It is wrong to mistreat animals because in doing so one develops habits that lead to the mistreatment of humans. It is humans who are of concern, not animals.

These passages from Aristotle, Thomas Aquinas, and Kant are sufficient to illustrate an anthropocentric strand that runs through the Western intellectual tradition. Aquinas and Kant speak of animals. Aristotle mentions plants as well. The reasoning of all three suggests that everything in the natural world, if it is of value, derives its value indirectly from the value of humans. Before we turn from this sampling of historical views, I should mention that there have also been clear minority opinions. Francis of Assisi is famous for demonstrating his more caring attitude toward animals through his life and writings, in which he refers to other aspects of creation as brothers and sisters. Jeremy Bentham, one of the founders of classical utilitarianism in the early modern period, argued that the pleasure and pain of animals should be considered equally with that of humans. Arthur Schopenhauer appeals to Asian thought for a more compassionate attitude toward animals and says he finds Kant's view that animals are to be treated kindly, merely as practice for how we treat humans, "revolting and abominable." In spite of these exceptions, however, the predominant and most influential strand in Western ethics has been clearly anthropocentric.

THE ANTI-ANTHROPOCENTRIST CHALLENGE

In spite of this prdominance, anthropocentrism has come under significant attack among contemporary environmental philosophers. Without going into detail, I will mention just four considerations that contribute to the anti-anthropocentric challenge. First, although anthropocentrism need not support environmentally harmful behavior, it has often done so. Humankind, thinking that it was the only being on earth that was intrinsically valuable, has too often devalued nonhuman forms of life on our planet. Although this does not, in itself, prove that the problem is with anthropocentrism per se, it has led many concerned environmentalists to consider theories of value grounded in something broader and "deeper" than humanity.

Secondly, the rise of the science of ecology has highlighted the fact that humankind is a part of the natural world, not above the natural world. Humans, like every other living thing, are linked in an intricately balanced system of life that makes up the biosphere in which we live. With the rising appreciation for the important role other life-forms play in this web, many have begun to believe that each element in the web should be equally valued.

Thirdly, our increasing knowledge of animals has blurred the line between humans and animals. Rationality, consciousness, the use of language, the use of tools, and problem-solving have all been characteristics or abilities that were thought to

our increasing knowledge of animals has blurred the line between humans and animals. Rationality, consciousness, the use of language, the use of tools, and problem-solving have all been characteristics or abilities that were thought to set humans apart from other animals.

[4]Kant, Immanuel, Lectures on Ethics, Translated by Louis Infield (Harper and Row, 1963) 239-241.

set humans apart from other animals. But studies suggest that some nonhumans communicate with one another in quite sophisticated ways, reason out solutions to problems, and use tools. It has become harder and harder to deny that some other animals are conscious of themselves, experience emotions, and use reason. The point is not that humans are not, as a whole, uniquely developed in some of their capabilities. It is rather that recent scientific discoveries have made this uniqueness appear to be a matter of degree. Rather than evidencing humankind's distinctness, the discoveries suggest continuity between higher animals and humans. And if there is no clear distinction between higher animals and humans, it becomes difficult to see why we should value humans intrinsically, while we value other animals only extrinsically. This again has led many environmental thinkers to look for a new and broader basis for environmental value.

Finally, and related to the third point, some philosophers have come to believe that anthropocentrists are guilty of a kind of moral inconsistency similar to racism. Racists place a higher value on members of one race (their own), simply because they are part of that race, and a lower value on members of another race, simply because they are members of that other race. We condemn racism because we believe that race has no moral significance. Members of one race are not more valuable or morally significant than members of another race simply because of their race. Racists base moral value on a factor that is morally irrelevant. In doing so, they violate the principle that those who are equal in relevant respects should be treated equally. Some philosophers, particularly those who have taken up the cause of animals, are now arguing that treating humans as more valuable than all other animals is similar to the injustice of racism. Their argument begins by pointing out that anthropocentrists place a higher value on the lives and well-being of other humans than on the lives and well-being of nonhuman animals. To be morally consistent, this difference of moral valuation must be based on some morally relevant difference between humans and members of other species. Finding such a difference is more difficult than it at first appears. As we have already seen, scientific studies are fuzzing the line between humans and higher animals. The distinction between intrinsically valuable beings and extrinsically valuable beings demands more than a fuzzy line for its justification.

The distinction between intrinsically valuable beings and extrinsically valuable beings demands more than a fuzzy line for its justification.

Anti-anthropocentrists drive home their point by considering individual cases. As we have seen, historically anthropocentrists have identified characteristics closely associated with human rationality, such as the ability to use language or tools as crucial to setting humans apart as intrinsically valuable. Other animals, which allegedly do not have these characteristics, are not intrinsically valuable. Thus these other animals can be used in various ways for human benefit. In this light, experiments on higher nonhuman animals are seen as morally justified because they benefit humans. But the opponents of anthropocentrism raise an objection: what about *humans* who do not have the specified characteristics? What about humans who do not have the normal human rational abilities? These humans, whose rational capacities are impaired, do not have the characteristics that anthropocentrists claim give humans intrinsic worth. Do these individuals thus lack intrinsic worth? Are experiments on them for the benefit of other humans justified? If not, why not?

The natural answer is to say that such experiments would be very wrong, especially because individuals who are rationally impaired cannot give their consent. But why is it wrong to perform the experiment on these individuals and not wrong to perform the same experiment on a nonhuman with similar or even greater rational capacities? The natural answer is to say, "because they are human beings!" And this is exactly where the anti-anthropocentric argument hits home. Unless one can identify some characteristic that is essential and unique to a species and that is morally relevant, species is no more relevant morally than race. It is not clear that there is any empirically observable characteristic that sets all humans clearly apart from all other animals. The only thing we are left with is the fact that humans are humans. But to value one individual more highly than another simply because of its species is to assign value on the basis of a morally irrelevant feature.

Religious anthropocentrists who base their views on the Bible may think there is an easy reply. Can't we simply say that humans are valuable because they are made in the image of God? As a Christian philosopher myself, I am sympathetic with this stance. Yet, I am not entirely satisfied with this as an answer to the anti-anthropocentric challenge. As a philosopher, I want to ask what it means to be made in the image of God. If one has just said that the feature that sets humans apart from other animals is the image of God, then one cannot simply turn around and say that the image of God is our humanness. That would make the theists' response to the above objection clearly circular. The reasoning would go like this: Humans are valuable because they are made in the image of God. To be made in the image of God is to be human. That is, humans are valuable because they are human. This is exactly the speciesist claim the objector is charging us with making. It does not identify a morally relevant feature that humans have and nonhumans do not. Furthermore, there seem to be good reasons to want to look for some account of what it means to be made in the image of God that does not identify the image with the biological species *homo sapiens.* It is certainly a logical possibility that God could create other beings, on another planet for example, who are also made in the image of God, and who are not *homo sapiens.* If so, God's image is not a matter of biological species. Then what is it? Thus, the response that what sets humans apart is the image of God ultimately leads to a question parallel to the question raised by the anti-anthropocentrist. What is it that sets us apart? If it is the image of God, what is that?

CONCLUSION

In this essay, I have attempted to provide a taste of the anthropocentric perspective that predominates in Western philosophy and religion. I have also outlined four considerations that have led to a challenge to this perspective in recent years. It is my hope that this will provide an introduction that will help you reader further on both sides of this discussion. Before I close, however, I must also say that it is a misunderstanding to equate anthropocentrism with a crude attitude of unthinking exploitation of the environment. In fact, it is possible to make a strong case for environmental protection on the basis of an anthropocentric approach to value. As we learn more and more about the interconnectedness between human life and the natural world, we discover more fully why we should preserve our environment, even from an anthropocentric perspective. We are learning that if we really value human life and culture we must care for the natural environment in which humans live. In short, there are both problems and promise in the anthropocentric tradition with regard to our attitudes toward nature and the environment.

NAME ______________________________

Worksheet 10 — Philosophical Ethics and Anthropocentrism

1. Do you believe that lying is ever moral? Explain your answer.

2. Imagine that you have promised to go to dinner and a concert with your good friend this evening. Early in the afternoon, another friend calls and talks with you. It is obvious that your other friend is quite depressed and could use a friend to talk with this afternoon and evening. You wonder whether you should cancel your date and spend the time with your other friend. According to classical utilitarianism, how do you decide what is the morally best thing to do?

3. Imagine that you have promised your uncle as he was dying that you would give his son, your cousin, $10,000. After your uncle's death, you discover that your cousin has a significant drug addiction, and you think the money will help him feed the addiction. Do you believe you should give him the money? Explain. What do you believe a Kantian would say you should do? Explain.

4. **What is the primary question asked by virtue ethics? Can you think of any human character traits (virtues) that would enable a person to live a more environmentally friendly life?**

5. **In the reading, "The Anthropocentric Tendency in Western Ethics," it is suggested that one could argue that we should care for the environment from an anthropocentric perspective. Give an example of an anthropocentric argument for taking care of some aspect of the environment.**

Pulling It Together — Philosophical Ethics and Anthropocentrism

QUESTIONS FOR THOUGHT AND DISCUSSION

1. Make a list of reasons we should care for our natural environment. Which of these reasons are "anthropocentric" reasons?

2. Do you believe that "value" is ultimately based on humans? Why or why not? If so, in what way?

3. Aquinas and Kant both argued that the reason it is wrong to be cruel to animals is that it leads to cruelty to humans? Is this the only reason it is wrong to be cruel to animals? Explain your answer.

EXERCISE

Research Exercise: Ask a variety of people (your roommate, friends, family members, and so forth) whether it is wrong to beat a pet dog. If they say 'yes,' ask them why. Write down their reasons. Reflect on the list of reasons you collect and explain whether each is an anthropocentric reason, and come to class ready to discuss your findings.

Chapter

11

Sentientism: Expanding the Circle to Include Animals

INTRODUCTION

In the previous chapter I introduced prominent approaches to philosophical ethics. Some of these theories are undeniably "anthropocentric," placing humans at the center of ethics. And even theories that I would argue need not be anthropocentric have usually been applied in an anthropocentric way. As I suggested in Chapter 12, and as we consider more fully in this chapter, utilitarianism is a theory that many would argue reaches beyond anthropocentrism. Classical utilitarianism claims that the greatest good is pleasure or happiness, and the greatest evil is pain or unhappiness. The implication of this is that actions should be weighed morally according to the amount of pleasure and pain they produce. But whose pain? Whose pleasure? If one is going to stick to a pure utlilitarian theory of value, one must respond: "any pleasure and pain." For on utilitarian grounds there is no basis for saying that the pain and pleasure of one is more or less valuable than the pain and pleasure of another. The pain and pleasure of anyone is morally important. In saying this, however, I have used a fudge word: "anyone." When we hear the word "anyone," most of us think of human beings. But the argument I have just given goes farther. If pleasure or happiness is the greatest good, and pain or unhappiness is evil or bad, then any pain or pleasure should be considered, no matter what being experiences it. The pleasure and pain of any being that can experience pleasure and pain counts. In spite of the fact that there have been philosophers, most notably Rene Descartes, who did not think nonhuman animals experience pain, the evidence is overwhelming that many do in fact experience significant pleasure and pain. It would seem, then, that if one is to consistently apply the classical utilitarian principle of ethics, one must include animals that can experience pleasure within the moral community, within the circle of beings who are morally considerable.

The view that I have been describing was suggested by Jeremy Bentham (1748–1832), one of the classical utilitarians in a much quoted passage that is the first reading in this chapter. This application of utilitarian ethics to all sentient animals has been powerfully defended today by philosopher Peter Singer. Singer defends what he calls the "animal liberation" view. Quoting classical utilitarian Jeremy Bentham, he argues that equal consideration should be given to the interests of any being that can suffer. This view has come to be called "sentientism." In this context, we will define a "sentient being" as a being that has the capacity to experience pleasure and pain.

Unlike the classical utilitarians, Immanuel Kant was unambiguously anthropocentric in his ethics. Nevertheless, Tom Regan, another contemporary philosopher active in today's discussion of ethics, has defended a theory of ethics that reflects Kantian ethics in important ways. Kant had argued that human beings are set apart from all other living things by their capacity to reason. It was this, Kant thought, that placed human beings inside what I have called the moral community. Kant contended that because of their rationality human beings have inherent worth. This is what makes humans ends in themselves. It is morally wrong to use a being with inherent value as a mere tool or instrument to get what one wants. So Kant said that humans should always be treated as ends-in-themselves and never merely as means. Humans are rational, have inherent value, and should be treated as ends-in-themselves. Other animals are not rational and thus can be treated as means to human ends.

But can we really say that all humans are rational, and all nonhumans are nonrational? Just as there is evidence that nonhuman animals can experience pleasure and pain, there is significant evidence that some animals have rational capacities that humans have failed to recognize until fairly recently. On the other side of the line, there are humans who have very limited, or perhaps no, rational capacities. So if we tie inherent worth to having rational capacities, then it would appear that some nonhumans do have inherent value and some humans do not. Tom Regan responds not by defending rationality as the criterion of moral worth, but by arguing that we should change the criterion for determining inherent value: the ability of a being to experience life as the "subject-of-a-life." What gives human beings inherent value is that "they have a life." To say that a being "has a life" is not merely to say that it is alive; it is to say that this being's life is valuable to itself. Its value is not external, but inherent within itself. It is this that grounds human rights. We value ourselves, and we accordingly have rights to not be treated in certain ways. Consistency, however, requires that this criterion be applied to any beings who are "subjects-of-a-life." Thus, Regan extends a kind of modified Kantianism to many nonhuman animals, who, he argues, are inherently valuable and have rights just as we do. In the reading in this chapter, Regan defends his view by first criticizing various other views, including the views of Kant, Singer, and others.

In both the popular debates and the philosophical literature about the views included in this chapter, there is often a loose use of terms. People sometimes speak of "animal rights" activists. In many cases, the phrase refers to political activists who argue against the use of animals in experimentation, as food, and for various other human benefits. In this loose sense of the phrase, both Singer and Regan are animal rights advocates. They both argue actively against use of nonhumans merely for human benefit, even as food. In a more technical sense, Singer and Regan defend their stands on behalf of animals on different theories of ethics. Singer's approach is an application of utilitarianism that is perhaps best labeled "the animal liberation view." Regan's approach is more similar to Kantianism and can be called "the animal rights view." There is a similar ambiguity in the use of the term "sentientism." In one sense, it is only Singer's position that appeals directly to the concept of sentience. Nevertheless, one finds the term "sentientism" used as a label for both views. It is in this broader way that I will use the term in this chapter. Although I recognize that the concept of "sentient being" and the concept of a "being that is a subject-of-a-life" are not identical in meaning, there does seem to be significant overlap between the individuals to which the two concepts refer. In a way similar to our diagram of anthropocentrism that I provided in Chapter 12, we might diagram the sentientist view like this:

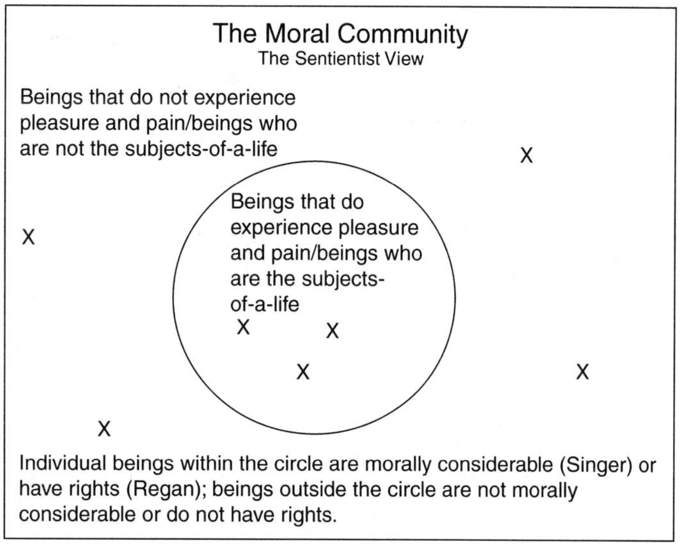

In this chapter, I have included a short reading from Jeremy Bentham to represent a utilitarian approach to animals and a longer reading by Tom Regan to represent a deontological approach. As you think about the material in this chapter, ask yourself what moral obligations you believe humans have toward animals.

Reading 11.1

Can They Suffer?

—*Jeremy Bentham*

Jeremy Bentham (1948–1832) was an English philosopher who worked to improve the conditions of all people. Convinced that other moral philosophies failed to take into consideration the interests of women, the weak, and the poor, Bentham developed a system of ethics known as utilitarianism. According to utilitarianism, the goals of maximizing pleasure and minimizing pain for all affected should guide decisions, both in the case of an individual choice and in the making of laws to guide society. In the following excerpt, Bentham applies his ethical theory beyond the domain of humankind. If the measure of good and evil is the amount of pleasure and pain that is produced by an action, then any being who can feel pain or experience pleasure is morally relevant. In defending this view, Bentham breaks with the anthropocentric approaches to ethics that have dominated Western philosophy and theology.

What other agents then are there, which, at the same time that they are under the influence of man's direction, are susceptible of happiness? They are of two sorts: (1) Other human beings who are styled persons. (2) Other animals, which, on account of their interests having been neglected by the insensibility of the ancient jurists, stand degraded into the class of *things.*

Under the Hindu and Mahometan religions, the interests of the rest of the animal creation seem to have met with some attention. Why have they not, universally, with as much as those of human creatures, allowance made for the difference in point of sensibility? Because the laws that are have been the work of mutual fear; a sentiment which the less rational animals have not had the same means as man has of turning to account. Why *ought* they not? No reason can be given. If the being eaten were all, there is very good reason why we should be suffered to eat such of them as we like to eat: we are the better for it, and they are never the worse. They have none of those long-protracted anticipations of future misery which we have. The death they suffer in our hands commonly is, and always may be, a speedier, and by that means a less painful one, than that which would await them in the inevitable course of nature. If the being killed were all, there is very good reason why we should be suffered to kill such as molest us: we should be the worse for their living, and they are never the worse for being dead. But is there any reason why we should be suffered to torment them? Not any that I can see. Are there any why we should *not* be suffered to torment them? Yes, several. The day has been, I grieve to say in many places it is not yet past, in which the greater part of the species, under the denomination of slaves, have been treated by the law exactly upon the same footing as, in England for example, the inferior races of animals are still. The day *may* come, when the rest of the animal creation may acquire those rights which never could have been withholden from them but by the hand of tyranny. The French have already discovered that the blackness of the skin is no reason why a human being should be abandoned without redress to the caprice of a tormentor. It may come one day to be recognized, that the number of the legs, the villosity of the skin, or the termination of the *os sacrum,* are reasons equally insufficient for abandoning a sensitive being to the same fate. What else is it that should trace the insuperable line? Is it the faculty of reason, or, perhaps, the faculty of discourse? But a full-grown horse or dog is beyond comparison a more rational, as well as a more conversable animal, than an infant of a day, or a week, or even a month, old. But suppose the case were otherwise, what would it avail? the question is not, Can they reason? nor, Can they talk? but, Can they suffer?

The question is not, Can they reason? nor, Can they talk? but, Can they suffer?

From The Principles of Morals and Legislation, *1789.*

Animal Rights, Human Wrongs

—*Tom Regan*

Tom Regan is a Professor Emeritus at North Carolina State University. In The Case for Animal Rights (1983), Regan articulated an ethical theory which affirms that not just humans, but other animals also, have moral rights. In numerous writings since, he has rearticulated and defended his animal rights philosophy. In the following article, Regan compares the animal rights view with both the anthropocentric Kantian view of ethics and the utilitarian view of animal liberation defended by Peter Singer. According to Regan, it is not a being's rational capacity in virtue of which it has rights, but the fact that it is a "subject-of-a-life." Since nonhuman animals are "subjects-of-a-life," they too must be recognized as having rights we often attribute only to humans.

In this essay, I explore the moral foundations of the treatment of animals. Alternative views are critically examined, including (a) the Kantian account, which holds that our duties regarding animals are actually indirect duties to humanity; (b) the cruelty account, which holds that the idea of cruelty explains why it is wrong to treat animals in certain ways; and (c) the utilitarian account, which holds that the value of consequences for all sentient creatures explains our duties to animals. These views are shown to be inadequate, the Kantian account because some of our duties regarding animals are direct duties to animals; the cruelty account because it confuses matters of motive or intent with the question of the rightness or wrongness of the agent's actions; and the utilitarian account because it could be used to justify identifiable speciesistic practices. I defend a fourth view. Only if we postulate basic moral rights in the case of humans, can we satisfactorily account for why it is wrong to treat humans in certain ways, and it is only by postulating that these humans have inherent value that we can attribute to them basic moral rights. Consistency requires that we attribute this same kind of value to many animals. Their having inherent value provides a similar basis for attributing certain basic moral rights to them, including the right not to be harmed. Possession of this right places the onus of justification on anyone who would harm these animals. I set forth conditions for such a justification which those who would abuse animals have failed to meet.

INTRODUCTION

At this moment workers on board the mother ship of a whaling fleet are disassembling the carcass of a whale. Though the species is officially protected by agreement of the member nations of the International Whaling Commission, it is not too fanciful to imagine the crew butchering a great blue whale, the largest creature ever known to have lived on the Earth—larger than thirty elephants, larger even than three of the largest dinosaurs laid end to end. A good catch, this leviathan of the deep. And, increasingly, a rare one. For the great blue, like hundreds of other animal species, is endangered, may, in fact, already be beyond the point of recovery.

But the crew has other things on their mind. It will take hours of hard work to butcher the huge carcass, a process now carried out at sea. Nor is butchering at sea the only thing in whaling that has changed. The fabled days of a real hunt, of an individual Ahab pitted against a treacherous whale, must remain the work of fiction now. Whaling is applied technology, from the use of the most sophisticated sonar to on-board refrigeration, from tracking helicopters to explosive harpoons, the latter a technological advance that expedites a whale's death. Average time to die: less than an hour. Here is one man's account of a whale's demise:

> ***The gun roars. The harpoon hurls through the air and the whale-line follows. There is a momentary silence, and then the muffled explosion as the time fuse functions and fragments the grenade. . . . There is now a fight between the mammal and the crew of the catching vessel—a fight to the death. It is a struggle that can have only one result. . . . Deep in the whale's vast body is the mortal wound, and even if it could shake off the harpoon it would be doomed. . . . A second harpoon buries itself just behind the dorsal fin . . . There is another dull explosion in the whale's vitals. Then comes a series of convulsions—a last despairing struggle. The whale spouts blood, keels slowly over and floats belly upward. It is dead.***[1]

For what? To what end? Why is this being done to the last remaining members of an irreplaceable species, certainly until recently, possibly at this very moment, by supposedly civilized men? For candle wax. For soap and oil. For pet food, margarine, fertilizer. For perfume.

In Thailand, at this moment, another sort of hunt, less technologically advanced, is in progress. The Thai hunter has hiked two miles through thick vegetation and now, with his keen vision, spots a female gibbon and her infant, sleeping high in a tree. Jean-Yves Domalain describes what follows:

> ***Down below, the hunter rams the double charge of gun-powder down the barrel with a thin iron rod, then the lead shot. The spark flashes from two flints, and the gun goes off in a cloud of white smoke. . . . Overhead there is an uproar. The female gibbon, mortally wounded, clings to life. She still has enough strength to make two gigantic leaps, her baby still clinging to the long hair of her left thigh. At the third leap she misses the branch she was aiming for, and in a final desperate effort manages to grasp a lower one; but her strength is ebbing away and she is unable to pull herself up. Slowly her fingers begin to loosen her grip. Death is there, staining her pale fur. The youngster flattens himself in terror against her bloodstained flank. Then comes the giddy plunge of a hundred feet or more, broken by a terrible rebound off a tree trunk.***[2]

The object of this hunt is not to kill the female gibbon, but to capture the baby. Unfortunately, in this case the infant's neck is broken by the fall, so the shots were wasted. The hunter will have to move on, seeking other prospects.

We are not dealing in fantasies when we consider the day's work of the Thai hunter. Domalain makes it clear that both the method of capture (killing the mother to get the infant) and the results just seen (the death of both) are the rule rather than the exception in the case of gibbons. And chimpanzees. And tigers. And orangutans. And lions. Some estimate that for every one animal captured alive, ten have been killed. Domalain further states that for every ten captured only two will live on beyond a few months. The mortality rate stemming from hunts that aim to bring animals back alive thus is considerable.

Nor do we romanticize when we regard the female gibbon's weakening grip, the infant's alarmed clutching, the bonds of surprise and terror that unite them as they begin their final descent. And for what? To what end? Why is this scene played out again and again? So that pet stores might sell "exotic animals." So that roadside zoos might offer "new attractions." So that the world's scientists might have "subjects" for their experiments.

---

[1]Captain W. R. D. McLaughlin, *Call to the South* (London: Harrap and Co., n.d.), quoted in Muriel the Lady Dowding. "Furs and Cosmetics: Too High a Price?" in Stanley and Roslind Godlovitch and John Harris, eds. *Animals, Men and Morals* (New York: Taplinger Publishing Co., 1972), p. 35.

[2]Jean-Yves Domalain, *The Animal Connection: The Confessions of an Ex-Wild Animal Trafficker* (New York: William Morrow and Co., 1977), p. 21. For additional information on endangered species, see Cleveland A mory, *Man Kind? Our Incredible War on Wildlife* (New York: Harper & Row Publishers, 1974) and Lewis Regenstein, *The Politics of Extinction* (New York: Macmillan Co., 1975).

Not far from here, perhaps at this moment, a rabbit makes a futile effort to escape from a restraining device, called a stock, which holds the creature in place by clamping down around its neck. Immediately the reader thinks of trapping in the wild—that the stock must be a sort of trap, like the infamous leg-hold trap—but this is not so. The stock is a handmaiden of science, and the rabbit confined by it is not in the wild but in a research laboratory. If we look closely, we will see that one of the rabbit's eyes is ulcerated. It is badly inflamed, an open, running sore. After some hours the sore increases in size until barely half the eye is visible. In a few days the eye will become permanently blind. Sometimes the eye is literally burned out of its socket.

This rabbit is a research subject in what is known as the Draize test, named after its inventor. This rabbit, and hundreds like it, is being used because rabbits happen not to have tear ducts and so cannot flush irritants from their eyes. Nor can they dilute them. The Draize test precedes routinely as follows concentrated solutions of a substance are made to drip into one of the rabbit's eyes; the other eye, a sort of control, is left untroubled. Swelling, redness, destruction of iris or cornea, loss of vision are measured and the substance's eye-irritancy is thereby scientifically established.

What is this substance which in concentrated form invades the rabbit's eye? Probably a cosmetic, a new variety of toothpaste, shampoo, mouthwash, talcum, hand lotion, eye cosmetic, face cream, hair conditioner, perfume, cologne. Why? To what end? In the name of what purpose does this unanesthetized rabbit endure the slow burning destruction of its eye? So that a researcher might establish the eye-irritancy of mouthwash and talc, tooth-paste and cologne.[3]

A final individual bids for our attention at this moment. A bobbie calf is a male calf born in a dairy herd. Since the calf cannot give milk, something must be done with it. A common practice is to sell it as a source of veal, as in veal Parmigiana. To make this commercially profitable the calf must be raised in highly unnatural conditions. Otherwise the youngster would romp and play as is its wont; equally bad, it would forage and consume roughage. From a businessman's point of view, this is detrimental to the product. The romping produces muscle, which makes for tough meat, and the roughage will contain natural sources of iron, which will turn the calf's flesh red. But studies show that consumers have a decided preference for pale veal. So the calf is kept permanently indoors, in a stall too narrow for it to turn around, frequently tethered to confine it further, its short life lived mostly in the dark on a floor of wood slats, its only contact with other living beings coming when it is fed and when, at the end, it is transported to the slaughterhouse.

Envision then the tethered calf, unable to turn around, unable to sit down without hunching up, devoid of companionship, its natural urges to romp and forage denied, fed a wholly liquid diet deliberately deficient in iron so as not to compromise its pale flesh but to keep it anemic. For what? To what end? In the name of what purpose does the calf live so? So that humans might have pale veal![4]

Multiply these cases by the hundreds, multiply them by the thousands, and we approach, perhaps, the magnitude of the death and suffering animals are enduring, at this moment, at the hands of human beings. The numbers are like distances in astronomy. We can write them down, compare them, add and subtract them, but, like light years stacked on light years, we lack the intellectual or imaginative wherewithal to hold them in steady focus. Item: two million whales killed in the last fifty years. Item: in the mid-nineteenth century there were upwards of two billion passenger pigeons in the United States; now there are none. Item: when white settlers set foot on North America, there were sixty million buffalo; now the largest wild herd numbers 600. Item: the bald eagle, the national symbol of the United States, and the kangaroo, Australia's national symbol, are both teetering on the edge of extinction.

---

[3]Especially important sources on the use of animals in research are Richard Ryder. *Victims of Science* (London: Davis-Poynter); Hans Ruesch, *Slaughter of the Innocent* (New York: Bantam Books, 1978): John Vyvyan, *The Dark Face of Science* (London: Michael Joseph, 1971); and E. A. Westacott, *A Century of Vivisection and Anti-Vivisection* (London: Daniel Co., 1949).

[4]For information on modern factory farming methods, see Ruth Harrison, *Animal Machines* (London: Stuart, 1964) and Peter Singer, *Animal Liberation* (New York: New York Review, 1975, New York: Avon Books, 1976).

Extinct or endangered species aside, consider these figures for legally imported wildlife brought into the United States, in one year (1970) alone, to supply the needs of the pet industry, zoos, and scientific researchers:

101,302 mammals, 85,151 of which were primates (mainly monkeys)

687,901 wild birds, not including canaries, parrots, and parakeets, millions of which are annually imported

572,670 amphibians (frogs, toads, salamanders)

2,109,571 reptiles[5]

In the United States alone for the year 1978, 400,000 dogs, 200,000 cats, 33,000 apes and monkeys, thousands of calves, sheep, ponies, horses, were used in research as well as rabbits, hamsters, guinea pigs, rats, birds, mice by the millions. Altogether an estimated 64 million animals were used in research. World-wide, educated estimates place the figure at about 200 million.[6]

As for slaughter, incomplete data suggest that, worldwide, the number of animals killed for food exceeds hundreds of millions for individual species, evolving upwards of one hundred million cattle, one hundred million sheep, two hundred million pigs, twenty-six million calves, and, in the United States alone, three billion chickens.[7]

But numbers prove nothing. They neither establish that something is right nor that it is wrong. They merely confirm what we know already. The use of animals as a primary food source and as subjects in scientific research knows no geographical boundaries. The extinction of unique species is not perpetrated by only a few political ideologies. Aside from economic considerations, and excluding the privileged status which pet animals may have as honorary members of the family, few things are regarded so cheaply as an animal's life.

Few things now call forth less collective compassion of humankind than the great suffering animals endure in the name of human interests.

But these numbers serve another function. They serve as an index of the complexity of any questioning of those activities which involve the pain or destruction of animals. Questions about morality here are not likely to have simple answers. They require patient examination, due attention to competing forces, an eye for subtlety and nuance. Difficult questions are like knots; they are not unraveled just by pulling hard on one idea. For example, there is no reason to believe that such ideas as "all life is sacred," or "we ought to reverence all life," or "all forms of life are good" will dissolve the complexity before us. For these ideas, simply stated, leave unanswered crucial questions that arise when the interests of one form of life come into conflict with those of another. What is needed is not a simple declaration that all life is sacred, etc. What is wanted is some rational way to resolve such conflicts. What is wanted are moral principles which resolve the conflict equitably.

Now, these moral principles will not tell us, for example, how the population of whales will be affected if whaling is stopped or how that will affect the population of tuna. By themselves, that is, moral principles will not tell us *the facts* about whaling practices, nor will they give us a means of predicting what the facts will be. What these principles *will* tell us is what sort of facts to look for. In other words, they will tell us what facts are *morally relevant* to reaching a rational decision. The task of identifying what these principles are is one of the distinctive tasks of moral philosophy. This task occupies most of my attention in what follows. But I also want to bring my results back to the rabbit, the calf, the whale, and the gibbon with which we began.

---

[5]As quoted in Regenstein, *Politics of Extinction*, p. 121.

[6]These figures are suggested by Ryder. See also his "Experiments on Animals" reprinted in Tom Regan and Peter Singer, eds., *Animal Rights and Human Obligations* (Englewood Cliffs: Prentice-Hall, 1975).

[7]These figures are based on estimates furnished for the year 1968 by the United Nations Food and Agriculture Organization and quoted in Nathaniel Altman, *Eating for Life: A Book About Vegetarianism* (London: Theosophical Publishing House, 1973), p. 59.

THE KANTIAN ACCOUNT

It is a commonplace to say that morality places some limits on how animals may be treated. We are not to kick dogs, set fire to cats' tails, torment hamsters or parakeets. Philosophically, the issue is not so much *whether* but *why* these acts are wrong.

An answer favored by many philosophers, including Thomas Aquinas and Immanuel Kant, is that people who treat animals in these ways develop a habit which, in time, inclines them to treat humans similarly.[8] People who torment animals will, or are likely to, torment people. It is this spillover effect that makes mistreating animals wrong. We are not concerned directly with the ill-treatment that the animals themselves receive. Rather, our concern is that this bodes ill for humankind. So, on this Kantian account, the moral principle runs something like this: don't treat animals in ways that will lead you to mistreat human beings.

On this Kantian account, the moral principle runs something like this: don't treat animals in ways that will lead you to mistreat human beings.

One need have no quarrel with this principle itself. The real quarrel lies with the grounds on which this principle is allegedly based. Peter Singer argues that there is a close parallel between this view and those of the racist and sexist, a view which, following Richard Ryder, he denominates speciesism.[9] The racist believes that the interests of others matter only if they happen to be members of his own race. The speciesist believes that the interests of others matter only if they happen to be members of his own species. Racism has been unmasked for the prejudice that it is. The color of one's skin cannot be used to determine the relevance of an individual's interests. Singer and Ryder both argue that neither can the number of one's legs, whether one walks upright or on all fours, lives in the trees, the sea or the suburbs. Here they recall Bentham.[10] There is, they argue forcefully, no rational, unprejudiced way to exclude the interests of nonhuman animals just because they are not the interests of human beings. Because the Kantian account would have us think otherwise, we are right to reject it.

THE CRUELTY ACCOUNT

A second view about constraints on how animals may be treated involves the idea of cruelty. The reason we are not to kick dogs is that we are not to be cruel to animals and kicking dogs is cruel. It is the prohibition against cruelty which covers and conveniently sums up our negative duties to animals, duties concerning how animals are *not* to be treated.

The prohibition against cruelty can be given a distinctively Kantian twist. This happens when the grounds given are that cruelty to animals leads people to be cruel to other people. John Locke suggests, but does not clearly endorse, this view:

> ***One thing I have frequently observed in Children, that when they have got possession of any poor Creature, they are apt to use it ill: They often torment, and treat very roughly, young Birds,***

---

[8]Relevant selections from both St. Thomas and Kant are included in Regan and Singer, *Animal Rights and Human Obligations.* What I call the Kantian account is critized further in my "Exploring the Idea of Animal Rights" in D. Paterson and R. Ryder, eds., *Animal Rights: a Symposium* (London: Centaur Press, 1979). Kant's views are criticized at length by Elizabeth Pybus and Alexander Broadie, "Kant's Treatment of Animals," *Philosophy* 49 (1974): 375–83. I defend Kant against their objections in my "Broadie and Pybus on Kant," *Philosophy* 51(1976): 471–72. Broadie and Pybus reply in their "Kant on the Maltreatment of Animals," *Philosophy* 53(1978): 560–61. At present I am persuaded that Broadie and Pybus are correct in arguing that Kant cannot account for the idea that animals themselves can be maltreated.

[9]Singer, *Animal Liberation,* and Ryder, *Victims of Science.*

[10]The famous passage from Bentham reads as follows (from *The Principles of Morals and Legislation* (1789), chap. 17, Sect. 1, reprinted in Regan and Singer, *Animal Rights and Human Obligations*): "The day has been, I grieve to say in many places it is not yet past, in which the greater part of the species, under the denomination of slaves, have been treated by the law exactly upon the same footing as, in England for example, the inferior races of animals are still. The day may come, when the rest of the animal creation may acquire those rights which never could have been withholden from them but by the hand of tyranny. The French have already discovered that the blackness of the skin is no reason why a human being should be abandoned without redress to the caprice of a tormentor. It may come one day to be recognized, that the number of the legs, the villosity of the skin, or the termination of the os sacrum, are reasons equally insufficient for abandoning a sensitive being to the same fate. What else is it that should trace the insuperable line? Is it the faculty of reason, or, perhaps, the faculty of discourse? But a full-grown horse or dog is beyond comparison a more rational, as well as a more conversable animal, than an infant of a day, or a week, or even a month, old. But suppose the case were otherwise, what would it avail? The question is not, Can they reason? nor, Can they talk? but, Can they suffer?"

Butterflies, and such other poor Animals, which fall into their Hands, and that with a seeming kind of Pleasure. This I think should be watched in them, and if they incline to any such Cruelty, they should be taught the contrary Usage. For the Custom of Tormenting and Killing of Beasts, will, by Degrees, harden their Minds even towards Men; and they who delight in the Suffering and Destruction of Inferior Creatures, will not be apt to be very compassionate, or benign to those of their own kind.[11]

Locke's position suggests the speciesism which characterizes the Kantian account and will not do for the same reasons. However, Locke's understanding of what cruelty is—tormenting a sentient creature or causing it to suffer, "with a seeming kind of Pleasure"—seems correct and has important implications. Many thinkers, including many persons active in the humane movement, champion the prohibition against cruelty to animals because it is wrong to be cruel to the animals themselves. This way of grounding the prohibition against, cruelty, which I call "the cruelty account," deserves our critical attention.

It is difficult to overestimate the importance the idea of preventing cruelty has played, and continues to play, in the movement to secure better treatment for animals. Entire societies are devoted to this cause, the Society for the Prevention of Cruelty to Animals (SPCA) in the United States and the Royal Society for the Prevention of Cruelty to Animals (RSPCA) in great Britain being perhaps the best known examples. I do not wish to deny the importance of preventing cruelty nor to deprecate the crusading work done by these organizations, but I must conclude that to stake so much on the prevention of cruelty both clouds the fundamental moral issues and runs a serious risk of being counterproductive.

Cruel is a term of moral appraisal used to refer either to the character of a person or to an individual action. Persons are cruel who are inclined to delight in or, in Locke's phrase, to take "a seeming kind of Pleasure" in causing pain. An individual action is cruel if one takes pleasure in making another suffer. It is clear that someone's being cruel is distinct from someone's causing pain. Surgeons cause pain. Dentists cause pain. Wrestlers, boxers, football players cause pain. But it does not follow that they are cruel people or that their individual actions are cruel. To establish cruelty we need to know more than that someone caused pain; we also need to know the state of mind of the agent, whether he/she took "a seeming kind of Pleasure" in the pain inflicted. It is faulty to reason in this way:

An individual action is cruel if one takes pleasure in making another suffer.

Those who cause pain are cruel. Surgeons (football players, etc.) cause pain. Therefore, surgeons (football players, etc.) are cruel.

But just as clearly, it is faulty to reason in the following way:

Those who cause pain are cruel. Those who experiment on animals (or kill whales, or raise veal calves in isolation, etc.) cause pain. Therefore, those who treat animals in these ways are cruel.

Those who are inclined to march under the banner of anti-cruelty must soon recognize the speciousness of this line of reasoning, if their thought, however well intentioned, is not to cloud the issues.

Once cruelty is understood in the way Locke saw that it should be, we can understand why more is needed. Take the case of the use of animals in the Draize test. Increasingly people want to object morally to this, to say it is wrong. However, if this required establishing cruelty, the weight of the evidence would be on the side of the experimenters and against the objectors, for there is no adequate evidence for believing that people who administer the Draize test are cruel people or that they are cruel when they administer this test. Do they take "a seeming kind of Pleasure" in causing the animals pain? That they cause *pain* to the animals is certain. But causing pain does not establish cruelty. Except for a few sadists in the scientific community, there is good reason to believe that researchers are no more cruel than are most persons.

---

[11]John Locke, *Some Thoughts Concerning Education,* 5th ed. (London, 1905). See also James Axtell, ed., *The Educational Writings of John Locke* (Cambridge: Cambridge University Press 1968), sec. 116, pp. 225–26.

Does this mean that using animals in the Draize test is right? Precisely not. Rather, to ask whether this is right is logically distinct from, and should not be confused with, asking whether someone is cruel. Cruelty has to do with a person's state of mind. The moral rightness or wrongness of a person's actions is different. Persons can do what is right or wrong whatever their state of mind. Researchers using the Draize test can be doing what is wrong, whether or not they enjoy causing animals to suffer. If they do enjoy this, we shall certainly think less of them as persons. But even if they enjoy the pain it will not follow that the pain is unjustified, any more than it will follow that the pain is justified if they feel sorry for the animals or feel nothing at all. The more we are able to keep in view how the morality of what a person does is distinct from his/her state of mind, distinct from the presence or absence of taking pleasure in pain; the better the chances will be for significant dialogue between vivisectors and anti-vivisectionists.

To charge vivisectors with cruelty *can* only serve to call forth all their defenses, because the charge will be taken as a denunciation of *what they are* (evil people) rather than of *what they do*. It will also give them an easy way out. After all, *they* are in priviledged position to know their own mental states; *they* can take a sober moment and see whether in fact they do take a "seeming kind of Pleasure" in causing pain. If, as will usually be the case, they find that they honestly do not, then they can reply that they are not cruel (evil) people. So we see now where the well-intentioned efforts of those defending animals can be and often are counterproductive. If it's cruelty they are charged with, and they are not cruel, then they can come away with a feeling that their hands are clean. They win, and the litany of accusations about cruelty is so much water off their backs. It is no good trying to improve the lot of animals by trying to convince persons who are not cruel that they are.

Some will complain that my argument is "too picky." They might say that cruelty has been interpreted too narrowly: what is meant is treating animals badly in ways they don't deserve, harming or wronging them. In practice this is what anti-cruelty charges often come to. But then this is the way the charges should be made, lest they be misunderstood or be counterproductive. To ask for more care in the charges leveled is not to strain at gnats. It is to begin to make the charges more difficult to answer. Perhaps a name like "The Society for the Prevention of Maltreatment of Animals" is not as euphonious as "The Anti-Cruelty Society," but a lack of euphony is a price those laboring for animal welfare should gladly pay.

THE UTILITARIAN ACCOUNT

Utilitarians give a different account of the constraints regarding how animals ought to be treated. The utilitarian account, or one version of it, involves two principles.[12] The first is a principle of equality. This principle declares that the desires, needs, hopes, etc. of different individuals, when these are of equal importance *to* these individuals, *are* of equal importance or value no matter who the individuals are, prince or pauper, genius or moron, white or black, male or female, *human or animal*. This equality of interests principle seems to provide a philosophical basis for avoiding the grossest forms of prejudice, including racism, sexism and, following Ryder and Singer, speciesism. Whether it succeeds is an issue which we shall take up below.

Since animals have interests, their interests must be taken into account, and because their interests are frequently as important to them as comparable interests are to human beings, their interests must be given the same weight as comparable human interests.

The second principle is that of utility itself. Roughly speaking, according to this principle, we are to act so as to bring about the greatest possible balance of good over evil, for example, the greatest possible balance of satisfaction over dissatisfaction, taking the interests of everyone affected into account *and* counting equal interests equally. Now, since animals have interests, their interests must be taken into account, and because their interests are frequently as important to them as comparable interests are to human beings, their interests must be given the same weight as comparable human interests. It is because kicking dogs and setting fire to cats' tails run counter to the principles of equality and utility that, on this utilitarian account, they are wrong.

---

[12]The utilitarian position I consider is the one associated with Bentham and forcefully presented by Peter Singer. That Singer is a utilitarian is made unmistakably clear in his "The Fable of the Fox and the Unliberated Animals" *Ethics* 88 (1978): 119–25.

Granted, this is a very rough sketch; nonetheless, it enables us to understand the main features of the utilitarian account, and also to find points of resemblance and contrast between it and the other accounts so far described. Like the Kantian account, but unlike the cruelty account, the utilitarian account emphasizes results or consequences, but unlike the Kantian account, and resembling the cruelty account, the utilitarian account recognizes the moral status of animals in their own right. We are not to measure morality by the speciesist yardstick of human interest alone. Finally, unlike the cruelty account, but in concert with the Kantian, the utilitarian does not conflate the morality of an act with the mental state of the agent. The utilitarian can be as opposed to cruelty as anyone else, but within the utilitarian theory right and wrong are determined by consequences, not feelings and intentions: the ordinary moral constraints placed on how we may treat animals are accounted for because they are necessary if we are not to violate the equality of interests principle *or* if we are to succeed in bringing about the greatest possible balance of good over bad.

The utilitarian account has much to recommend it. How far can it take us in challenging the way in which animals are routinely treated, for example, as subjects in scientific research? Peter Singer, a utilitarian whose work has well deserved influence, holds that utilitarianism leads to far-reaching consequences here. Singer argues that we become vegetarians *and* that we oppose much (even if not quite all) research involving animal subjects. Singer's main argument is that the intensive rearing of animals as well as their routine use in experimentation violates the equality of interests principle. The animals involved, we have reason to believe, have an interest in not being made to suffer, and this interest, we have further reason to believe, is as important to them as is the comparable interest in the case of human beings. This being so, Singer contends, it is wrong to do to animals what we would not do to humans. It cannot be right to raise animals intensively or use them in research if we would morally oppose doing these things to human beings. We do condemn cannibalism and the coerced use of humans in research and we must, Singer argues, morally condemn the comparable treatment of animals. We have a moral obligation to become vegetarians and oppose much, if not quite all, vivisection.

As clear and powerful as this argument is, I do not believe that Singer succeeds in making a fully convincing case. He shows that animals are treated differently than human beings, but not that this differential treatment violates either the equality of interests principle or the principle of utility. Consider the equality of interests principle first. We can count equal interests equally, no matter whose interests they are, and still treat individuals quite differently. For example, I might correctly regard my son's and my neighbor's son's interests in receiving a medical education as being equal and yet help only my son. I treat them differently but I do *not* necessarily count their equal interests differently, and neither do I thereby do anything that is in any obvious sense morally reprehensible. I have duties to my son which I do not have to the children of others.

The general point is this: the differential treatment of individuals with equal interests does not by itself violate the equality of interests principle. Singer has to give an *argument* which shows *more* than that they are treated differently. What argument does he give and how adequate is it? Singer proceeds by asking whether we would do to humans what we allow to be done to animals.[13] For example, would a researcher use an orphaned, profoundly retarded human baby in the sort of painful experiment in which he is willing to use a more intellectually and emotionally developed animal? If the researcher says no, Singer charges him with speciesism, with violating the equality of interests principle. The animal's interest in avoiding pain is just as important to it as is the infant's interest to him/her.

This argument begs the question. It assumes that by treating the involved individuals differently, we count their equal interests differently. As I have explained, however, this is not always true. Whether it is true in any particular case, therefore, is something which must be established, not simply assumed on the basis of differential treatment. Singer, I believe, assumes just this, and thus begs the question.

Singer's argument has a further deficiency, which involves the principle of utility. First, Singer does not show that the differential treatment of animals runs counter to the utilitarian objective of bringing about

---

[13]See Singer, *Animal Liberation,* expecially pp. 78–83.

the greatest possible balance of good over evil. To show this Singer would have to give an elaborate, detailed description, not only of how animals are treated, a part of the task which he does complete with great skill, but an analysis of what, all considered, are the consequences for everyone involved. He would have to inquire how the world's economy depends on present levels of productivity in the animal industry, how many people's lives are directly and indirectly involved with the maintenance or growth of this industry, etc. Even more, he would have to show in detail what would probably be the consequences of a collapse or slowdown of the animal industry's productivity.

Secondly, Singer needs to make a compelling case for the view that *not* raising animals intensively or *not* using them routinely in research leads to better consequences, all considered, than those which now result from treating animals in these ways. Singer is required to show that better consequences *would* result, or at least that it is *very probable* that they would. Showing that it is possible or conceivable that they might is insufficient. It comes as a disappointment, therefore, that we do not find anything approaching this kind of required empirical data. What we find, instead, are passages where he bemoans (rightly, I believe) the fact that animals are fed protein-rich grains which could be fed to malnourished human beings.[14] The point, however, is not whether these grains *could* be fed to the malnourished; it is whether we have solid empirical grounds for believing that they *would* be made available to and eaten by these people, if they were not fed to animals, *and* that the consequences resulting from this shift would be better, all considered. I hope I am not unfair to Singer in observing that these calculations are missing, not only here, but, to my knowledge, throughout the body of his published writings.

This, then, is the first thing to note regarding Singer and the principle of utility: *he fails to show, with reference to this principle, that it is wrong* to treat animals as they are now being treated in modern farming and scientific research. The second thing to note is that, for all we know and so long as we rely on the principle of utility, the present treatment of animals might actually be justified. The grounds for thinking so are as follows.

On the face of it, utilitarianism seems to be the fairest, least prejudicial view available. Everyone's interests count, and no one's counts for more or less than the equal interests of anyone else. The trouble is, as we have seen, that there is no necessary connection, no preestablished harmony between respect for the equality of interests principle *and* promoting the utilitarian objective of maximizing the balance of good over bad. On the contrary, the principle of utility might be used to justify the most radical kinds of differential treatment between individuals or groups of individuals, and thus it might justify forms of racism and sexism, for these prejudices can take different forms and find expression in different ways. One form consists in not even taking the interests of a given race or sex into account at all; another takes these interests into account but does not count them equally with those of the equal interests of the favored group. Another does take their interests into account equally, but adopts laws and policies, engages in practices and customs which give greater opportunities to the members of the favored group, because doing so promotes the greatest balance of good over evil, all considered.

Thus, forms of racism or sexism, which seem to be eliminated by the utilitarian principle of equality of interests, could well be resurrected and justified by the principle of utility. If a utilitarian here replies that denying certain humans an equal opportunity to satisfy or promote their equal interests on racial or sexual grounds must violate the equality of interests principle and so, on his position, is wrong, we must remind him that differential treatment is not the same as, and does not entail, violating the equality of interests principle. It is quite possible, for example, to count the equal interests of blacks and whites the same (and thus to honor the equality principle) and still discriminate between races when it comes to what members of each race are permitted to do to pursue those interests, on the grounds that such discrimination promotes the utilitarian objective. So, utilitarianism, despite initial appearances, does not provide us with solid grounds on which to exclude all forms of racism or sexism.

Similarly with speciesism. The same kind of argument can show a possible utilitarian justification of an analogous speciesism. We count the equal interests of animals and humans equally; it just so happens

[14]Ibid., chap. 4.

that the consequences of treating animals in ways that humans are not treated, such as intensively raising animals, but not humans, are better, all considered, than are other arrangements. Thus, utilitarianism might provide a basis for speciesist practices. Whether it *actually* does depends on whether the consequences are better, all considered, if animals continue to be treated as they are. Since Singer fails to provide us with empirical data showing that the consequences would be better if we changed, it follows that, for all we know, the present speciesist way of treating animals might actually be justified, given his version of utilitarianism.

ANIMAL RIGHTS

Our results to this point are mainly negative. I have thus far argued (1) that the moral principles we seek cannot refer to the agent's state of mind, to whether the agent takes a "seeming kind of Pleasure" in causing animal suffering. (2) These principles cannot refer only to consequences that harm or benefit human beings, since this prejudicially leaves out of account the harms and benefits to the animals themselves. (3) These principles cannot refer only to the utilitarian objective of maximizing the balance of good over evil, even if animal harms and benefits are taken into account. What is wanted, then, is an account which avoids each of these shortcomings. This account is to be found, I believe, by postulating the existence of animal rights. Indeed, I believe that only if we postulate human rights can we provide a theory which adequately guards humans against the abuses which utilitarianism might permit.

Various analyses of the concept of a right have been proposed. We will bypass the nooks and crannies of these competing analyses so as to focus attention on the role that moral rights play in our thinking about the status of the individual, relative to the interests of the group. Here the truth seems to lie where Ronald Dworkin sees it: the rights of the individual trump the goals of the group.[15]

What does this mean? It means that the moral rights of the individual place a justifiable limit on what the group can do to the individual. Suppose a group of people stand to gain enjoyment by arranging for others to be harmed. Imagine, for example, the Romans enjoying how the Christians go up against lions. Such a group does wrong because they allow their interests to override the individual's moral rights. This does not mean that there are no circumstances in which an individual's rights must give way to the collective interest. Imagine that Bert has inadvertently swallowed the microfilmed code which we must have in order to prevent a massive nuclear explosion in New Zealand. Bert sits safely in Tuscon, Arizona. We explain the situation but Bert refuses to consent to our request that we operate, retrieve the code, and prevent the explosion. He cites his right to determine what is to be done to his body. In such a case it is not implausible to say that Bert's right must give way to the collective interests of others.

Individual rights then normally, but not always, trump collective interests. To give a precise statement of the conditions which determine which ought to prevail is very difficult indeed, but the following conditions, which deal only with the right not to be harmed, at least seem to incorporate necessary conditions for overriding this right.[16]

Individual rights then normally, but not always, trump collective interests.

An individual's right not to be harmed can justifiably be overridden only if—

a. we have very good reason to believe that overriding the individual's fight by itself will prevent, and is the only realistic way to prevent, vastly greater harm to other innocent individuals; or

b. we have very good reason to believe that allowing the individual to be harmed is a necessary link in a chain of events which collectively will prevent vastly greater harm to innocent individuals, *and* we

---

[15]Ronald Dworkin, *Taking Rights Seriously* (Cambridge: Harvard University Press, 1977).

[16]The present statement of these conditions deviates somewhat from my earlier effort in "The Moral Basis of Vegetarianism," *Canadian Journal of Philosophy* 5 (1975): 181–214. I believe the inclusion of conditions (b) and (c) marks an improvement over the earlier formulation. However, a fuller statement has to include more than simply the idea of *preventing* vastly greater harm; for example, *reducing* already existing harm also has a place.

have very good reason to believe that this chain of events is the only realistic way to prevent this vastly greater harm; or

c. we have very good reason to believe that it is only if we override the individual's right that we can have a reasonable hope of preventing vastly greater harm to other innocent individuals.

There is much that is vague in these conditions, e.g., "vastly greater harm," "innocent individuals," "reasonable hope." At present, however, we will have to make do with them as they stand. Even so we can see that these conditions attempt to do justice to the complexity of conflicts of interest. In particular, they attempt to explain how, in a principled way, we might justify overriding an individual's right not to be harmed even though *just* doing this will not guarantee the prevention of vastly greater harm. Condition (b) brings this out—harming an individual is only one part of a more complex series of events which we have very good reason to believe will prevent vastly greater harm, or because—as (c) brings out—we simply do not know how things will turn out, but do have very good reason to believe that we have no reasonable hope of preventing some catastrophe unless we allow an individual to be harmed. Possibly some will find these conditions too liberal. Condition (c) in particular might seem too lenient. Even (b) might go too far. I am not certain what to say here and beg to leave this issue unresolved, except to say that the case for not harming animals is proportionately greater the more one is inclined to restrict the above set just to condition (a). For reasons that will become clearer, however, even the more liberal view, that harm can not be justified unless one of the three conditions is met, is sufficient to make a strong case against our routine abuse of animals.

These conditions share an extremely important feature. Each specifies what we must know or have good reason to believe if we are justified in overriding an individual's right not to be harmed. Each requires anyone who would harm an individual to show that this does not involve violating the individual's right. Part of the importance of the question, whether animials have rights, specifically, the right not to be harmed, now comes into clear focus. *If* they have this right, then it will be violated whenever animals are harmed and condition (a), (b), (c) is not satisfied. Moreover, the onus of justification is always on those who harm animals to explain how they are not violating the right of animals not to be harmed, if animals have this right. So, the question continues to press itself upon us. Do animals have the right not to be harmed?

The question continues to press itself upon us. Do animals have the right not to be harmed?

This is not an easy question to answer. One is reminded of Bentham's observation that the idea of moral rights is "nonsense on stilts." Bentham meant this in the case of *human* moral rights. One can only speculate regarding what he might have thought concerning the moral rights of *animals*! So, how is one to procede? The circuitous path we must cautiously travel, I think, is in broad outline as follows.[17]

We begin by asking about our reasons for thinking that human beings have the moral right not to be harmed; then we go on to ask whether, given these reasons, a case can be made for saying that animals have this right as well. Let us go back to the idea that individual human beings have this right and that, except in extreme cases, this right trumps collective interest. Why? What is there about being a human being to which we can point and say, "*That's* why you must not harm the individual even if the group benefits?"

The heart of the answer lies, I believe, in thinking that human beings have a certain kind of value, inherent value. By this I mean that each human being has value logically independently of whether he/she is valued by anyone else (or, what perhaps comes, to the same thing, whether he/she is the object of anyone else's interest).[18] The view that human beings have inherent value implies that the kind of value properly attributable to them is not exclusively instrumental. Humans have value not just because, and not just so long as, they are good for something. They have value distinct from their utility and skill.

---

[17]See my "An Examination and Defense of One Argument Concerning Animal Rights," *Inquiry* 22 (1979): 189–219.

[18]Whether sense can be made of including irreversibly comatose human beings in the class of beings having inherent value is a troublesome question indeed. I consider this issue, perhaps not very adequately, in the essay referred to in footnote 17.

If this is true, we can explain, in general terms reminiscent of Kant, what is involved in mistreating human beings. Humans are mistreated if they are treated as valuable only if they forward the interests of other beings. To treat a human being thus is to show a lack of proper respect for the sort of value humans have. In Kant's terms, what has value in itself must always be treated as an end, never merely as a means. However, this is precisely what we are doing if we harm an individual so that others might gain pleasure or profit; we are treating the individual merely as a means, as valuable only to the extent he/she contributes to the collective interest.

Now, *if* we accept the postulate that human beings have inherent value, we can press on and ask how rights enter the picture. They enter in being grounded in inherent value. In other words, it is individuals who have inherent value who have moral rights, and it is *because* they have value of this kind that they have a moral right not to be treated in ways that deny their having this kind of value. Rather than rights being connected with *the value of consequences* which affect individuals for good or ill, rather than rights being justified by the utility of recognizing them, rights are based on *the value of individuals.* In the case of the right not to be harmed, then, what we can say is that individuals who have inherent value have the right not to be harmed, which precludes treating them merely as a means. This would fail to treat these individuals with that respect to which, because of the kind of value they have, they are entitled.

Now, certainly the foregoing is not a definitive account of the view that individuals having inherent value have basic moral rights, in particular the right not to be harmed. One omission is especially conspicuous. What is there about being a human being that underlies this inherent value? Any answer is controversial, and a sustained defense of the answer proposed here is not possible.[19] But here is the answer I would give: human beings not only are alive; *they have a life.*[20] What is more, we are the subjects of a life that is better or worse for us, logically independently of anyone else's valuing us or finding us useful.

I do not mean that others cannot contribute to or detract from the value of our lives. On the contrary, the great goods of life (love, friendship, and, in general, fellow feeling) and its great evils (hatred, enmity, loneliness, alienation) all involve our relationships with other persons. What I mean, rather, is that our being *the subject* of a life that is better or worse for us does not depend logically on what others do or do not do. This fact, I believe, provides the illumination we seek. Humans have inherent value because we are ourselves the subjects of a life that is more or less valuable to us. In sum:

> ***Human beings have inherent value because, logically independently of the interest of others, each individual is the subject of a life that is better or worse for that individual. Because of the type of value that human beings have, it is wrong (a sign of disrespect and a violation of rights) to treat humans as if they had value merely as a means (e.g., to use humans merely to advance the pleasures of the group). In particular, to harm human beings for the sake of the profit or pleasure of any group is to violate their right not to be harmed.***

The question now arises whether this same line of argument can be developed in the case of animals? It can, at least in the case of those animals who are the subjects of a life that is better or worse for them, logically independently of whether they are valued by anyone else. And there can be no rational doubt that there *are* numerous species of animals of which this is true, including the great blue whale, gibbons, and the others with which we began. They too have a distinctive kind of value in their own right, if we do; therefore, they too have a right not to be treated in ways that fail to respect this value, if we do. And, like humans, this right of theirs will be overridden unjustifiably if they are harmed merely to advance the profits or pleasures of others.

They too have a distinctive kind of value in their own right, if we do; therefore, they too have a right not to be treated in ways that fail to respect this value, if we do. And, like humans, this right of theirs will be overridden unjustifiably if they are harmed merely to advance the profits or pleasures of others.

---

[19]I do not believe it is absurd to think of natural objects which lack consciousness, or collections of such objects, as having inherent value, in the sense in which I use this expression. An *X* has inherent value if it has value logically independently of anyone's valuing *X*. I do not say this is easy to clarify or to defend, and it may be wrongheaded. At present, however, I believe it is a view that must be held, if we are to develop an environmental ethic, as distinct from an ethic for the use of the environment.

[20]The distinction between being alive and having a life is one James Rachels frequently makes. See, for example, his "Euthanasia" in Tom Regan, ed., *Matters of Life and Death* (New York: Random House, 1980). Rachels does not, so far as I am aware, relate this distinction to the idea of inherent value.

CONCLUSION

Two final philosophical points are in order, before I bring the results of my argument to bear on how animals are treated in the world at large. First, it is important to realize that I have not *proven* that animals have rights, or even that *human* beings have rights. Rather, I have argued that if humans have rights, so do many animals. More particularly, I have argued for what appears to be the most promising line of argument for explaining human rights, the view that we have inherent value, and that this can rationally be extended to animals of some kinds. So, while I admit that I have not proven that animals (or humans) have rights, I hope at least to have made clear the direction in which future argument ought to proceed. Erecting pointers, to be sure, is not the same as constructing proofs, but pointers are the best I can do here.

Second, the history of moral philosophy teaches us that utilitarianism dies hard. Just when one thinks it has been forced off the stage for good, one finds it loitering in the wings, awaiting yet another curtain call. The utilitarian can be counted on to say that there is nothing introduced by the idea of rights for which he cannot account.[21] One has only to see that the utilitarian objective is promoted if we recognize a strict obligation not to harm individuals except in extreme cases, *and* that, furthermore, utility is promoted by saying that individuals have the right not to be harmed, this invocation of a right functioning as an especially forceful way of conveying the idea that we ought not to harm individuals.

I am not convinced by this attempt to resurrect utilitarianism, and here I raise my final and most fundamental objection to it. The utilitarian is in no position to say that he knows that the utilitarian objective is promoted by talk of individuals having rights. But even if it is true that talk of rights helps promote the utilitarian objective, and for this reason such talk ought to be encouraged and honored, there can only be a *contingent* connection between any right, such as the right not to be harmed, and the fact that respecting this right forwards the utilitarian objective. The most that the utilitarian can say is that recognizing the right not to be harmed *as a matter of fact* fits in with forwarding his goal of maximizing the balance of good over evil.[22] The utilitarian must also accept that things could have been (and could become) otherwise. He must accept the possibility that it could have been or might become all right to harm individuals if this ever happened to forward the utilitarian objective. But neither the wrongness of harming individuals nor the right not to be harmed can change in the ways utilitarian theory implies they can. They are not contingent upon *utility*. Neither depends on the value of consequences. Instead, each depends on *the value of individuals*.

Let us put this in perspective before applying it. Earlier we said that before making an informed judgment about the morality of whaling or the use of the Draize test we must know both facts and moral principles. Otherwise, we cannot know which facts are morally relevant; and without this preliminary knowledge, we do not know what moral judgments to make. To determine what these principles are, we said, is one of the distinctive tasks of moral philosophy. Three positions were examined and found wanting: the Kantian account, the cruelty account, and the utilitarian account. We then considered an account ascribing rights to animals, a position which meets the objections which were fatal to the views examined earlier. Unlike the Kantian account, the rights account insists upon the moral status of animals in their own right; unlike the cruelty account, the rights account does not confuse the morality of acts with the mental states of agents; and unlike utilitarianism, this account closes the door to the justification of prejudices which merely happen to bring about the best consequences and emphasizes the value of individuals as distinct from the value of consequences. This emphasis on the value of individuals becomes prominent now as we turn at last to the task of applying the rights account to the whale, the veal calf, and the others.

It would be grotesque to suggest that the whale, the rabbit, the gibbon, the bobbie calf, the millions of animals brought so much pain and death at the hands of humans are not harmed, for harm is not restricted

---

[21]It is possible that Mill meant to give rights a utilitarian basis. On this see David Lyons, "Human Rights and the General Welfare," *Philosophy and Public Affairs* 6(1977): 113–29, reprinted in David Lyons, ed., *Rights* (Belmont, California: Wadsworth Publishing Co., 1979). The principal objection to this enterprise is the third objection I raise against utilitarianism here.

[22]I do not believe utilitarianism is alone in implying that the duty not to harm an individual (or the individual's right not to be harmed) are *contingent* moral truths, which *might* have been otherwise (or *might* become otherwise). Certain aspects of Kant's theory as well as ethical egoism arguably imply this as well. This is absolutely fatal to these theories, a point I argue in my "Utilitarianism, Vegetarianism, and Animal Rights," forthcoming in *Philosophy and Public Affairs*.

to human beings. They are harmed, harmed in a literal, not a metaphorical sense. They are made to endure what is detrimental to their welfare, even death. Those who would harm them, therefore, must justify doing so. Thus, members of the whaling industry, the cosmetics industry, the farming industry, the network of hunters-exporters-importers must justify the harm they bring animals in a way that is consistent with recognizing the animals' right not to be harmed. To produce such a justification it is not enough to argue that people profit, satisfy their curiosity, or derive pleasure from allowing animals to be treated in these ways. These facts are not the morally relevant ones. Rather, what must be shown is that overriding the right of animals not to be harmed is justified because of further facts. For example, because we have very good reason to believe that overriding the individual's right prevents, and is the only realistic way to prevent, vastly greater harm to other innocent individuals.

Let us ask the whaling industry whether they have so justified their trade. Have they made their case in terms of the morally relevant facts? Our answer must be: no! And the cosmetic industry? No! The farmers who raise veal calves? No! The retailer of exotic animals? No! A thousand times we must say: no! I do not say that they cannot possibly justify what they do. The individual's right not to be harmed, we have argued, almost always trumps the interests of the group, but it is possible that such a right must sometimes give way. Possibly the rights of animals must sometimes give way to human interests. It would be a mistake to rule this possibility out. Nevertheless, the onus of justification must be borne by those who cause the harm to show that they do not violate the rights of the individuals involved.

We allow then that it is *possible* that harming animals might be justified; but we also maintain that those harming animals typically fail to show that the harm caused is *actually* justified. A further question we must ask ourselves is what, morally speaking, we ought to do in such a situation? Reflection on comparable situations involving human beings will help make the answer clear.

Consider racism and sexism. Imagine that slavery is an institution of the day and that it is built on racist or sexist lines. Blacks or women are assigned the rank of slave. Suppose we are told that in extreme circumstances even slavery might conceivably be justified, and that we ought not to object to it or try to bring it down, even though no one has shown that it is actually justified in the present case. Well, I do not believe for a moment that we would accept such an attempt to dissuade us from toppling the institution of slavery. Not for a moment would we accept the general principle involved here, that an institution actually is justified because it might conceivably be justified. We would accept the quite different principle that we are morally obligated to oppose any practice which appears to violate rights unless we are shown that it really does not do so. To be satisfied with anything less is to cheapen the value attributable to the victims of the practice.

Exactly the same line of reasoning applies in the case where animals are regarded as so many dispensible commodities, models, subjects, etc. We ought not to back away from bringing these industries and related practices to a halt just because it is *possible* that the harm caused to the animals *might* be justified. If we do, we fail to mean it when we say that animals are not mere things, that they are the subjects of a life that is better or worse for them, that they have inherent value. As in the comparable case involving harm to human beings, our duty is to act, to do all that we can to put an end to the harm animals are made to endure. The fact that the animals themselves cannot speak out on their own behalf, the fact that they cannot organize, petition, march, exert political pressure, or raise our level of consciousness—all this does not weaken our obligation to act on their behalf. If anything, their impotence makes our obligation the greater.[23]

We can hear, if we will but listen, the muffled detonation of the explosive harpoon, the sharp crack of the Thai hunter's rifle, the drip of the liquid as it strikes the rabbit's eye, the bobbie calf's forlorn sigh. We can see if we will but look, the last convulsive gasps of the great blue whale, the dazed terror of the gibbons' eyes, the frenzied activity of the rabbit's feet, the stark immobility of the bobbie calf. But not at this moment only. Tomorrow, other whales, other rabbits will be made to suffer; tomorrow, other gibbons, other calves will be killed, and others the day after. And others, stretching into the future. All this we know with certainty. All this and more, incalculably more, *will* go on, if we do not act today, as act we must. Our respect for the value and rights of the animals cannot be satisfied with anything less.

---

[23]For a more complete list of recent philosophical work relating to the topics discussed in the present essay, see Charles Magel and Tom Regan, "Animal Rights and Human Obligations: A Select Bibliography," *Inquiry* 22(1979): 243–47.

NAME ______________________

Worksheet 11

Sentientism: Expanding the Circle to Include Animals

1. **In the brief reading from Jeremy Bentham, what is the crucial question to ask in deciding what beings we have a moral responsibility to treat well?**

2. **Tom Regan criticizes several other perspectives before defending his own view. Briefly summarize each view and his criticism of it on the chart below.**

The Kantian account **Describe the position:**	**Summarize Regan's critique:**
The cruelty account **Describe the position:**	**Summarize Regan's critique:**
The utilitarian account **Describe the position:**	**Summarize Regan's critique:**

3. **Briefly summarize Regan's "animal rights account."**

4. **According to Regan, what is it that gives a being (human being or animal) rights?**

5. **List four implications of Regan's view for the way we live.**

Pulling It Together

Sentientism: Expanding the Circle to Include Animals

QUESTIONS FOR THOUGHT AND DISCUSSION

1. What obligations do you believe humans have toward animals? Why do we have these obligations?

2. Do you believe that it makes sense to speak of animals as having 'rights'? Why or why not?

3. Do you believe that humans are more valuable than other animals? Why or why not? If you believe they are, explain what it is that you believe sets humans apart and bestows greater value on them.

4. Imagine that you are on a mountain hike and you come across a bear which is attacking a child. Do you believe it would be morally justified to kill the bear if it was necessary to save the child's life. Why or why not?

5. Do you believe that it is moral to test medicines and drugs on non-human animals? Why or why not? Would it be moral to perform that same experiments on humans?

EXERCISES

1. Many zoos and wild animal parks such as the San Diego Zoo and the San Diego Zoo Wild Animal Park keep wild animals in captivity. Conduct research in the library and on the internet to discover what sentientists, such as Singer and Regan, say about zoos and wild animal parks. Write a one page essay in which you briefly explain their views on zoos and their argumens for their views.

2. Write an essay in which you discuss the arguments concerning the use of non-human animals for medical research. Is such experimentation morally justified? Present arguments on differing sides of this question, including arguments by sentientists and others. After presenting different positions, explain which position you find the most plausible and why.

Chapter 12

Biocentrism: Expanding the Circle to Include All Living Things

INTRODUCTION

Men and women in today's consumer, fast-food, instant-gratification culture find the views of sentientists such as Singer and Regan radical. Yet in this chapter we consider a perspective that pushes the circle of the moral community out even farther. According to biocentrism, all life is valuable, not just human life, or even just sentient life. Value is not confined to beings who are subjects-of-a-life. Biocentrism affirms that all living beings have inherent value.

In this chapter, we will focus primarily the views of Albert Schweitzer (1875–1965). Schweitzer lived a fascinating, multifaceted life that will be summarized later. While serving as a medical missionary in Africa, Schweitzer developed a unique philosophy of life. For the truly ethical person, he writes, "all life is sacred, including forms of life that from a human point of view may seem to be lower than ours." Although Schweitzer writes in passionate prose, his view rests less on rational argumentation and more on the mystical experience of the unity of life, similar to ideas and experiences encountered in Eastern traditions

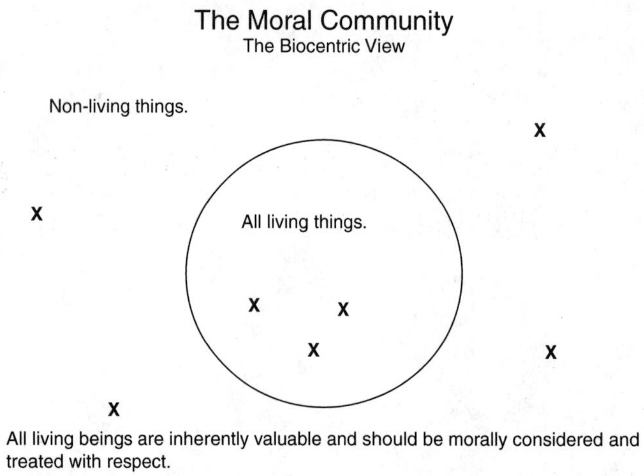

The Moral Community: The Boicentric View

explored earlier in this book. Schweitzer devoted much of his later life to the cause of the philosophy of "reverence for life." My essay on his life and ideas provides a taste of his philosophy.

A second biocentric perspective is that of Paul Taylor, a living philosopher who has worked out what he calls an "ethics of respect for nature." Although I have no included a reading from Taylor in this book, it will be helpful to briefly summarize his ideas here. Far from being a naïve idealist, who would suggest that humans could live life without killing another living being, Taylor recognizes our need for food and the fact that there are times when destroying life is a necessity. He does not react by rejecting the value of life, however, but by working out a detailed system of principles by which human beings can respect all life, even when it is necessary to take it.

His argument is, in a sense, quite like that of Tom Regan, and could be seen as an even more radical extension of the basic Kantian idea that some beings have inherent worth that should be respected (although Taylor does not speak of this respect in terms of "rights"). Taylor argues that it is not reason (Kant) or being the subject-of-a-life (Regan) that gives a being inherent worth, but life itself. Life, he says, is characterized by teleology. Living things are teleological systems. That is, they do not just operate, but they develop and grow toward a goal. Because of this, each living thing can be benefited and harmed. A young oak tree, for example, grows toward the goal of developing into a mature oak tree. But this growth can be helped or hindered, the tree can be benefited or harmed, by many factors in its environment. Humans can clearly benefit or harm the tree. The point is that if the tree can be benefited and harmed, it must have "a good of its own." Thus Taylor argues that each living thing has its own good and has inherent worth, and should thus be treated with respect.

Albert Schweitzer and Paul Taylor represent two approaches to ethics that place life itself at the center. After reading the essay on Schweitzer, I encourage you to read further in the articles and books by both Schweitzer and Taylor. Ask yourself, "what insights do these thinkers contribute to my environmental perspective?"

Albert Schweitzer and Reverence for Life: An Introduction

Albert Schweitzer was born into an Alsatian (German-French) family that had been Christian ministers and musicians for several generations. As a young child Albert Schweitzer had already become famous for his ability to play the organ. As a young man he studied theology, served as a pastor and seminary administrator, and published The Quest of the Historical Jesus, a book that rocked the field of biblical studies. By his early 30s, he had works in both theology and music for which he would be remembered long after his lifetime. But all of this was overshadowed by the humanitarian mission that propelled the remainder of his life. Even while his books on theology and music were beginning to be published, Schweitzer developed a desire to become a doctor and use his medical skills to serve the people of Africa as a missionary. This led him, in 1905, to begin studies in medicine. Receiving his M.D. degree in 1911, and marrying soon after, Schweitzer moved with his wife in 1912 to Africa to found a hospital in the village of Lambaréné in French Equatorial Africa, in an area now known as Gabon. His work in Africa was interrupted in 1917, during World War 1, when he and his wife were placed in a French internment camp. Being released in 1918, Schweitzer spent six years in Europe preaching, lecturing, and performing organ concerts. But in 1924, he returned to Africa and spent most of the remaining years of his life at work caring for African people as a doctor, pastor, and hospital and village administrator. In 1953, Schweitzer was award the Nobel Peace Prize for the year of 1952. By the 1960s, under his leadership, the hospital had grown to include 70 buildings, with a capacity for serving 500 patients. Schweitzer constantly channeled money earned from his publications, concerts, and lectures into his work in Lambaréné. The Albert Schweitzer Hospital continues to serve the people of Africa, currently hosting a leading medical research unit that specializes in infectious diseases prevalent in sub-Saharan Africa.

While in Africa, Schweitzer developed a philosophy of life and ethics, radical for both his time and ours, which he called "Reverence for Life." In the following excerpt from one of his works, Schweitzer describes an experience that was seminal in the development of his ideas:

> ***At sunset of the third day, near the village of Igendja, we moved along an island set in the middle of the wide river. On a sandbank to our left, four hippopotamuses and their young plodded along in our same direction. Just then, in my great tiredness and discouragement, the phrase, "Reverence for Life," struck me like a flash. As far as I knew, it was a phrase I had never heard nor ever read. I realized at once that it carried within itself the solution to the problem that had been torturing me. Now I knew that a system of values that concerns itself only with our relationship to other people is incomplete and therefore lacking in power for good. Only by means of reverence for life can we establish a spiritual and humane relationship with both people and all living creatures within our reach. Only in this fashion can we avoid harming others, and, within the limits of our capacity, go to their aid whenever they need us.***

Only by means of reverence for life can we establish a spiritual and humane relationship with both people and all living creatures within our reach.

> ***It also became clear to me that this elemental but complete system of values possessed an altogether different depth and an entirely different vitality from one that concerned itself only with human beings. Through reverence for life, we come into a spiritual relationship with the universe. The inner depth of feeling we experience through it gives us the will and the capacity to create a***

spiritual and ethical set of values that enable us to act on a higher plane, because we then feel ourselves truly at home in our world. Through reverence for life, we become, in effect, different persons. I found it difficult to believe that the way to a deeper and stronger ethic, for which I had searched in vain, had been revealed to me as in a dream. Now I was at last ready to write the planned work on the ethics of civilization.

Schweitzer went on to write about his new philosophy in several important works. "Reverence for Life" was not just a speculative philosophy of life, but an implicit worldview with practical implications, as is illustrated in the two short readings that follow.

THE GROUNDING OF THE REVERENCE FOR LIFE IN CONCIOUSNESS

Based on his reflections and experiences, Albert Schweitzer developed a new approach to ethics. This new ethics is not grounded in looking at the world, or in attempting to think rationally, but in thinking deeply about one's own conscious experience. Schweitzer believed that if one is truly honest with oneself, one will discover the necessity of reverence for all life.

If man wishes to reach clear notions about himself and his relation to the world, he must ever again and again be looking away from the manifold, which is the product of his thought and knowledge, and reflect upon the first, the most immediate, and the continually given fact of his own consciousness. Only if he starts from this can he arrive at a thinking worldview.[1]

To discover the foundation for a clear understanding of one's relationship to the world, Schweitzer is saying, one must look away from the variety of experiences and examine what is present to us in consciousness in general, that is, in every act of conscious experience. If we turn away from the variety of forms our conscious experience takes and look at the nature of our experience itself, he believes, we will discover the key to a clear understanding of ourselves and the world. What will we discover?

The most immediate fact of man's consciousness is the assertion: "I am life which wills to live, in the midst of life which wills to live," and it is as will-to-live in the midst of will-to-live that man conceives himself during every moment that he spends in meditating on himself and the world around him.[2]

When we examine the nature of our consciousness, Schweitzer is suggeseting, each of us will discover that all of his or her life is based on a fundamental will to live and we experience ourselves not as isolated beings, but as a being among other beings who share with us this will to life. None of us can think seriously about ourselves and the world in which we live without assuming that we and many other beings share a fundamental drive to continue to live. This drive, Schweitzer calls, "the will-to-live."

Can a person deny that he or she has the will-to-live? Schweitzer recognizes that some individuals, and even some philosophies of life, seem to attempt to deny the will-to-live, but he argues that this attempt involves a "self-contradiction." Even these individuals who speak as if they deny this fundamental desire to live are constantly making exceptions and are full of inconsistencies in their thinking and acting. Schweitzer's argument is based on the assumption that each individual, even if he or she denies it, has already made a commitment to live.

This assumption comes to the front when Schweitzer turns to consider the person who consciously affirms the will-to-live.

If man affirms his will-to-live, he acts naturally and honestly. He confirms an act which has already been accomplished in his instinctive thought by repeating it in his conscious thought. . . To affirm life is to deepen, to make more inward, and to exalt the will-to-live.[3]

---

[1]Albert Schweitzer, Out of My Life and Thought: An Autobiography, translated by C. T. Campion, (New York: Henry Holt and Company, 1933), 186.
[2]Albert Schweitzer, Out of My Life and Thought: An Autobiography, translated by C. T. Campion, (New York: Henry Holt and Company, 1933), 186.
[3]Albert Schweitzer, *Out of My Life and Thought: An Autobiography*, translated by C. T. Campion, (New York: Henry Holt and Company, 1933), 187.

In affirming one's will-to-live, one is deepening one's commitment to an act that has "already been accomplished" simply by living one's life. It is in the nature of life itself, one might say, to have this desire, this drive, to continue life. And thus one who reflects upon his situation carefully and responds honestly will affirm that this will-to-live is in fact at the center of his or her life.

But honesty also brings another realization. The fundamental conscisouness that is at the core of all experience is not just of one's own will-to-live, but that of others living beings as well. Going against the grain of philosophers who follow in the footsteps of Rene Descartes and see one's foundational self-consciousness to be of oneself alone, Schweitzer contends that at the heart of all of our experience is a recognition that we live "in the midst of" other beings who share with us the will-to-live. In a manner reminiscent of Hinduism, Schweitzer links one's experience of the "will-to-live" in oneself with to the experience of the universal will-to-live that is in all living beings.

> ***In my will-to-live the universal will-to-live experiences itself otherwise than in its other manifestations. In them it shows itself in the process of individualizing which, so far as I can see from the outside, is bent merely on living itself out to the full, and in no way on union with any other will-to-live. The world is a ghastly drama of will-to-live divided against itself. One existence makes its way at the cost of another; one destroys the other. One will-to-live merely exerts its will against the other, and has no knowledge of it. But in me the will-to-live has come to know about other wills-to-live. There is in it a yearning to arrive at unity with itself, to become universal.***[4]

But in me the will-to-live has come to know about other wills-to-live. There is in it a yearning to arrive at unity with itself, to become universal.[4]

In an unreflective life, the world seems as a ghastly battle between wills. One will seeks to live at the expense of others. But, Schweitzer says, when one looks inward one sees that each will-to-live is only a manifestation of a universal will-to-live that is shared by all. To reflect upon oneself and others, and to come to terms with that it is this universal will-to-live that one experiences in one's own self.

At the same time the man who has become a thinking being feels a compulsion to give every will-to-live the same reverence for life that he gives to his own.

Schweitzer suggests that this move to reflective recognition of the universal will-to-live brings with it a basic ethical affirmation as well.

> ***At the same time the man who has become a thinking being feels a compulsion to give every will-to-live the same reverence for life that he gives to his own. He experiences that other life in his own. He accepts as being good: to preserve life, to promote life, to raise to its highest value life which is capable of development; and as being evil: to destroy life, to injure life, to repress life which is capable of development. This is the absolute, fundamental principle of the moral, and it is a necessity of thought.***[5]

Thus, for Schweitzer, the foundations of ethics are grounded in the reflective realization that at the core of one's life and consciousness is an affirmation of life, and that this affirmation is shared by others. It is, so to speak, the fundamental value. Whether it is found in oneself or in another living being, it is the same will-to-live and equally valuable. Thus, Schweitzer believes, the thinking person is committed to a particular concept of good and a particular concept of evil. Whatever preserves, promotes, and nourishes life, wherever it is found, is good. Whatever destroys, injurers or represses life, wherever it is found, is evil. This is the basis for ethics. This is reverence for life.

---

[4]Albert Schweitzer, *The Philosophy of Civilization* (Buffalo: Prometheus Books, 1987), quoted in Susan J. Armstrong and Richard G. Botzler, *Environmental Ethics: Divergence and Convergence* (New York: McGraw-Hill, Inc, 1993), 344.

[5]Albert Schweitzer, *Out of My Life and Thought: An Autobiography*, translated by C. T. Campion (New York: Henry Holt and Company, 1933), 187–8.

LIVING WITH A PUZZLING AND HORRIBLE NECESSITY

Schweitzer is not naïve. He is well aware that one could not live without harming and destroying other living beings. The world," he writes, "offers us the horrible drama of the Will-to-Life divided against itself."[6] As the thinking person becomes self-aware, he or she experiences a solidarity with all life, but it is a solidarity he or she

> ***cannot completely bring about because man is subject to the puzzling and horrible law of being obliged to live at the cost of other life, and to incur again and again the guilt of destroying and injuring life. But as an ethical being he strives to escape whenever possible from this necessity, and as one who has become enlightened and merciful to put a stop to this disunion of the Will-to-Live so far as the influence of his existence reaches. He thirsts to be permitted to preserve his humanity, and to be able to bring other existences release from their sufferings.***[7]

The man who is guided by the ethics of reverence for life stamps out life only from inescapable necessity, never from thoughtlessness. He seizes every occasion to feel the happiness of helping living things and shielding them from suffering and annihilation.[9]

Although we cannot escape our situation, neither is it morally right to acquiesce in it. Whereas we cannot avoid causing suffering and death, we can seek to minimize the amount of suffering and death that we cause. Whereas we cannot protect and nurture all life, we can protect and nurture as life as much as we are able. With this as our ideal, Schweitzer suggests, we "reach a power of reflection which enables us to distinguish between what is essential in civilization and what is not."[8] We will, in short, form a new set of priorities by which to live our lives. The highest priority will be on life.

But how will this work out in a human life? Schweitzer suggests that the ethical person must judge each situation individually, making the "horrible," and sometimes "arbitrary" decisions with seriousness.

> ***To the truly ethical man, all life is sacred, including forms of life that from the human point of view may seem to be lower than ours. He makes distinctions only from case to case, and under pressure of necessity, when he is forced to decide which life he will sacrifice in order to preserve other lives. In this deciding from case to case, he is aware that he is proceeding subjectively and arbitrarily, and that he is accountable for the lives thus sacrificed.***
>
> ***The man who is guided by the ethics of reverence for life stamps out life only from inescapable necessity, never from thoughtlessness. He seizes every occasion to feel the happiness of helping living things and shielding them from suffering and annihilation.***[9]

LIVING WITH REVERENCE FOR LIFE

Schweitzer gives examples of implications of the philosophy of reverence for life. A farmer, he writes, who has cut "down a thousand flowers in his meadow in order to feed his cows" should nevertheless be careful on the way home not to destroy a single flower along the way thoughtlessly. "For then he sins against life without being under the compulsion of necessity."[10]

---

[6]Albert Schweitzer, *Out of My Life and Thought: An Autobiography*, translated by C. T. Campion (New York: Henry Holt and Company, 1933), 188.
[7]Albert Schweitzer, *Out of My Life and Thought: An Autobiography*, translated by C. T. Campion (New York: Henry Holt and Company, 1933), 189.
[8]Albert Schweitzer, *Out of My Life and Thought: An Autobiography*, translated by C. T. Campion (New York: Henry Holt and Company, 1933), 189.
[9]Albert Schweitzer, *The Teachings of Reverence for Life*, translated by Richard and Clara Winston (New York: Holt, Rinehart and Winston, 1965), 47–48.
[10]Albert Schweitzer, *The Teachings of Reverence for Life*, translated by Richard and Clara Winston (New York: Holt, Rinehart and Winston, 1965), 48.

In another example, Schweitzer considers the use of animals in scientific experiments. He does not deny that such experiments might justified "in order to apply the knowledge gained to the alleviation of human ills." But with each experiment, the scientist must consider the case. No general principle that justifies medical experiments is enough to justify all experiments.

> ***In each individual case they must ask themselves whether there is a real necessity for imposing such a sacrifice upon a living creature. They must try to reduce the suffering insofar as they are able. It is inexcusable for a scientific institution to omit anesthesia in order to save time and trouble. It is horrible to subject animals to torment merely in order to demonstrate to students phenomena that are already familiar.***[11]

As we have seen before, whenever an action causes suffering and death to any living thing, it should be done only "under necessity." In this case, it seems that he believes that some experiments are necessary in order to alleviate other suffering and death, but that many that are not.

Just as it is wrong to cause unnecessary suffering in animals, it is wrong to sit by while animals suffer as the result of the actions of others. When we are aware that needless pain and death is being inflicted on animals, we have an ethical responsibility to act. It is wrong to merely pretend that we have not seen the plight of these living beings.

> ***When abuse of animals is widespread, when the bellowing of thirsty animals in cattle cars is heard and ignored, when cruelty still prevails in many slaughterhouses, when animals are clumsily and painfully butchered in our kitchens, when brutish people inflict unimaginable torments upon animals and when some animals are exposed to the cruel games of children, all of us share in the guilt.***[12]

The examples that Schweitzer gives are very interesting. His references to the farmer's cows and to slaughtering animals does not seem to suggest that he believes that use of animals in agriculture or even for food is wrong in general. His recognition of the "necessity" of experiments on animals in medical research seperates him from many of the animal activists today who argue for an end to animal experimentation. Yet all of these activities, while not ruled out, must be engaged in thoughtfully, being sure from "case to case, he is aware that he is proceeding subjectively and arbitrarily, and that he is accountable for the lives thus sacrificed."[13] In Schweitzer's worldview, there is a terrible necessity of death that one must never forget, and that one must always seek to escape as much as possible.

SCHWEITZER'S CRITICISM OF PREVIOUS ETHICISTS

Not surpringly, Albert Schweitzer has harsh criticism for philosophers and ethicists who have gone before him.

> ***The great fault of all ethics hitherto has been that they believed themselves to have to deal only with the relations of man to man. . . . A man is ethical only when life, as such, is sacred to him, that of plants and animals as that of his fellow-men, and he devotes himself helpfully to all life that is in need of help.***[14]

A man is ethical only when life, as such, is sacred to him, that of plants and animals as that of his fellow-men, and he devotes himself helpfully to all life that is in need of help.[14]

I close my exposition of Schweitzer's philosophy of reverence for life with a passage which illustrates both his criticism of other ethicists and the radical nature of the way of life which he challenges us to live.

> ***As the housewife who has scrubbed the floor sees to it that the door is shut, so that the dog does not come in and undo all her work with his muddy paws, so religious and philosophical thinkers have gone to some pains to see that no animals enter and upset their systems of ethics.***

---

[11] Albert Schweitzer, *The Teachings of Reverence for Life*, translated by Richard and Clara Winston (New York: Holt, Rinehart and Winston, 1965), 48.
[12] Albert Schweitzer, *The Teachings of Reverence for Life*, translated by Richard and Clara Winston (New York: Holt, Rinehart and Winston, 1965), 49.
[13] Albert Schweitzer, *The Teachings of Reverence for Life*, translated by Richard and Clara Winston (New York: Holt, Rinehart and Winston, 1965), 48.
[14] Albert Schweitzer, *Out of My Life and Thought: An Autobiography*, translated by C. T. Campion (New York: Henry Holt and Company, 1933), 188.

It would seem as if Descartes, with his theory that animals have no souls and are mere machines which only seem to feel pain, had bewitched all of modern philosophy. Philosophy has totally evaded the problem of man's conduct toward other organisms. We might say that philosophy has played a piano of which a whole series of keys were considered untouchable.

To the universal ethics of reverence for life, pity for animals, so often dismissed as sentimentality, becomes a mandate no thinking person can escape.

The time will come when public opinion will no longer tolerate amusement based on the mistreatment and killing of animals. The time will come, but when? When will we reach the point that hunting, the pleasure in killing animals for sport, will be regarded as a mental aberration? When will all the killing that necessity imposes upon us be undertaken with sorrow?

SOME CLOSING REFLECTIONS

Albert Schweitzer's philosophy of Reverence for Life represents an example of "biocentric ethics," ethics which places intrinsic value in life itself. The principle of ethics he suggests is radical. It does not rationalize the activities we already perform, as ethical principles sometimes seem to do, but challenges us to live a very different type of life, a life which respects all of life. Philosophers sometimes describe the view that it is life itself that is intrinsically valuable as "biocentrism." A more recent philosopher who has defended a biocentric approach to ethics, in a much more philosophically sophisticated manner, is the contemporary philosopher Paul W. Taylor. Taylor argues that all living things have ends, goals toward which they strive, and because of this they have goods of their own. Each living thing can be helped and harmed. On the basis of this core idea, Taylor develops a set of principles that outline an ethical system of "respect for nature." Schweitzer is, in a sense, a prophet. Taylor is a philosopher. And while they share the biocentric perspective on ethics, they are differences between them as well. Unlike Schweitzer, who we have seen advocates a "case by case" approach, Taylor seeks to develop principles to help an ethical person live with respect for all life. This essay is written to serve as an introduction to biocentric ethics. I would encourage you to read more in the Schweitzer and Talyor for a deeper understanding of biocentrism. Only with further study will you be in a position to really evaluate the ideas of these thinkers.

I do, however, hope that I have given you significant food for thought. In my teaching of environmental ethics, I find that few students take biocentrism seriously. And I will admit, that I find its message almost too radical to hear myself. But I have also been struck by an interesting inconsistency. While few take biocentrism seriously, many recite a twentieth century myth about native American culture with much approval. "The native Americans," they say, "never killed any more buffalo (or other animals) than they needed. And when they did they asked the victims for permission, respectfully offering sacrifices of thanksgiving, recognizing that these animals were their brothers and sisters."[15] What strikes me as odd is this: The biocentric perspective we have been considering is rejected out of hand. But in a sense, it seems to be saying no more than this American myth: We should only harm and kill the things of nature when it is necessary. To do otherwise is wrong. And when we find that we must, we should do so with reverence and sorrow for the life we take.

Why is it, I ask, that we blithely spout "native American" wisdom and reject biocentrism. I suggest two reasons. First, few of us seriously consider what biocentrists are really saying. We assume that they are advocating a life that *never* harms either plants or animals, a life that is patently impossible for us to live. And thus we reject it out of hand. To the extent that this is how we are reasoning, we are guilty of what philosophers call "the straw man fallacy." We reject biocentrism because we can destroy a similar-looking

[15]I call this a "myth" not because I don't believe there is an element of truth in it, but because like many other descriptions of America's indigenous peoples it is too generalized to be entirely accurate. It is recited not in order to present a true description of the past but to make a point about the human relationship to nature. Unfortunately, in the process it fails to give due respect to the particular tribes and ways of life that are represented among the peoples of whom it speaks.

position we have constructed out of straw. As anyone who has learned the lessons of a logic class should know, this is not a good way to reason. Second, in telling the native American myth, we are thoughtlessly engaging in a ritual of political correctness. Just as many (unfortunately) engage in rituals in the church for which they do not understand the meanings, using words they neither understand nor intend to live out, we tell this nice story about those who came before us in America without really intending to take it seriously. Oh, I'll admit in both cases we have a point. We take our kids to church and engage in its rituals in hopes that we can pass on standard American "morals." And we tell the myth of the buffalos to reinforce in our kids the old American proverb, "Waste not, want not." In the meantime, we miss the real teachings of Jesus and we fail to really learn from the native Americans. Biocentrism is a challenging position. Let us all think about it carefully. Whether we accept it in the end, I hope that it can help us to become more ethical men and women.

NAME ______________________

Worksheet 12

Biocentrism: Expanding the Circle to Include All Living things

1. **According to Albert Schweitzer, what basic attitude or desire is central to conscious life?**

2. **Why does Schweitzer believe one should care about other living things?**

3. **After reading about Schweitzer, how would you characterize the attitude he believed one should have about the necessary killing of other living beings.**

Pulling It Together

Biocentrism: Expanding the Circle to Include All Living Things

QUESTIONS FOR THOUGHT AND DISCUSSION

1. Do you believe that plants have value in themselves? Why or why not?

2. What considerations do you believe support the importance of preserving plant life? Analyze these reasons. Are they anthropocentric? sentientist? biocentric?

EXERCISE

Philosopher Paul Taylor defends a more sophisticated biocentric theory of ethics. An intro to his view is found in "The Ethics of Respect for Nature" in the journal *Environmental Ethics,* Fall 1981. In the Fall 1982 issue of the same journal, Gene Spitler wrote a criticism of Taylor's view, entitled "Justifying a Respect for Nature." Then the Fall 1983 issue of *Environmental Ethics* includes an article entitled, "In Defense of Biocentrism," in which Taylor responds to Spitler's critique. Go to the library and find these article and read them carefully. This is a typical philosophical dialuge between two philosophers. After reading the three articles, write a 5 to 10 page research paper in which you outline the arguments and responses that are part of the dialogue.

Chapter 13

Ecocentrism: Valuing the Land as a Whole

INTRODUCTION

In this section, we traced an expanding circle of moral concern. Most standard philosophical ethics, at least in the Western tradition, focuses on human beings. Theories that place ultimate moral value in humans alone we call examples of anthropocentrism, human-centered ethics. Sentientism pushes the circle of moral consideration out to include not just humans, but other sentient beings as well. Because many non-human animals are sentient in the relevant sense, sentientists argue that we must consider these animals when we determine what is morally right and wrong. We have considered two types of argument for this more inclusive approach to ethics. With biocentrism, the circle is expanded even further, to include all living things. Because every living thing has a good of its own, we must respect each instance of life. The idea of an expanding circle is a helpful way to see the progression through these theories.

In this chapter we consider an approach to environmental ethics inspired by the work of Aldo Leopold. In his essay, "The Land Ethic," Leopold writes, "The land ethic simply expands the boundaries of the community to include soil, waters, plants, and animals, or collectively: the land." Leopold's approach to this "land ethic" is summarized in the first reading in this chapter. At first glance, Leopold's words suggest that his proposed ethic simply expands the circle to one step further to include nonliving, as well as living, beings. This would indeed be an inclusive ethic! On closer inspection, however, it is clear that Leopold is not suggesting that we simply expand the categories of individual beings that are included in the moral community, but rather that we shift our focus from the individuals to the community as a whole. In other words, it turns out that the word "collectively" is very important. As an ecologist, Leopold knows that elements in what he calls a "biotic community" cannot be understood apart from the community as a whole.

To translate this into more common terms, the function and role of the parts of an ecosystem, the water, the animals, the soil, cannot be fully understood if they are isolated and studied solely on their own. They must be studied in the context of the system or community of which they are a part. Only then can one see the real value they have to the system as a whole. So Leopold is suggesting that we think of ethics in these same terms. We must think, first of all, of a biotic community (a community of life), as a whole. Once we do this, we can understand the value of the parts. In short, Leopold's ethics calls for a shift from what philosophers have called individualist ethics to a holistic ethic. When we make this shift, we realize that it is not just humans, or animals, or even living things that are valuable, but water and soil as well.

In a frequently quoted passage, Leopold writes, "A thing is right as it tends to preserve the beauty, integrity, and stability of the biotic community; wrong as it tends otherwise." The meaning of this principle has been much discussed in environmental philosophy. J. Baird Callicott is a contemporary philosopher who has consistently sought to build an environmental ethics on the foundation suggested by Leopold. In the second reading in this chapter, James D. Heffernan explores the meaning of Leopold's famous principle and Callicott's interpretation of it. Aldo Leopold and those inspirerd by him give us lots of food. As you read this chapter, try to grasp the meaning of, arguments for, and implications of a holistic approach to ethics. What aspects of Leopold's ideas do you find most plausible? Why?

Aldo Leopold and The Land Ethic: An Introduction

One cannot explore the field of environmental ethics for long without coming across the name of Aldo Leopold. Leopold's ideas have shaped ideas and policies in such diverse fields as game management, forest conservation, and environmental ethics. His works are read in literature and environmental ethics courses around the world. Who was this man? The Aldo Leopold Foundation's Web page describes him as "a conservationist, forester, philosopher, educator, writer, and outdoor enthusiast."[1]

Aldo Leopold was born on January 11, 1887, in Burlington, Iowa. From an early age, Leopold would go with his father and siblings on walks, exploring the fields and woods near his home. His high school years were spent at a private boarding school in New Hampshire, where his record was not stellar, allegedly due to his love for the outdoors.[2] He went on to study for a Master's degree as one of the first students in the School of Forestry at Yale University. From 1909 to 1928, Leopold worked for the U.S. Forest Service. His first assignments were in Arizona and New Mexico, where he worked successfully for the establishment of the nation's first designated wilderness area. In 1924, Leopold was transferred to the Forest Service's Products Laboratory in Madison, Wisconsin. After several years, Leopold left the Forest Service to become a private game consultant. In 1933, he began a career on the faculty of the University of Wisconsin, serving as the initial chair of the University's game management program in the Department of Agricultural Economics and later as chair of the University's Department of Wildlife Management. During his career at the University, Leopold and his wife purchased an old farm on the Wisconsin River. Restoring the health of the land on the farm became a family project for Leopold, his wife Estella, and their five children. On April 21, 1948, while helping fight a grass fire on neighboring farmland, Leopold died of a heart attack.[3]

Although his life was cut short, his influence is great. As a forest and game manager, Leopold experienced the natural world and the human impact on our natural environment first-hand. Hours spent planting trees and restoring prairies on his farm gave Leopold time to reflect on the land community of which he was a part. As a result of these thoughts, Leopold penned his most influential work, published posthumously in 1949, under the title of The Sand County Almanac. His down to earth, yet prophetic, writing has remained relevant for over fifty years, and it continues to inspire those concerned with environmental issues today. Although Leopold was not an academic philosopher, his essay, "The Land Ethic," included in The Sand County Almanac, planted seeds for important developments in environmental philosophy.

---

[1]The Aldo Leopold Foundation. "Aldo Leopold," http://www.aldoleopold.org/About/AldoLeopold.pdf (accessed July 25, 2009).
[2]The Wilderness Society. "Aldo Leopold," http/wilderness.org/content/aldo-leopold (accessed July 24, 2009).
[3]The Aldo Leopold Foundation. "Aldo Leopold," http://www.aldoleopold.org/About/AldoLeopold.pdf (accessed July 25, 2009), and The Wilderness Society. "Aldo Leopold," http/wilderness.org/content/aldo-leopold (accessed July 24, 2009).

EXTENDING ETHICS TO INCLUDE THE LAND COMMUNITY

As a scientist and conservationist, Leopold pioneered the shift toward a more ecological perspective. Unlike the "classical sciences," which studied natural phenomena by breaking them down into their constituent parts and studying the parts, ecology emphasizes the need to study the whole to understand the parts. Thus, it considers things systematically, holistically. Individual organisms and species cannot be understood apart from the role they play in the larger habitat in which they live. One sees an organism in the context of its community. Leopold saw "the land" through ecological lenses. Many of the essays in *A Sand County Almanac* communicate his unique perspective through stories and observations of nature. But in the "The Land Ethic," Leopold takes a more straight-forward line of attack to advocate for a new approach to ethics. It is an ethics based squarely on the concept of community. All ethics, he argues, have been based on the idea of community. And through time, humankind has recognized that the membership of this community must be expanded. Leopold's reflections on the land led him to conclude that the time had come for an even wider, more inclusive ethic. He wrote,

> ***All ethics so far evolved rest upon a single premise that the individual is a member of a community of interdependent parts. His instincts prompt him to compete for his place in that community, but his ethics prompt him also to co-operate (perhaps in order that there may be a place to compete for). The land ethic simply enlarges the boundaries of the community to include soils, waters, plants, and animals, or collectively: the land.***[4]

Leopold anticipates a question: Don't we already declare our love for the land of the free and the home of the brave? His answer is, in fact, "no." At least not if we by "land," we mean the fertile topsoil that we allow to flow down the rivers, or the plant or animal species that we systematically destroy. The soil, the water, the plants, the animals are not respected as members of our ethical community. Leopold calls for this to change.

> ***A land ethic of course cannot prevent the alteration, management, and use of these "resources," but it does affirm their right to continued existence, and, at least in spots, their continued existence in a natural state. In short, A land ethic changes the role of Homo sapiens from conqueror of the land-community to plain member and citizen of it. It implies respect for his fellow-members, and also respect for the community as such.***[5]

A land ethic changes the role of Homo sapiens from conqueror of the land-community to plain member and citizen of it. It implies respect for his fellow-members, and also respect for the community as such.[5]

THE CONCEPT OF THE LAND

In order to develop the idea of the land community further, Leopold explains the land as an energy circuit. Energy travels through what he calls "the land pyramid." This pyramid consists of several layers, all of which are important for the flow of energy in the circuit.

> ***Plants absorb energy from the sun. This energy flows through a circuit called the biota, which may be represented by a pyramid consisting of layers. The bottom layer is the soil. A plant layer rests on the soil, an insect layer on the plants, a bird and rodent layer on the insects, and so on up through various animal groups to the apex layer, which consists of the large carnivores.***[6]

The pyramid structure is not based on origins or status in the system, but on food supply. "Each successive layer depends on those below it for food and often for other services, and each in turn furnishes food and services to those above. Proceeding upward, each successive layer decreases in numerical abundance."[7]

---

[4]Aldo Leopold. "The Land Ethic," in *A Sand County Almanac* (Oxford University Press, 1949). 203–204.
[5]Leopold, 204.
[6]Leopold, 215.
[7]Leopold, 215.

Humans, writes Leopold, share an intermediate layer with other omnivores, such as bears, raccoons and squirrels. Within this pyramid, there is a complex web of food chains, with each species having a part in numerous chains. "The pyramid is a tangle of chains so complex as to seem disorderly, yet the stability of the system proves it to be a highly organized structure."[8] When the system is healthy, the structure is very stable. Each elements of the system is both nourished by many other parts and a source of nourishment for many other elements. It is this entire pyramid that Leopold refers to as "the land."

> ***Land, then, is not merely soil; it is a fountain of energy flowing through a circuit of soils, plants, and animals. Food chains are the living channels which conduct energy upward; death and decay return it to the soil. . . . The velocity and character of the upward flow of energy depend on the complex structure of the plant and animal community, much as the upward flow of sap in a tree depends on its complex cellular organization. Without this complexity, normal circulation would presumably not occur. Structure means the characteristic numbers, as well as the characteristic kinds and functions, of the component species. This interdependence between the complex structure of the land and its smooth functioning as an energy unit is one of its basic attributes.***[9]

Leopold concludes his explanation of his concept of the land with the following summary.

> ***This thumbnail sketch of land as an energy circuit conveys three basic ideas:***
>
> ***(1) That land is not merely soil.***
>
> ***(2) That the native plants and animals kept the energy circuit open; others may or may not.***
>
> ***(3) That man-made changes are of a different order from evolutionary changes, and have effects more comprehensive than is intended or foreseen.***
>
> ***These ideas, collectively, raise two basic issues: Can the land adjust itself to the new order? Can the desired alterations be accomplished with less violence?***[10]

Because the changes humans make are more major than the small changes that happen in the evolutionary course of nature, we must ask whether we are right to make these changes.

It is here, that ethical questions are raised. Human beings make alternations to the land. But what are the effects of these alterations for the land community? Plants and animals that are native to a biotic community (the land pyramid in a specific place) enable the energy to flow through the circuit. Humans make frequent changes to the plant and animal life in a particular place. What impact will that have on the biotic community? Will it become unstable? Will the land be able to adjust and remain healthy? Because the changes humans make are more major than the small changes that happen in the evolutionary course of nature, we must ask whether we are right to make these changes.

THE LAND ETHIC

How are we to think about this expanded ethic? And what is the prospect for its development. Leopold recognizes obstacles in the way of a shift to an ethic that thinks of the land in ethical terms, yet he proposes an principle that has spawned significant discussion among environmental ethicists. First, he discusses the prospects.

> ***It is inconceivable to me that an ethical relation to land can exist without love, respect, and admiration for land, and a high regard for its value. By value, of course, I mean something far broader than mere economic value; I mean value in the philosophical sense.***[11]

---

[8]Leopold, 215.
[9]Leopold, 218.
[10]Leopold, 218.
[11]Leopold, 223.

"Unfortunately," Leopold writes, "our culture, both in its economic system and its educational system stands in the way of truly developing a land ethic. They do not encourage us to value the land. First, typical men and women in today's world are "separated from the land by many middlemen, and by innumerable gadgets."[12] This is as true today as it was when Leopold wrote it, if not more so. Second, men and women are "bored stiff" when they are set loose to spend a day in the wild. If we could create food, clothing, or building materially through synthetic methods, most of us would not care. We have little room for nature in our lives. Third, Leopold argues, the development of true love for the land is impeded by the fact that many a farmer sees the land as "an adversary, or a taskmaster that keeps him in slavery." Finally, in spite of extensive systems of higher education, true training in ecological understanding among those who are being educated is very scarce.

After this litany of woes, Leopold writes, "The case for a land ethic would appear hopeless but for the minority which is in obvious revolt against these 'modern' trends." Toward the end of his essay, Leopold suggests a principle for these minority rebels who seek change.

> ***The "key-log" which must be moved to release the evolutionary process for an ethic is simply this: quit thinking about decent land-use as solely an economic problem. Examine each question in terms of what is ethically and esthetically right, as well as what is economically expedient. A thing is right when it tends to preserve the integrity, stability, and beauty of the biotic community. It is wrong when it tends otherwise.***[13]

A thing is right when it tends to preserve the integrity, stability, and beauty of the biotic community. It is wrong when it tends otherwise.[13]

THE CHALLENGE OF EXPANDING ON LEOPOLD'S LEGACY

As in the areas of forestry, game management, and ecology, so in environmental ethics, Aldo Leopold is an original and seminal thinker. His legacy confronts us with a challenge. How will we comes to terms with his argument and interpret his proposal. In particular, what exactly is he suggesting? One possibility is that Leopold is suggesting a new foundation for all ethical thought. The two sentences I have italicized in the quotation are stated very much like the basic principles of other philosophers. John Stuart Mill, for example, proposed that ethics be based on "the greatest happiness principle, which Mill states as follows: "Actions are right in proportion as they tend to promote happiness, wrong as they tend to produce the reverse of happiness."[14] For Mill all ethical actions are to be judged as right or wrong on the basis of this principle. So one interpretation is that Leopold is offering an alternative basic principle, suggesting a principle by which all human actions are to be judged as right or wrong. If Leopold is recommending a principle such as this, he is making a truly radical suggestion.

At first glance, it looks as if Leopold is simply pushing the boundaries of ethics farther out to include more individuals. In recent environmental ethics, there have been philosophers who have argued that we must expand the boundaries of our ethical concern to include at least certain animals. Mammals, for example, should be considered to be morally relevant. Thus when I weigh what is right or wrong, I must take into account how the actions I am considering will affect these individuals. Both Tom Regan and Peter Singer, in different ways, have argued that it is unethical for us to continue in disregard for animals. Others have argued that we should recognize that even other living beings, including plants, have inherent moral value. Albert Schweitzer, in the early twentieth century, developed a philosophy of "reverence for life," which argued that we should value the life in all living things. In a much more philosophically sophisticated way, Paul W. Taylor has defended a variation of this biocentric perspective. Both the animal advocates,

---

[12]Leopold, 224.

[13]Leopold, 224–225.

[14]John Steward Mill, *Utilitarianism*, 1863.

sometimes called sentientists, and the biocentrists are arguing that the boundaries of the moral community must be expanded. At first, it seems that Leopold is simply calling for us to push the boundary out to include nonliving things, such as soil, water, and rocks, as well.

If one thinks carefully about the principle suggested by Leopold, however, one realizes that he has something different in mind. Rather than simply arguing for expanded boundaries, Leopold is taking ethics in a different direction. Individualism has had a strong grip on Western ethics. Each of the theories I have referred to in the previous paragraph is an individualistic theory of ethics. Each identifies the primary focus of ethics on individual beings. What is right or wrong is determined by the effects of actions upon, or by the rights of, individuals. The differences between the theories concern what type of individuals are morally considerable. Anthropocentric approaches say individual humans must be considered. Sentientsts say some individual animals must also be included. Biocentrists argue that all individual living things are morally considerable. For all of these theories, one must look at the well-being of the individuals to determine what is right or wrong. Leopold's principle, if that is what we are to take it to be, turns away from individualism and places the primary focus of right and wrong, of good and evil, on the well-being of the whole, of the biotic community, consisting of the whole complex pyramid of life. Like the science of ecology, the land ethic is based on the holistic concept of community. It begins with the idea that what is right or wrong is determined by understanding the working and welfare of the entire community of life. The moral good, above all else, is a healthy, stable, beautiful biotic community.

It begins with the idea that what is right or wrong is determined by understanding the working and welfare of the entire community of life. The moral good, above all else, is a healthy, stable, beautiful biotic community.

Thus, it is right to call the Land Ethic a holistic. Holism has interesting implications. Take an example from Leopold's own life and practice. Leopold was a hunter. In one famous essay, "Thinking like a Mountain," Leopold writes critically of thoughtless killing of wolves by hunters. But in general, he is not opposed to hunting. Tom Regan would argue that hunting deer is clearly a violation of the rights of individual deer. Peter Singer would argue that hunting deer causes more pain and suffering than happiness, when we take into consideration not only human pain and pleasure, but that of the deer and its relatives as well. The land ethic, on the other hand, asks what impact hunting of deer has on the stability, beauty, and integrity of the biotic community as a whole. And the answer to this question could go either way. In general, human beings are part of the land pyramid, and they are omnivores like others on their layer. There seems nothing wrong with fulfilling this role in the community as a whole. In addition, there may be times when an overpopulation of deer is clearly a threat to the health of a biotic community in a particular place, in which case it may be that thinning out the deer population through hunting would strengthen the biotic community. It is the impact of an action on the health, beauty, and stability of the whole community that is the key, not the impact on the individuals involved.

If this is what Leopold is suggesting, it is a radical suggestion. But it raises concerns as well. For example, in the previous paragraph, I have suggested that there may be times when it is appropriate to allow the reduction of the deer population through hunting if an overpopulation of deer threatens the stability or integrity of the biotic community. Could the same be said about human population? Like deer, humans are members of the biotic community. Humans clearly threaten the health and stability of biotic communities. Is it right to reduce their population through allowing violence against humans? When I ask this in my classes, I usually get a few students who say, "sure, why not?" But I seriously doubt many would be willing to see the suggestion realized. How, then, can we adapt a holistic environmental ethic and meet these concerns? One way to avoid this implication would be to privilege human individuals, a move in the direction of anthropocentrism. Presumably, this move would be based on a pluralistic approach to ethics that incorporates Leopold's holistic ethical principle, but also includes other principles as well, some of which would call for respect for human rights or human happiness. A second avenue to explore might the possibility that some species are somehow more valuable to the community, or more valuable in some other way, than members of other species, and that humans are among these. Still another idea might be to suggest that each species is justified in developing its own mores to insure its own health and survival, and that within the human context, harming of humans is morally wrong. All of these are but stabs in various directions; none of them is completely faithful to the holism suggested by Leopold and his followers.

In conclusion, this essay is merely an introduction to the Land Ethic. I have offered neither a complete explanation and defense nor a complete critical analysis. It would be wrong to dismiss either Aldo Leopold's Land Ethic or other versions of ecocentric holism merely on the basis of this summary. There are several ethicists who have worked hard to develop an ethic based on Leopold's ideas and to respond to objections. The most prominent of these is J. Baird Collicott. I encourage you to read one or two of Collicott's books or articles to get a better grasp of the possible interpretations of an ethic based firmly on Leopold's ideas. Leopold's contribution to ethics, like his contribution to conservation and ecology, are valuable and suggestive. It is my hope that this essay has challenged you to struggle with the question of how his ideas can be appreciated and incorporated into our attitudes toward "the land," which is our home.

The Land Ethic: A Critical Appraisal

*—James D. Heffernan**

In the previous reading, we have discussed how Aldo Leopold's "land ethic" has been interpreted as proposing a new basic principle of ethics. In this reading, philosopher James D. Heffernan evaluates this proposal. In the process, Heffernan seeks to clarify what Leopold means by "the integrity, stability, and beauty of the biotic community," and how Leopold intends that we should preserve them. Heffernan raises a key question: "is Leopold proposing a new standard of right and wrong that will replace or supplement the old?" In the end, he finds the latter option more acceptable. If this is so, then Leopold's insights may be helpful not in replacing other ethical theories, but in enriching them.

Aldo Leopold's "Land Ethic" centers on the maxim: "A thing is right when it tends to preserve the integrity, stability, and beauty of the biotic community. It is wrong when it tends otherwise." I contribute to the critical appraisal of this maxim by providing answers to the following questions: (1) what is referred to by the phrase "the integrity, stability, and beauty of the biotic community"? (2) What "things" tend to preserve or threaten the integrity, stability, and beauty of the biotic community? (3) Are the integrity, stability, and beauty of the biotic community goods such that preserving them is right and failing to do so wrong?

There are flood and drouth
Over the eyes and in the mouth,
Dead water and dead sand
Contending for the upper hand.
The parched eviscerate soil
Gapes at the vanity of toil,
Laughs without mirth.
This is the death of earth.
T. S. Eliot[1]

I

No maxim is more often cited in discussions of environmental ethics than Aldo Leopold's: "A thing is right when it tends to preserve the integrity, stability, and beauty of the biotic community. It is wrong when it tends otherwise."[2] Yet relatively little has been done by way of sustained critical appraisal of this principle.[3] In this essay I contribute to the task of appraisal by answering three questions: (1) what is referred to by the

---

[1]T. S. Eliot, "Four Quartets: Little Gidding," in *Complete Poems and Plays: 1909–1950* (New York: Harcourt, Brace & World, 1952), p. 140.

[2]Aldo Leopold, *A Sand County Almanac and Sketches Here and There* (New York: Oxford University Press, 1968), pp. 224–55.

[3]The reader will find some perfunctory criticism of Leopold's maxim in John Passmore's *Man's Responsibility for Nature: Ecological Problems and Western Traditions* (New York: Charles Scribner's Sons, 1974), p. 116. A response to Passmore's criticism as well as further analysis of Leopold's maxim can be found in J. Baird Callicott's article "Elements of an Environmental Ethic: Moral Considerability and the Biotic Community" (*Environmental Ethics* 1 (1979): 71–81). Callicott's "Animal Liberation: A Triangular Affair," in *Environmental Ethics* 2 (1980): 311–38, adds some further analysis. Finally, Holms Rolston's "Is There an Ecological Ethic?" in *Ethics* 85 (1975): 93–109, is an important contribution as well. I am indebted to Rolston for reminding me of several of these earlier discussions.

phrase "the integrity, stability, and beauty of the biotic community"? (2) What "things" tend to preserve or threaten the integrity, stability, and beauty of the biotic community? (3) Are the integrity, stability, and beauty of the biotic community goods such that preserving them is right and failing to do so wrong?

II

In his essay "The Land Ethic" Leopold asserts: "All ethics so far evolved rest upon a single premise: that the individual is a member of a community of interdependent parts."[4] What is distinctive about the land ethic, he continues, is that it "enlarges the boundaries of the community to include soils, waters, plants, and animals, or collectively: the land."[5] Since Leopold includes the nonliving environment in his notion of the *biotic community,* he can be understood to be referring to what contemporary ecologists call an *ecosystem.* For example, G. Tyler Miller, Jr. in his *Living in the Environment* defines an ecosystem as a "community of living things interacting with one another and with their physical environment (solar energy, air, water, soil, heat, wind, and various essential chemicals)."[6] Some examples of ecosystems are forests, ponds, lakes, rivers, grasslands, deserts, even the entire planet Earth. This last example, the totality of terrestrial ecosystems, is often called the *biosphere.*[7]

Of the triad of characteristics—integrity, stability, and beauty—only stability seems to have found a secure place in ecological literature. In his *Fundamentals of Ecology,* Eugene Odum defines *stability* as the "tendency for biological systems to resist change and to remain in a state of equilibrium."[8] Miller defines it as the "ability to withstand or to recover from externally imposed changes or stresses."[9] Moreover, Miller maintains that stability implies "persistence of structure over time."[10] In similar fashion Leopold speaks of "land health" as the "capacity of the land for self-renewal."[11] Leopold also suggests a connection between structure and stability:

> ***Structure means the characteristic numbers, as well as the characteristic kinds and functions, of the component species. This interdependence between the complex structure of the land and its smooth functioning as an energy unit is one of its basic attributes.***[12]

The characteristic structure of an ecosystem seems to be what Leopold means by its integrity, for he suggests that the "stability (of an ecosystem) depends on its integrity." He seems also to think of the complex structure of an ecosystem as its beauty; for he refers to "the incredible intricacies of the plant and animal community" as "the intrinsic beauty of the organism called America."[13]

Hence, when Leopold talks of preserving the "integrity, stability, and beauty of the biotic community" he is referring to preserving the characteristic structure of an ecosystem and its capacity to withstand change or stress. Moreover, maintaining the characteristic structure of the ecosystem, its objective beauty, is the key to preserving its stability.

III

The claim that the stability of an ecosystem is a function of its structural complexity is related to a hypothesis frequently discussed in ecological literature, viz., the "diversity-stability" hypothesis. The initial plausibility of this hypothesis is explained in the following fashion by Miller:

> ***Intuitively it seems that species diversity (the number of different species and their relative abundance) and food-web complexity should help stabilize ecosystems. With so many different species and ecological***

---

[4]Leopold, *Sand County Almanac,* p. 203.
[5]Ibid., p. 204.
[6]G. Tyler Miller, Jr., *Living in the Environment,* 2d ed. (Belmont, California: Wadsworth, 1979), p. 43.
[7]Ibid.
[8]Eugene P. Odum, *Fundamentals of Ecology,* 3rd ed. (Philadelphia: W. B. Saunders, 1971), p. 140.
[9]Miller, *Living in the Environment,* p. 82.
[10]Ibid.
[11]Leopold, *Sand County Almanac,* p. 221.
[12]Ibid., p. 216.
[13]Ibid., p. 174.

niches, risk is widely spread; the system should have more ways to respond to environmental stress, and it should be more efficient in capturing and using matter and energy. The diversity of negative feedback controls should tend to keep the ecosystem functioning smoothly. A complex food-web should also promote stability. If one species is eliminated many predators can shift to another food source. In other words, it seems intuitively obvious that it is better not to have all of one's eggs in the same basket. [14]

However, if the diversity-stability hypothesis is taken to mean that the more diverse an ecosystem is the more stable it is, there has been some disconfirming evidence. One of the most diverse ecosystems yet investigated is the tropical rainforest; yet it also appears to be the most fragile, in that clearing a sufficiently large area for agriculture or too frequently clearing a small area leads to complete breakdown. David Ehrenfeld in *The Arrogance of Humanism* reports that mathematical modeling of the diversity-stability hypothesis has shown that "the most diverse systems ought to be the most delicate; they were the ones at greatest risk of collapse following human-induced change."[15] Moreover, according to Ehrenfeld, there is direct evidence from conservation work that "the diverse, 'mature' communities were almost always the first to fall apart under heavy human-imposed stress and were always the most difficult to protect."[16] This kind of evidence leads ecologists like Miller to treat the diversity-stability hypothesis with caution. Miller says: "The idea that diversity leads to ecosystem stability may be valid in some types of ecosystems, especially if stress is not enough to wipe out the dominant species. But we should be wary of applying this idea to all situations,"[17] John Passmore expresses similar reservations in his *Man's Responsibility for Nature:*

There are, then, two principles which seem to be untenable: the first, that it is always better to increase the diversity of an ecosystem; the second, that it is never better to do so. All that can be properly said is that in modifying the degree of diversity there are always inherent dangers, biological dangers, and there is also the real risk of destroying the 'character' of a landscape, the complex set of relationships which constitute its attractiveness. [18]

It might be objected, however, that Leopold's ethical directive implies no commitment to the diversity-stability hypothesis unless *diversity* is understood to mean "characteristic diversity." It seems plausible to claim that the stability of an ecosystem is a function of its characteristic diversity. The characteristic diversity of an ecosystem cannot be raised (e.g., by introducing exotic species) or lowered (e.g., by removing indigenous species) without disturbing its integrity; and that in turn may, in Leopold's judgment, threaten its stability.

It also needs to be pointed out that stability is a relational property. Ecosystems are stable relative to the characteristic climactic, geological, and biological fluctuations in which they have evolved. No ecosystem is stable relative to all possible fluctuations. For example, no terrestrial ecosystem could withstand the stress induced by the burning out of the sun. Along the same lines, ecosystems that are stable relative to characteristic fluctuations or stresses in which they have evolved may not be stable relative to human-induced stress no matter how diverse they are, simply because shifts in the characteristic diversity induced by high technology or large population influxes are not the kinds of stress to which even the most diverse ecosystems have evolved a resistance.

Thus, we have a reply to the second question raised above: what "things" tend to preserve or threaten the integrity, stability, and beauty of the biotic community? They are the things that preserve or threaten the characteristic structure and characteristic diversity of ecosystems, or things that involve uncharacteristic stresses. Examples of such things are rife: depletion of the ozone layer in the atmosphere, eutrophication of lakes and other bodies of water, elimination of important species by pesticides or herbicides, introduction of certain exotic species, and desertification.

---

[14]G. Tyler Miller, Jr. *Living in the Environment,* p. 87.
[15]David W. Ehrenfeld, *The Arrogance of Humanism* (New York: Oxford University Press, 1978), p. 195.
[16]Ibid.
[17]G. Tyler Miller, Jr., *Living in the Environment,* p. 88.
[18]John Passmore. *Man's Responsibility for Nature* (New York: Scribner's, 1974), p. 120.

Leopold's principle proscribes such "things" and any activities that lead to such changes. Proscribed actions would probably include the use of fluorocarbons in aerosal sprays, the dumping of industrial wastes in fresh water, the use of broad-spectrum pesticides and herbicides in agriculture, and the over-grazing of pasture lands. Presumably to refrain from activities that threaten ecosystem stability is the best way to preserve it.

IV

Now that we have examined the principal concepts of Leopold's maxim we can deal with the last question: are the integrity, stability, and beauty of the biotic community goods such that preserving them is right and failing to do so wrong?

In general a thing may be either good in itself, i.e., intrinsically good, or good because it leads to something good in itself, i.e., instrumentally good. There is relatively little difficulty in showing that the integrity, stability, and beauty of the biosphere are instrumentally good for human beings and other forms of life. Ecology has acquainted us with the various ways in which the continuance of life on this planet depends on the integrity and stability of the biosphere. For example, J. E. Lovelock has argued that if the proportion of oxygen in the atmosphere were to be increased to twenty-five percent by destruction of the biological controls on oxygen, the planet would be plagued by virtually uncontrollable lightning-initiated fires that would destroy almost everything flammable in a relatively short time.[19] The integrity and stability of the biosphere, then, are instrumentally good in that they constitute the conditions within which life can continue.

With regard to lesser ecosystems the task of establishing that their integrity and stability is instrumentally good for human life is not so easy. The modification of ecosystems in agriculture certainly is a case in which the destruction or at least radical reduction in the integrity and stability of a wild ecosystem leads to good for human life. The flooding of a wild river canyon for a reservoir is an instance in which the integrity and stability of riparian and other habitat are destroyed for the human goods of flood control, hydroelectric power, irrigation water, and boating. Sometimes the preservation of individual ecosystems may be good for human beings either economically (habitat for valuable species) or scientifically (natural laboratory) or aesthetically (especially beautiful places) or recreationally (places for hiking). But one would be hard pressed to show that the preservation of ecosystems always leads to something good for humans. Often, the opposite is the case, that the destruction of the integrity and stability of an ecosystem leads to good for humans. For example, the city of Los Angeles derives the benefit of water from the destruction of Mono Lake and numerous people derive recreational benefits from the destruction of desert habitats by trail bikes and other off-road vehicles.

Recognizing this last difficulty a number of philosophers have been led to attempt to establish that the integrity, stability, and beauty of an ecosystem are intrinsic goods. This is perhaps what Leopold meant by "value in the philosophical sense":

> ***It is inconceivable to me that an ethical relation to land can exist without love, respect, and admiration for land, and a high regard for its value. By value, I of course mean something far broader than mere economic value; I mean value in the philosophical sense.***[20]

Some philosophers have thought that to say that the integrity, stability, and beauty of an ecosystem are intrinsic goods is incomprehensible. For such philosophers the concept of intrinsic goods (or bads) is intimately bound up with concept of benefits (or harms). For example, William Frankena in a critique of reverence-for-life ethics says the following:

> ***The difficulty about it, to my mind, is that I can see no reason, from the moral point of view, why we should respect something that is alive but has no conscious sentiency and so can experience no***

[19] J. E. Lovelock, *Gaia: A New Look at Life on Earth* (New York: Oxford University Press, 1979), pp. 70–71.
[20] Leopold, *Sand County Almanac*, p. 223.

pleasure or pain, joy or suffering, unless perhaps it is potentially a consciously sentient being, as in the case of a fetus. Why, if leaves and trees have no capacity to feel pleasure or to suffer, should I tear no leaf from a tree? Why should I respect its location any more than that of a stone in my driveway, if no benefit or harm comes to any person or sentient being by my moving it?[21]

Frankena seems here to be committed to a principle something like the following: it is only possible to benefit or harm a consciously sentient being or one that is potentially so.

Peter Singer takes a similar position in his *Animal Liberation:*

> ***The capacity for suffering and enjoyment is a prerequisite for having interests at all, a condition that must be satisfied before we can speak of interests in a meaningful way. It would be nonsense to say that it was not in the interests of a stone to be kicked along the road by a schoolboy. A stone does not have interests because it cannot suffer. Nothing that we can do to it could possibly make any difference to its welfare. A mouse, on the other hand, does have an interest in not being kicked along the road, because it will suffer if it is.***[22]

Thus far then we seem to be left with the dilemma: either a creature is consciously sentient (at least potentially) or it is incapable of being benefited or harmed. Since ecosystems are not consciously sentient, they are incapable of being benefited or harmed.

Recently Kenneth Goodpaster has argued contra Frankena and Singer that it is possible to benefit or harm anything that is alive:

> ***There is no absurdity in imagining the representation of the needs of a tree for sun and water in the face of a proposal to cut it down or pave its immediate radius for a parking lot. . . . In the face of their obvious tendencies to maintain and heal themselves, it is very difficult to reject the idea of interests on the part of trees (and plants generally) in remaining alive.***[23]

Similar remarks can be made about ecosystems and the biosphere as a whole.

Peter Singer is aware that it is possible to speak of the "needs" or "interests" of plants, but considers it to be a loose way of talking with absurd consequences:

> ***While there may be a loose sense of the term in which we can say that it is in the interests of a tree to be watered, this attentuated sense of the term is not the sense covered by the principle of equal consideration of interests. All we mean when we say that it is in the interests of a tree to be watered is that the tree needs water if it is to continue to live and grow normally; if we regard this as evidence that the tree has interests, we might almost as well say that it is in the interest of a car to be lubricated regularly because the car needs lubrication if it is to run properly. In neither case can we really mean (unless we impute consciousness to trees or cars) that the tree or car has any preference about the matter.***[24]

However, Singer fails to note that a car has no interest in running properly because it is really we who have the interest in its running properly. The "interests" of machines that have no autonomous goals are really our interests. But plants, for example, certainly have an autonomous interest in staying alive that is not really our interest. They grow, develop, nourish themselves, and reproduce autonomously, spontaneously, without our programming them to do so. Indeed a plant's interest in growth, development, and reproduction may conflict with our interests in uncluttered gardens or root-free drainage systems.

Currently existing machines (as opposed to those imagined by science-fiction writers) do not have the interests in growth, development, nourishment, and reproduction that living things have. Nor do

---

[21] W. K. Frankena, "Ethics and the Environment," in K. E. Goodpaster and K. M. Sayre, eds., *Ethics and the Problems of the 21st Century* (Notre Dame: University of Notre Dame Press, 1979) p. 11.

[22] Peter Singer, *Animal Liberation: A New Ethics for Our Treatment of Animals* (New York: Avon Books, 1997), p. 8 (author's emphasis).

[23] Kenneth E. Goodpaster, "On Being Morally Considerable," *The Journal of Philosophy* 75 (1978): 319.

[24] Peter Singer, "Not for Humans Only: The Place of Nonhumans in Environmental Issues," in K. E. Goodpaster and K. M. Sayre, *Ethics*, pp. 194–95.

they exhibit the interest in stability that ecosystems manifest. These differences between living things and ecosystems, on the one hand, and machines, on the other, suggest, according to Goodpaster, that "the core of moral concern lies in respect for self-sustaining organization and integration in the face of pressures toward high entropy."[25]

At this point, then, it seems appropriate to claim, contra Frankena and Singer, that living things and even ecosystems, are things that have interests and, hence, may be benefited or harmed by having their interests nurtured or thwarted, even though these things are not sentient creatures. Notice that in arguing that it is not possible to harm a nonsentient being Frankena talks about plucking a leaf from a tree or not respecting its location. But these examples are both things not (at least, not necessarily) harmful to trees. Plucking a leaf and transplanting may both be done without harming a tree, whereas girdling or uprooting may not.

If those things that can be benefited or harmed are those things that have intrinsic value, then ecosystems and their inhabitants have intrinsic value. If this is true, then preserving the integrity, stability, and beauty of an ecosystem is right. Not to do so is wrong. But this last claim raises a further problem.

If those things that can be benefited or harmed are those things that have intrinsic value, then ecosystems and their inhabitants have intrinsic value. If this is true, then preserving the integrity, stability, and beauty of an ecosystem is right. Not to do so is wrong.

The problem is this: is Leopold proposing a new standard of right and wrong that will replace or supplement the old? For example, he may be suggesting: "Hitherto we have used benefit or harm to human beings as the intrinsic good to be done or evil to be avoided, but I now propose to replace benefit and harm to human beings by benefit and harm to ecosystems as the intrinsic good to be done or evil to be avoided." Or he may be suggesting: "Hitherto we have used benefit or harm to human beings as the sole intrinsic good to be done or evil to be avoided, but I now propose to supplement benefit and harm to human beings with benefit and harm to ecosystems as another intrinsic good to be done or evil to be avoided."

The problem is this: is Leopold proposing a new standard of right and wrong that will replace or supplement the old?

If the former interpretation were the correct one it would have morally unacceptable consequences, for it would imply that the standard against which our actions are measured would be benefit and harm to ecosystems. On this interpretation, then, feeding starving peoples on the African Sahel would be wrong precisely insofar as their continued existence leads to deterioration of the ecosystems they inhabit through their practices of wood gathering and dung burning. Agriculture, except of the most primitive kind, would be wrong precisely insofar as cultivation threatens the integrity and stability of wild ecosystems. Indeed, almost the only right actions on this interpretation of Leopold's principle would be the cessation of most human projects and the setting up of wilderness preserves.

In his recent article "Animal Liberation: A Triangular Affair"[26] J. Baird Callicott appears to support this interpretation of Leopold. He argues that Leopold's land ethic may be superior to other environmental ethics in that it, like ecology, is holistic. In so far as this is simply asserting that wholes like ecosystems have intrinsic value, what I have said thus far in this paper is in agreement. However, Callicott seems to mean more: not only is the integrity, stability, and beauty of an ecosystem *a* good; it is, he asserts, "*the summum bonum.*" He thinks this is an advantage:

> ***An environmental ethic which takes as its summum bonum the integrity, stability, and beauty of the biotic community is not conferring moral value on something else besides plants, animals, soils, and waters. Rather, the former, the good of the community as a whole, serves as a standard for the assessment of the relative value and relative ordering of its constitutive parts and therefore provides a means of adjudicating the often mutually contradictory demands of the parts***

---

[25]Goodpaster. "On Being Morally Considerable," p. 323.

[26]J. Baird Callicott, "Animal Liberation: A Triangular Affair," *Environmental Ethics* 2 (1980): pp. 221–38.

considered separately for equal consideration. . . . Animals of these species, which like the honey bee, function in ways critically important to the economy of nature, moreover, would be granted a greater claim to moral attention than psychologically more complex and sensitive ones, say, rabbits and moles, which seem to be plentiful, globally distributed, reproductively efficient, and only routinely integrated into the natural economy.[27]

It is difficult to know how far Callicott wishes to take this view. One wonders whether human beings "function in ways critically important to the economy of nature" in the sense of contributing significantly to preserving the integrity and stability of the biosphere. In view of their increasing abuse of the environment to satisfy their desires, it is hard to avoid the conclusion that they do not. But if the integrity, stability, and beauty of the biotic community is the *summum bonum*, the best thing we can do is to find some ecologically sound way of disposing of the human race or at least of drastically reducing the human population. Perhaps, however, this is unfair to Callicott. Perhaps, he is only asserting, in an exaggerated fashion, that the integrity, stability, and beauty of the biotic community are intrinsic values to be considered alongside the intrinsic value of human life. This is indicated by his remarks appropos of Edward Abbey's misanthropy:

Edward Abbey in his enormously popular Desert Solitaire bluntly sates that he would sooner shoot a man than a snake. Abbey may not be simply depraved; this is perhaps only his way of dramatically making the point that the human population has become so disproportionate from the biological point of view that if one has to choose between a specimen of Homo sapiens and a specimen of a rare even if unattractive species the choice would be moot.[28]

This may be nothing more than a dramatic presentation of a familiar dictum that we need not take extraordinary measures to preserve human life. For example, suppose an aged, senile person can only be kept alive by constant injection of a serum prepared from an endangered plant species. One can imagine deciding here not to use the serum. On the other hand, suppose a population of children can be saved from a virulent plague only by a serum prepared from eggs of the endangered California condor. Here, one would probably prepare the serum.[29]

This line of thought suggests that the more acceptable interpretation of Leopold's maxim is the second one introduced above in which ecosystemic good is to be weighed along with human good in deciding the rightness or wrongness of actions. An indication that Leopold intended his maxim to be taken in this way is the following:

A land ethic of course cannot prevent the alteration, management, and use of these 'resources', (i.e., soil, water, plants, and animals), but it does affirm their right to continued existence, and, at least in spots, their continued existence in a natural state.[30]

The second interpretation of Leopold's maxim avoids the morally unacceptable implications of the first interpretation suggested above, but it does so at a price. On the second interpretation there are numerous goods to be considered alongside human good, i.e., the good of ecosystems and the good of their nonhuman components, and these goods can conflict with one another. Perhaps then Leopold should be interpreted as proposing a new rule of conduct, but a rule having exceptions, because on occasion it can conflict with other rules, e.g., "A thing is right when it tends to preserve the life of an innocent human being. It is

---

[27]Ibid., pp. 324–25.

[28]Ibid., p. 326.

[29]A case in which there was an apparent conflict between the human interest in health and animal interests was described several years ago: Nicholas Wade, "New Vaccine May Bring Man and Chimpanzee into Tragic Conflict," *Science* 200 (1978): 1027–30. An American pharmaceutical firm applied for a permit to import chimpanzees, an endangered species, as research subjects to test a vaccine for the hepatitis B vaccine. The strongest objection was that more than four times as many chimpanzees would die in providing research subjects than would actually be used to test the vaccine because of the technique used for capturing them in the wild. Wade argued: "Since the vaccine seems usually innocuous, and since the disease is only rarely fatal, it would perhaps be more just if the larger population could find some way of solving its problem that was not to the detriment of the smaller." Paul and Ann Ehrlich, who discuss this case in their recent book, *Extinction* (New York: Random House, 1981), pp. 60–61, report that the permit was not granted.

[30]Leopold, *Sand County Almanac*, p. 204.

wrong when it tends otherwise." A classical example of conflict between rules is the conflict between the rule regarding promise keeping and that of preserving human life: a person to whom I have promised a gun has decided to use the gun to hold up a store. Should I give the person the gun and thus keep my promise or withhold the gun and thus protect human, lives?

On the second interpretation there are numerous goods to be considered alongside human good, i.e., the good of ecosystems and the good of their nonhuman components, and these goods can conflict with one another.

Some philosophers have tried to deal with exceptions to rules by claiming that they are *prima facie* rules of conduct, not *actual* rules of conduct.[31] They are rules to be followed unhesitatingly, just in case no other rules apply to the situation. This condition is represented by saying that rules of conduct have an implicit *ceteris paribus* clause ("other things being equal"). So, for example, provided that in doing so I commit no wrong, it is right to keep a promise. And, providing that I am not avoiding a supervening moral wrong, it is wrong not to keep a promise.

Some philosophers have attempted to provide ranking systems, e.g., claiming that preventing a death is more important than keeping a promise. However, others have been persuaded by the complexity of actual situations to forego the attempt to rank these rules and claim that such rules indicate wrong-making (or right-making) features that may be outweighed in various circumstances by other such features. Thus, if an act involves breaking a promise, that is a moral stroke against it; but that may be outweighed by other strokes for or against it, i.e., that it involves saving a life. On this interpretation Leopold may be viewed as offering a *prima facie* rule of conduct: "Provided that in doing so I commit no greater wrong, a thing is right when it tends to preserve the integrity, stability, and beauty of an ecosystem."

Now this interpretation of Leopold's maxim rids it of the absurd consequence of condoning the refusal to feed a people that is destroying its ecosystem, but it does so, perhaps at the price of rendering it innocuous or unhelpful in moral decision making. Because moral rules, on this interpretation, do not prescribe or proscribe an act on all occasions or because a set of moral rules does not come with a precise ranking, they are considered useless. This, however, is like arguing that geometrical relationships are useless in construction because no physical structure precisely embodies them. Such rules can guide moral thinking without determining it. For example, guided by the rule that there is *prima facie* wrong involved in interfering with the integrity, stability, and beauty of an ecosystem, I disapprove of the use of ORVs in desert landscapes because, although recreation is important, it can be gotten in other ways. I do not disapprove of agriculture (at least not in a blanket way) since it is not clear that humans can be fed without it. However, even in agriculture I am on the lookout for ways of raising food that are less destructive of ecosystems than current energy-intensive methods. I do suppose this means that one might permit or acquiesce in the destruction of ecosystems to save human lives.

But this does not mean that saving human lives always outweighs preserving the integrity and stability of ecosystems. I think most people would find it absurd to turn the Appalachian or the John Muir trails into asphalt highways on the grounds that it contributes to the preservation of human life by making rescue operations easier. To say that the integrity and stability of ecosystems and human lives are intrinsic values does not imply that everything must be sacrificed to preserve them.

To say that something is intrinsically valuable is to say that, *ceteris paribus,* to preserve it is right, to destroy it wrong. But in life, as in ecology, *ceteris non paribus.* In certain circumstances it may not be wrong to destroy an ecosystem and in certain circumstances it may not be wrong not to satisfy a human interest. The human interest in ORV recreation perhaps should be sacrificed to the integrity, stability, and beauty of the Mojave desert, whereas the stability of the tall grass prairie may have to yield to the reduced stability of the cornfield. The human interest in the convenience of fluorocarbon aerosals should yield to the

[31]Cf. W. K. Frankena, *Ethics,* 2d ed. (Englewood Cliffs: Prentice-Hall. 1973), pp. 24–28.

interests of the biosphere, whereas I find it difficult to recognize as an ethic one that condones not giving aid to a country simply on the grounds that is destroying its own ecosystems.

Perhaps we can venture a general principle at this point: the survival interests of human beings ought to outweigh those of the rest of the biotic community and the survival interests of the rest of the biotic community ought to outweigh the non-survival interests of human beings. This helps decide some cases. For example, the human interest in attractive shoes should not outweigh the survival interests of alligators, for clearly an interest in *attractive* shoes is not a survival interest, but the interest in some kind of food for native Africans may for a time outweigh the preservation of rainforest ecosystems, for the interest in food is a survival interest in human beings. Of course, widespread destruction of rainforests could lead to an upset in the integrity of the biosphere and this lends some urgency to finding ways of feeding people or reducing their numbers that do not involve destruction of the rainforests.

I am not sanguine about the possibility of providing a rule so precisely formulated that all cases of conflict can be easily resolved, but even an imprecise rule provides guidance, without determination, in many cases, i.e., those cases in which it is clear that the human interest at stake is or is not a survival interest and/or the ecosystemic interest is or not a survival interest.

To say that something is intrinsically valuable is to say that, ceteris paribus, to preserve it is right, to destroy it wrong. But in life, as in ecology, ceteris non paribus. In certain circumstances it may not be wrong to destroy an ecosystem and in certain circumstances it may not be wrong not to satisfy a human interest.

V

In this essay I have provided a critical analysis of Aldo Leopold's land ethic as summarized in the maxim: "A thing is right when it tends to preserve the integrity, stability, and beauty of the biotic community. It is wrong when it tends otherwise." Current ecological literature suggests a rephrasing of this maxim as follows: "A thing is right when it tends to preserve the characteristic diversity and stability of an ecosystem (or the biosphere). It is wrong when it tends otherwise." Those things that threaten the stability of an ecosystem tend not to be minor changes in the structure, i.e., in the "characteristic numbers, as well as the characteristic kinds and function, of the component species," but rather major ones. Ecological science cannot supply us with a simple way of characterizing stability-threatening activities, not, for instance, through the "diversity-stability hypothesis." Nonetheless, in practice numerous examples of such activities are easy to find.

Against those philosophers who contend that the stability of an ecosystem is not an intrinsic value, since it is not possible to harm or benefit anything not at least potentially sentient, I have argued, following Goodpaster, that it is possible to harm or benefit anything that exhibits "self-sustaining organization and integration in the face of pressures toward high entropy." Lastly I have argued that it is more appropriate to interpret Leopold as offering an additional prima facie rule of conduct rather than a single new standard of right and wrong. The latter way of interpreting the maxim has, in my estimation, immoral consequences, whereas the former way allows recognition of the intrinsic value of the stability of ecosystems without the consequence that it is never permissible to sacrifice the stability of ecosystems. While this interpretation does not make our duties toward ecosystems precisely calculable, that seems only reasonable. "Weighing the alternatives" has always seemed to me a metaphor for hard thinking rather than something we can do precisely.

NAME ______________________________

Worksheet 13

Ecocentrism: Valuing the Land as a Whole

1. **As Aldo Leopold uses the concept, what is "the land"?**

2. **What basic principle does Leopold suggest for the Land Ethic? (Write it out here.)**

3. **What does it mean to say that the Land Ethic is "holistic"? In what sense are other ethical theories individualistic?**

The following Web site contains a brief essay by Aldo Leopold in which he describes an experience he had earlier in life: http://www.eco-action.org/dt/thinking.html. Read the essay and answer the following questions:

4. **What is the point that Leopold is making with this story?**

5. **How does it illustrate the Land Ethic?**

Questions on "The Land Ethic: A Critical Appraisal," by James D. Heffernan.

6. What criticisms does Heffernan raise against the way in which some have interpreted the Land ethic?

7. After reading these essays, are there any elements of Leopold's thinking that you agree with and find helpful in building your own environmental perspective? Explain.

Pulling It Together

Ecocentrism: Valuing the Land as a Whole

QUESTIONS FOR THOUGHT AND DISCUSSION

1. What is a "holistic" approach to ethics? What implications does such an approach have for individual beings? for human beings?
2. Do you believe that hunting is morally permissible? Why or why not?
3. When do you believe that it is morally permissible to limit the rights of human beings in order to protect the beauty, integrity and stability of the environment? Explain your answer.

EXERCISES

1. Compare and contrast one of the approaches to environmental value (anthropocentrism, sentientism, biocentrism, or ecocentrism) to one or more of the religious perspectives we studied previously in the class. Point out as many similarities and differences as you can.
2. Explain the basic ideas of anthropocentrism, sentientism, biocentrism and ecocentrism. What aspects of the environment does each focus on? How might a person argue for environmental action on the basis of each approach?
3. Pick two of the approaches (anthropocentrism, sentientism, biocentrism and ecocentrism) and compare how they would evaluate the activities of hunting. Would each approve? In what cases, if any would hunting be morally permissible or impermissible according to each of these theories? Explain?

Chapter

14

Summary and Concluding Thoughts from a Theocentric Perspective

Toward a Theocentric Approach to Environmental Value

—*Roger P. Ebertz*

Over the past forty years there has been a growing conviction in many circles that humankind should be concerned about its natural environment. But why? The claim that we should be concerned implies that the environment is in some sense valuable. But in what sense? Why is the environment valuable? In this paper I will briefly explore four general approaches to environmental value being discussed in the contemporary philosophical literature. Following others in the discussion, I will call these four approaches anthropocentrism, sentientism, biocentrism, and ecocentrism. After critically considering each of these general views, I will suggest a "theocentric" approach to environmental value and argue that this approach provides a rich framework for bringing numerous insights concerning environmental value into a coherent perspective.

THE ANTHROPOCENTRIC TRADITION

Deeply embedded in Western thought is the view that humans are intimately connected with, if not central to, earthly value. A common thread that runs from Greek philosophy through much of medieval thought and into the modern worldview affirms that humankind stands above the rest of the natural world in such a way that nonhuman elements are valuable, if they are valuable, because of their value to humans. In slightly more technical terms, this approach, sometimes called "anthropocentrism," claims that humans alone, of earthly things at least, are intrinsically valuable. All other elements of the natural world, including other animals and plants, are extrinsically (or instrumentally) valuable in as much as they are of value to humans. Aristotle, for example, wrote that "plants exist on account of animals . . . and the other animals exist for the sake of man."[1] Aristotle does not stand alone among the Greeks on this theme. With Plato and many others, he believes that human reason sets humans above other animals.

It is not surprising that a similar view is found in the work of Thomas Aquinas. Thomas writes that a rational creature (human) has "dominion over its actions," and thus God provides for it "for its own sake." Other things are cared for "not for their own sake, but as being directed to other things." Thus, according to Aquinas, only rational beings are valuable in themselves. And for Aquinas, the category of rational beings does not include nonhuman animals. Thus regarding the treatment of animals, he writes, "they are intended for man's use in the natural order. Hence it is no wrong for man to make use of them, either by killing or in any other way whatever."[2] I will readily admit that Thomas does not represent the only medieval attitude toward the nonhuman world. The more positive view of St. Francis of Assisi is well known. But these passages from Aquinas do illustrate that an anthropocentric thread is alive and well in medieval thought.

In the modern period, we find a very similar argument in Immanuel Kant, who has this explanation of our duties to animals: "Animals are not self-conscious and are there merely as a means to an end. That end is man.

---

[1]Aristotle. "From Politics," in *Reflecting on Nature: Readings in Environmental Philosophy*, Lori Gruen and Dale Jamieson, eds. (Oxford: Oxford University Press, 1994), 19–20.

[2]Thomas Aquinas. "Differences between Rational and Other Creatures," in *Environmental Ethics: Divergence and Convergence*, ed. Susan J. Armstrong and Richard G. Botzler (New York: McGraw-Hill, Inc., 1993), 278–280.

We can ask, 'Why do animals exist?' But to ask, 'Why does man exit?' is a meaningless question."[3] Anthropocentrism is common among us as well. Even our language about the environment reflects it. We speak, of preserving our natural resources, speaking of both nonliving and living things. The very expression "natural resources" suggests that the physical world is a repository of "resources" for humans to use. And this includes animals as well as minerals. The tendency to view the value of the rest of the natural world as derivative from human value runs deep in our tradition.

But lest we draw too harsh a conclusion from this fact, we must avoid a common assumption. It is a misunderstanding to equate anthropocentrism with a crude attitude of unthinking exploitation of the environment. In fact, it is possible to make a strong case for environmental protection on the basis of an anthropocentric approach to value. As we learn more and more about the interconnectedness between human life and the natural world, we discover more fully why we should preserve our environment, even from an anthropocentric perspective. We are learning that if we really value human life and culture, we must care for the natural environment in which humankind lives. Anthropocentrism need not be antienvironmental.

Yet, environmental activists and philosophers frequently criticize anthropocentrism for several reasons. First, although anthropocentrism need not support environmentally harmful behavior, it has often done so. Humankind, thinking that it was the only thing on earth that was intrinsically valuable, has too often devalued nonhuman forms of life on our planet. Second, the rise of the science of ecology has emphasized that humans are part of the natural world, not above it. Some believe that as a part of the web of life humans should be valued equally with other living things, not above them. Third, the more we know about animals, the less obvious it is that there is a clear, nonarbitrary distinction between our own species and others. Other animals have been shown to solve problems through the use of reason, they appear clearly conscious; some animals have been discovered to use tools and speak to one another in their own languages. All of these traits and abilities were once thought to set humans apart as unique. The point is not that humans are not, as a whole, uniquely developed in some of their capabilities. It is rather that recent scientific discoveries have made this uniqueness appear to be a matter of degree, with some other animals sharing traits we had formerly thought of as human, and some humans lacking these same traits. And if there is no clear distinction between higher animals and humans, it becomes difficult to see why we should value humans intrinsically while we value other animals only extrinsically.

Finally, anthropocentrists have been charged with a kind of moral inconsistency: speciesism. Some philosophers, particularly those who have taken up the cause of animals, are now arguing that treating humans as more valuable than all other animals is similar to the injustice of racism. Their argument begins by pointing out that anthropocentrists place a higher value on the lives and well-being of humans than on the lives and well-being of nonhuman animals. To be morally consistent, this difference of moral valuation must be based on some morally relevant criterion that humans satisfy and other beings do not have. There seems to be no clear criterion. As we have already seen, scientific studies are fuzzing the line between humans and higher animals. The distinction between intrinsically valuable beings and extrinsically valuable beings demands more than a fuzzy line for its justification.

Theists will be tempted to appeal to the "image of God" as that which sets humans apart. As a Christian, this is appealing. But what is the image of God? Historically, theologians have identified features such as the ability to reason and the ability to communicate as characteristics we have because we are made in the image of God. These, however, are the same types of features we have been discussing above. Whatever specific feature one might point to, it seems that if it appears relevant to moral worth, then it also turns out that some nonhumans have that feature to a greater extent than some humans. One might then be tempted to say that the image of God is our humanness. In the current context, however, this will not suffice. The theistic anthropocentrist would be arguing that humans are more valuable because they are

---

[3]Immanuel Kant. "Duties to Animals," in *Environmental Ethics: Divergence and Convergence*, ed. Susan J. Armstrong and Richard G. Botzler (New York: McGraw-Hill, Inc., 1993), 285–286.

made in the image of God, and that to be made in the image of God is to be human. In short, humans are more valuable because they are human. This turns out to be no argument or criterion at all, merely a bald claim, exactly the speciesist claim the objector is charging us with making. It does not identify a morally relevant feature that humans have and nonhumans do not.

CONTEMPORARY NONANTHROPOCENTRIC ALTERNATIVES

Out of dissatisfaction with anthropocentrism, recent environmental philosophers have defended several nonanthropocentric approaches to environmental value. Each of these views expands the circle of beings it recognizes as intrinsically worthy of moral consideration, the moral community. Anthropocentrism views the moral community as coterminous with the human community. The three general views we will consider in this section seek to widen the circle to include some nonhumans.

Sentientism

The first of these views I will refer to as sentientism.[4] The argument I summarized earlier charging anthropocentrism with speciesism is most frequently leveled by sentientists. Those I am lumping together in this group agree that the circle of beings that are worthy of moral consideration include some higher animals. There are two distinct ways contemporary philosophers defend this claim. The first approach to extending the circle builds upon utilitarian ethics and argues, in effect, that animal pain and pleasure should be considered in the utilitarian calculus just as human pain and pleasure are. The second approach extends the Kantian idea of inherent worth to some animals. Both conclude that higher animals should in some sense be treated equally with humans.

The first of these two approaches is best represented by Peter Singer, a professional philosopher and outspoken defender of animal interests. Singer levels the charge of speciesism against anthropocentrists, and against much of our human conduct. Singer rejects mere species difference as the criterion of moral considerability, and seeks to uncover a criterion that is morally relevant. Singer believes that "sentience"—the ability to experience pleasure and pain—is just such a criterion. He argues that all pain and pleasure, including that of nonhuman animals, should be taken into consideration in determining which actions are morally permissible. Singer turns to the classical utilitarian Jeremy Bentham for inspiration. In a famous passage, Bentham compares our treatment of nonhuman animals with slavery and looks forward to the day when we will recognize that the key question for deciding who should be morally considered is not "Are they rational?" or "Can they talk?" but "Can they suffer?"[5] The key points in Singer's reasoning are these: The basic principle of ethics is that actions should aim at the maximum satisfaction of interests. (This is a version of what Mill called the Greatest Happiness Principle.) Furthermore, in determining what actions maximize interest satisfaction, every individual's interests should be considered equally. (Singer calls this the Principle of Equal Consideration of Interests.) Finally, the individuals whose interests should be considered equally include any animal that can experience pleasure and pain.

> ***The capacity for suffering and enjoying things is a prerequisite for having interests at all, a condition that must be satisfied before we can speak of interests in any meaningful way. . . . If a being suffers, there can be no moral justification for refusing to take that suffering into***

---

[4]The term "sentientism" best fits the views of utilitarians such as Singer, who argue that sentience, the ability to experience pleasure and pain, is the criterion of moral considerability. Although others, such as Regan, do not make sentience itself the criterion in quite the same way, there is a general similarity between the two views that makes it helpful to refer to both views using one term. The label "animal rights" theorist is more appropriate to Regan than to Singer, because according to Regan's view, animals have inherent value and thus rights. For Singer, on the other hand, the basic concept is that of interests, not rights.

[5]Peter Singer, "All Animals Are Equal," in *Applied Ethics*, Peter Singer, ed. (Oxford: Oxford University Press, 1986), 221.

consideration. No matter what the nature of the being, the principle of equality requires that its suffering be counted equally with the like suffering—insofar as rough comparisons can be made—of any other being.[6]

Because all sentient creatures have the interest in avoiding pain and experiencing pleasure, all sentient animals should be morally considered. In this way Singer argues that the moral community, the circle of morally considerable beings, is wider than is usually assumed in our ethics and everyday actions.

It is important to avoid one misunderstanding. Singer is not saying that nonhuman animals should be treated exactly like humans. The principle of equal consideration of interests does not imply that because it is wrong to slap a baby, it is equally wrong to slap a horse. After all, babies and horses do not experience a slap in the same way. The pain a human baby would experience is significantly greater than the pain a horse would experience from a comparable slap. Similarly, Singer's view need not imply that the taking of the life of a nonhuman animal is equivalent to taking the life of a human being. The psychological capacities of a human and the ability to make plans for the future may make a human's interests in continuing to live greater than the similar interest of some others animals. Thus equal consideration of interests may allow for the choice of a human life over an animal life in some circumstances. But even here, it is not merely because one of the individuals is human that she is saved rather than an animals. It is rather because the two individuals have relevantly different interests. Like utilitarianism in general, the principle Singer is defending does not imply that all individuals should be treated equally. Rather it implies that the *interests* of all the individuals involved should be equally considered. Different levels and intensities of interests in different individuals may justify different treatment.

Another well-known defender of the view I am calling sentientism is Tom Regan. Like Singer, Regan accuses anthropocentrists of speciesism. But Regan argues for a different moral theory and identifies a different criterion of moral worth. Rejecting utilitarianism, Regan defends what he calls "the rights view" of ethics. At the heart of this view is the idea that individuals have inherent value. In contrast to utilitarianism, which Regan accuses of making individuals "mere receptacles"[7] of satisfactions and frustrations, the rights view recognizes that beings with inherent value "all possess it equally, and all have an equal right to be treated with respect, to be treated in ways that do not reduce them to the status of things, as if they exist as resources for others." Just as Singer's reasoning builds upon utilitarianism, Regan's is clearly akin to the Kantian tradition. But as I have indicated, Regan rejects anthropocentrism, accusing it of arbitrarily limiting morally considerable beings to the human species. Like Singer, Regan wants to find a morally relevant criterion of inherent worth. Such things as the ability to read and the ability to solve highly abstract problems are not the factors that give us moral worth. It is true that animals do not have these abilities. But it is also true that many *humans* lack these abilities. What is it, Regan asks, that all humans do share in virtue of which they are inherently worthy of respect? In answer to this question, Regan writes,

> ***The really crucial, the basic similarity is simply this: we are each of us the experiencing subject-of-a-life, each of us a conscious creature having an individual welfare that has importance to us whatever our usefulness to others.***[8]

According to Regan, what sets us apart is that each of us is a subject. We experience life from the inside. We have desires, memories, and expectations. For each of us there are many things—pleasures and pains, hopes satisfied and dreams frustrated—that either add to or subtract from the quality of our lives. We are not merely objects. We are "subjects-of-a-life." And thus we should not be treated as objects. We should be treated as beings with inherent value.

---

[6]Singer, 221.

[7]Tom Regan. *The Case of Animal Rights* (Berkeley: University of California Press, 1983) 205.

[8]Tom Regan. *In Defense of Animals* (Oxford: Basil Blackwell, 1985) 44, 45. The passages quoted are on pages 44 and 45 of the excerpt found in Louis Pojman, ed. *Environmental Ethics: Readings in Theory and Application,* Jones and Bartlett, 1994.

It is on the basis of this criterion of moral considerability that Regan argues that some animals also have inherent worth. Regan does not maintain that all nonhuman animals have inherent value. He isn't committed, for example, to the claim that mosquitos have equal rights with humans, or even rights at all. For it is not clear that mosquitos experience life as subjects. But Regan does think that many animals—in particular the other mammals—are subjects-of-a-life. They can be satisfied or frustrated. They can experience pleasure or pain. They can live lives of either high or low quality. Thus, Regan argues, they have rights. And because each being with inherent moral value has it equally, this means that nonhumans with inherent value have rights equal to those of humans.

In sum, both Singer and Regan argue that the moral community includes all animals that are the subjects of experiences of pleasure and pain. Both of their arguments imply that radical changes are necessary in our lifestyles. Both Singer and Regan defend vegetarianism, condemn animal agriculture and hunting, and criticize zoos. There are some important differences between the two views. For the most part, these differences reflect differences between utilitarianism and Kantianism. Singer allows for the interests of one individual to be weighed against the interests of another. Regan, on the other hand, emphasizes that no inherently valuable being should be used for the sake of another. Yet both agree that the mere species of an individual is never an adequate reason for treating that individual as more worthy than another. Finally, both Singer and Regan identify a criterion that picks out a wider group of animals as morally considerable. Although the two identify the criterion in different ways, it is roughly the same animals that each identifies as worthy of moral consideration.

It is tempting, I think, for some to dismiss the sentientist position too quickly. But these philosophers can contribute some very important insights concerning environmental value. First, sentientists remind us that it is not easy to identify an empirical criterion, other than species, that clearly divides humans from nonhumans in a morally relevant way. Many of the characteristics that we might naturally appeal to in support of our attempt to establish the moral superiority of humans can be found to a degree in some nonhuman individuals. This makes the distinction far less clear. Second, sentientists do point to features of higher animals that do seem morally relevant. Both the ability to experience pleasure and pain and the presence of self-consciousness in a creature do seem to be features that are indeed morally relevant to deciding how we should treat that object. One need not be a sentientist in the sense we have been considering to be sympathetic with the claim that animal pain is in some sense bad in itself and that cruelty to animals is immoral for reasons other than its effects on humans. Sentientists point us to features of higher animals that help to explain why such cruelty is wrong.

But I do think there are two features of sentientism that count against its plausibility as an account of value. First, it makes little room for degrees of value. This is especially true of Regan's "rights view." Here some higher nonhumans have equal rights with those of humans. In many cases our intuitions would suggests that, even if nonhuman animals are intrinsically valuable, humans are nevertheless more valuable. Yet it is difficult to see how Regan's view can accommodate this. Of course, Regan can say that our intuitions are simply wrong here and that we should yield to the reasoning he has presented. Nevertheless, these intuitions do tend to diminish the overall plausibility of the theory. A second, and perhaps more important, difficulty with sentientism is its failure to account for an important difference between the way we regard human actions and the way we regard actions of nonhumans. There *is* a continuity between higher animals and humans that is relevant to how we treat the nonhumans. But the continuity is along a spectrum with a morally important difference between the ends of the spectrum. At the upper end, and I do think the world "upper" is appropriate here, individuals are not only considered to be morally valuable. They are also held to be morally responsible. We expect those who fall at this end of the spectrum, including most humans, to be the *subjects,* as well as the objects, of moral consideration. We believe they are capable of and responsible for taking other morally valuable beings into consideration. At the lower end we do not hold individuals morally responsible. And although the line is not clear—I grant that we do not hold all humans morally accountable for their actions—there is something that distinguishes the two ends of the spectrum. It is this difference that sentientists do not seem to provide an account of in their theory of environmental value. It is not entirely implausible that this ability to consider others has implications for how one should be valued.

Biocentrism

Another general approach to environmental value called "biocentrism" argues that the circle of moral accountability is even wider. Defenders of this view contend that *all* living things are intrinsically valuable. Perhaps the most well-known proponent of this approach is the famous scholar, organist, missionary doctor, and author, Albert Schweitzer. From one's own realization of one's will to live, Schweitzer suggests, a person can come to appreciate his or her unity with every other living thing in its will to live. "Ethics consists," he wrote, "in my experiencing the compulsion to show to all will-to-live the same reverence as I do to my own." This provides the clue to what Schweitzer calls the "basic principle of the moral": "It is good to maintain and to encourage life; it is bad to destroy life and obstruct it."[9]

In more recent years, a similar "biocentric outlook on nature" has been defended by philosopher Paul Taylor. Taylor exposits this outlook in terms of four central beliefs. First, the biocentric outlook sees humans as members of the community of life on equal terms with all other living beings. Second, humans and other living things are all "integral parts of a system of interdependence" in such a way that the survival of any element of this system depends on the health of its physical environment and its relationships to other elements of the system. In these first two points, Taylor draws upon the insights that have emerged in recent years from the science of ecology . Third, Taylor says that the biocentric outlook recognizes all living things as "teleological centers of life."[10] Here, like Singer and Regan, Taylor identifies a characteristic that he believes marks some entities off from others as having moral worth. Each living thing, Taylor suggests, directs its growth and activities toward goals. It has "a good of its own" toward which it aims.[10] This "good is realized to the extent that (the organism) is strong and healthy."[10] Each plant, each animal, can be benefited or harmed. Taylor should not be read as saying that all living things are conscious, or that they have desires or even interests. He is appealing to something more basic to life than the features identified by the sentientists. Every living organism, whether sentient or not, whether conscious or not, is a "teleological center of life." And Taylor believes that this fact has important implications for our attitude and behavior toward fellow members of the Earth community. Having a good of its own gives an individual moral worth. He writes, "The good of each is to be accorded some value and so acknowledged as having some weight in the deliberations of all rational agents."[10]

Building on the fact that each individual organism has a good of its own, Taylor identifies the fourth basic biocentrist belief as the claim "that humans are not inherently superior to other living things."[9] Whereas Regan extends the Kantian idea of the inherent worth of all rational beings to all subjectively conscious beings, Taylor argues that the circle of inherent value is even wider. Because all living things have their own good, we have a duty, he contends, to recognize them as having inherent worth.[10] To paraphrase Taylor in Kantian terms, he is saying that all living things should be treated as ends in themselves, and never merely as means. He writes, "Insofar as we regard any organism, species population, or life community as an entity having inherent worth, we believe it must never be treated as a mere object or thing whose entire value lies in being instrumental to the good of some other entity. The well-being of each is judged to have value in and of itself"[11] As for Regan, so for Taylor, the possession of inherent moral value is an all or nothing thing. Thus Taylor concludes that we must reject the anthropocentric bias that says that humans are morally superior to animals and recognize that as interdependent living things in the community of life, we are all of equal value. Although we cannot consider them here, we should note that Taylor develops a quite elaborate set of principles based on the biocentric outlook. Like the views of the sentientists, the view proposed by Taylor has radical implications for our lives.

As with the sentientists, many will no doubt dismiss Taylor's view as far too radical for their intuitions. Yet I think Taylor has identified a very significant feature of living things. They are natural, goal-directed

---

[9]Albert Schweitzer. *Civilization and Ethics*, A. Nash, trans. (London: Black, 1923) 343. The passages quoted are found on page 343 of the excerpt in Susan J. Armstrong and Richard G. Botzler, eds. *Environmental Ethics: Divergence and Convergence,* McGraw-Hill, Inc. 1993.

[10]Paul Taylor, *Respect for Nature*, ed. Marshall Cohen, *Studies in Moral, Political and Legal Studies* (Princeton, NJ: Princeton University Press, 1986) 100.

[11]Paul Taylor. "The Ethics of Respect for Nature," *Environmental Ethics* 3, no. Fall (1981): 197–218.

systems. And it does seem that this feature is conceptually related to important moral concepts. Living systems can be harmed and benefited. Activities in the world can be good or bad "for them." I find these observations very interesting.

Yet I do think there are important problems with Taylor's proposal. Much that was said about sentientism could be said about Taylor's biocentrism as well. Like the sentientists, Taylor downplays the fact that humans are not only teleological centers of life, but moral agents. Taylor does recognize that this distinction is important, but he uses it for other purposes. In places he carefully distinguishes the morally proper response to nonhuman behavior from the proper response to similar human behaviors. He argues, for example, that when organisms suffer and die because of factors that are not the result of human behavior, we should not intervene but let nature take its course.[12] But when organisms suffer or die because of human actions, Taylor calls for various kinds of restitution.[13] Similarly, Taylor argues for a "Principle of Fidelity" that is violated by any activities, such as hunting and trapping, that are based on deceiving animals with the intent to harm them. Hunters who sneak up on animals in order to kill them are tricking the animals into thinking the hunter is not there, or that the hunter does not intend to harm them. Taylor says this violates the human duty of fidelity toward animals. But on the other hand, many nonhuman animals camouflage themselves and sneak up stealthily to overtake their prey. In these cases, Taylor defends a hands-off policy and a respect for nature. Thus he applies a set of moral standards to humans that he does not apply to nonhumans.

In fact Taylor explicitly recognizes the distinction. But he argues that this does not imply that humans are morally superior. He writes,

> ***One cannot validly argue that humans are morally superior beings on the ground that they possess, while others lack, the capacities of a moral agent. The reason is that, as far as moral standards are concerned, only beings that have the capacities of a moral agent can meaningfully be said to be either morally good or morally bad. . . . Just as animals and plants can be neither good nor bad scientists, engineers, critics, or Supreme Court justices, so they can be neither good nor bad moral agents. More precisely, it is meaningless to say either that they are morally inferior to humans or that humans are morally superior to them.***[14]

Taylor's argument here misses the point by confusing two issues. On the one hand one might ask whether one being is morally better (good) or worse (bad) than another. Taylor is right that this is a category that can be applied only to moral agents. Thus it would be nonsensical to say that plants or animals are either morally better or morally inferior to humans in this sense. But it is a different issue to ask whether one being is *more valuable* than another. A being need not be a moral agent to be valuable. But the capacity to be a moral agent does seem to be relevant to considerations of moral value. In seeking to develop a theory of environmental value, we are looking for characteristics that are conceptually related to moral value by which we can distinguish more valuable entities from less-valuable entities. I am willing to consider the sentientist suggestion that consciousness of pleasure is such a feature, and the biocentrist suggestion that goal-directedness is such a feature, because these features have conceptual links to the notions of good and bad. But I also believe that the capacity to be a moral agent is another such feature of which an adequate theory of moral value must give an account. This is one respect in which I find biocentrism less than adequate.

Ecocentrism

We turn now to a third nonanthropocentric approach to value that has been defended in the recent philosophical literature. This view often takes its inspiration from the progressive wildlife manager and creative thinker, Aldo Leopold. In his essay "The Land Ethic," Leopold suggests that, through history, ethics has

---

[12]Paul Taylor. *Respect for Nature*, Marshall Cohen, ed. *Studies in Moral, Political and Legal Studies* (Princeton, NJ: Princeton University Press, 1986) 100.

[13]Taylor, (1986) 177.

[14]Taylor, (1986) 186ff.

evolved from focusing on the relationships between individuals, to focusing on relationships between the individual and society. Now, he says, the time has come to expand the focus once again to include the relationship between the individual and his or her natural environment. Leopold writes,

> ***All ethics so far evolved rest upon a single premise: that the individual is a member of a community of interdependent parts. . . . The land ethic simply enlarges the boundaries of the community to include soils, waters, plants, and animals, or collectively: the land.***[15]

It is natural to see Leopold's suggestion as a continuation of the trend toward widening the circle of moral considerability that we have seen in the sentientists and the biocentrists. And there is a clear sense in which this is correct. Leopold includes nonliving as well as living things in the community.

But there is also a very important difference between the view inspired by Leopold and the others we have considered above. Leopold's suggestion is not to apply the concept of inherent worth or intrinsic moral considerability to additional *individuals.* Rather, he transfers the locus of value *from the individuals to the community as a whole.* Thus at the end of the essay, Leopold suggests the following principle: "A thing is right when it tends to preserve the integrity, stability, and beauty of the biotic community. It is wrong when it tends otherwise."[16]

This principle has become the cornerstone of the ecocentric approach to environmental value. Rather than put humans at the center (as in anthropocentrism), or individual living things at the center (as in sentientism and biocentrism), the ecocentric view places ecosystems at the center of value. It is the entire ecosystem that is valuable in itself, with each part playing a role in the whole. One of the most able defenders of ecocentric ethics is J. Baird Callicott. Callicott brings out the holism of the view very well in the following passage.

> ***An environmental ethic which takes as its summum bonum the integrity, stability, and beauty of the biotic community is not conferring moral standing on something else besides plants, animals, soils, and waters. Rather, the former, the good of the community as a whole, serves as a standard for the assessment of the relative value and relative ordering of its constitutive parts and therefore provides a means of adjudicating the often mutually contradictory demands of the parts considered separately for equal consideration.***[17]

There is a sense, then, in which the ecocentric approach moves in a direction opposite of some of the other views we have considered. Callicott seems to imply, for example, that individual members of the biotic community have only instrumental or extrinsic value. It is the community *as a whole* that is the locus of inherent worth or intrinsic value. Each part is valuable in as much as it contributes to the whole.

Thus Leopold, in his application of the "land ethic," reached conclusions quite different from those of the sentientists or the biocentrists. Whereas both of these other views condemned hunting, Leopold did not. Because what is crucial to ethics is the health of the entire ecosystem, in some situations hunting may be morally required to protect a local ecological system from overpopulation by certain animal species.[18] Similarly, while sentientists defend vegetarianism, Callicott speculates that if humankind were all to become vegetarians, the results would probably be "ecologically catastrophic."[19] Thus a vegetarian lifestyle is not called for by the biocentric principle. As Callicott summarizes, "In every case the effect upon ecological systems is the decisive factor in the determination of the ethical quality of actions."[20]

Ecocentrism has a number of strengths. It recognizes, first of all, that individuals cannot be understood as isolated units, but must be seen as parts of a larger whole. Closely related to this is the insight that

---

[15]Aldo Leopold. "The Land Ethic," in *A Sand County Almanac* (London: Oxford University Press, 1949), 203–204.

[16]Leopold, 224–225.

[17]J. Baird Callicott. *In Defense of the Land Ethic: Essays in Environmental Philosophy*, David Edward Shaner, ed. *SUNY Series in Philosophy and Biology* (Albany: State University of New York Press, 1989), 25.

[18]Callicott, 21.

[19]Callicott, 35.

[20]Callicott, 21.

the value of a system cannot always be reduced to the value of the individuals that make up that system. Ecocentrism also has a kind of clear simplicity. Whereas some other views result in a number of competing entities with inherent worth, Ecocentrism focuses value on the whole. In this regard, it shares an advantage held by Singer's utilitarianism. There is room for balancing all the individuals for the common good. Yet unlike Utilitarianism, Ecocentrism sees the value as residing in the ecosystem as a whole and not as being the mere addition of the satisfaction of individuals.

There are however, several considerations that I believe count against ecocentrism. First, because all members of the biotic community are valuable only inasmuch as they contribute to the health of the entire community, ecocentrism seems to imply a less than satisfactory account of the value of human life. If it is morally permissible, or even morally required, to reduce the population of deer in an area through the issuing of hunting licenses for the sake of the health of the biotic community, it is difficult to see why the same cannot be said concerning the overpopulation of an area by the human species.[21]

Finally, there is one additional question that can be raised concerning the adequacy of the ecocentric approach to environmental value. This is the question of who it is who values the biotic community as a whole. It is clear that, for anthropocentrism, the valuers are human beings. Similarly, for sentientism, all sentient animals could be understood as beings that value their own pleasure and satisfaction. It is less clear, but perhaps possible to extend this function of valuing to other individual organisms. But ecocentrism identifies the whole system as the locus of value. Does there have to be a valuer of this system to coherently speak of its intrinsic value? This is an issue on which philosophical intuitions seem to be split. Holmes Rolston III, for example, speaks "against the standard view that value requires a *beholder*," and suggests that value, as storied achievement, "requires only a *holder*, which can be an individual, but which can also be the historic system that carries value to and through individuals."[22] Yet is not the concept of value without a valuer paradoxical? I find myself asking whether an ecosystem, or the ecosphere as a whole, would be *valu-able* in any sense if there were no being that was *able* to *value* it. If, for example, the universe were a collection of bits of matter in which there had evolved no beings who could enjoy such things as the appearance of the reds and oranges of a beautiful sunset, and if there is no higher being to delight in the cosmos, I find it difficult to see how the interaction of energy and matter that results in the sunset is valuable in any sense. Similarly, if the cosmos is such that it has evolved complex systems of life, including many creatures that value various things, I wonder how there could be a more overriding value of the system as a whole if there is no higher valuer that is able to value the system as a whole. Although not all philosophers find this question troublesome, it is one factor that, for me at least, makes the ecocentric approach less than completely satisfying.

THE THEOCENTRIC POSSIBILITY

Against the background of the discussion concerning the nature of environmental value, I'd like to turn now to explore the possibility of a coherent, theistic alternative. It is not my purpose to give a systematic exposition of the biblical message concerning the environment. There are a number of others who have made significant contributions in this area. Rather, I want to reflect philosophically upon the concept

---

[21]It should be noted that Callicott does argue that ecocentric ethics need not have inhumane consequences, by suggesting that there are different levels of moral obligations. Just as one's obligations to one's family can override obligations to one's nation, the obligations of human morality should not be ignored for the sake of obligations to the biotic community. Thus, Callicott says, it would be perverse to advocate "visiting war, famine, or pestilence on human populations. . . in the name of the integrity, beauty, and stability of the biotic community." J. Baird Callicott. *In Defense of the Land Ethic: Essays in Environmental Philosophy,* State University of New York Press, 1989, 94. Nevertheless, it is difficult to see how Callicott can make this move without significantly compromising the ecocentric approach. Although one might be able to argue that human ethics should be taken seriously because in general it contributes to the biotic community, clearly philosophers like Callicott want to suggest that human-centered ethical systems do not always lead to the actions that are right from the ecocentric perspective. If there are no situations in which traditional ethics and ecocentrist ethics conflict, then it is difficult to see why we need a new ethic. If there are situations in which the two conflict, and Callicott is correct that the good of the biotic community is the "decisive" factor, then there will be situations in which ecocentrism has inhumane implications.

[22]Holmes Rolston, III. "Values Gone Wild," *Inquiry* 26, April 1987 (1987): 172–90. The quotation is found on page 64 of the reprint of this article in Susan J. Armstrong and Richard B. Botzler. eds. *Environmental Ethics: Divergence and Convergence,* McGraw Hill, Inc. 1993.

of value against the background of both the discussion above and a Judeo-Christian worldview. My goal will be to sketch out a theocentric theory of value that can bring together the insights of other approaches into a coherent whole.

What shape will a theistic approach to environmental value take? Many people, including many of the Christian students who take my environmental philosophy course, assume that the Christian perspective is anthropocentric. Like many theists in the history of Western thought, they quote the Scriptures and argue that God created the earth for humankind and gave it all to us to use and develop for our good. One possible approach for Christian environmentalism would be to accept this view of environmental value and to argue that, for our own good, we must take care of the world around us.

Some reflections on the Scriptures

Do the Scriptures really teach an anthropocentric view of environmental value? I don't think so. Let me briefly mention two features of the biblical stories that point away from anthropocentrism. First, a number of passages suggest that the nonhuman world has value to God quite apart from humankind. Second (and more disturbing), there are passages of Scripture that relativize the value of even human life. I cannot develop either of these themes in detail here. On the first theme, I will only remind you of two well-known passages from the book of Genesis. In the account of the creation in Genesis 1, we read over and over again that, as God created the cosmos, God looked at it and "saw that it was good." All but the last of these divine observations of the goodness occur *before* humankind is created. This suggests that the creation is good, that it is valuable, quite apart from any relationship it has to humankind. If so, anthropocentrism cannot be the whole story. It is true that at the end of the sixth day of creation, the day on which humankind is created, God sees that the final product is "very good." But even here it is not humankind that is said to be "very good." What God observes to be very good is not humankind in particular, but "all that God had made." It is the whole, wonderful cosmos that is in focus, not just human beings.

Another passage that suggests that nonhuman creation is valuable to God in ways that go beyond its value to humans is Genesis 7–9. In the story of Noah, it is noteworthy that God tells Noah to take both the clean and the unclean animals into the ark, seemingly a demonstration of God's direct concern for all the animals, whether they are good for humans or not. And after the flood waters subside, God puts a rainbow in the sky as a sign of "the everlasting covenant between God and every living creature of all flesh that is on the earth." There is no indication that God's relationship to the other animals is in any way mediated by humankind. The animals, it seems, are pictured as relating to God alongside humankind, not through it. These are just a few of many passages that suggest that all aspects of nonhuman creation are enjoyed by God, valued by God, and relate to God directly. We could add other passages in which the elements of creation are spoken of as belonging to and giving glory and praise to God.

But there is another strand in the stories of Scripture that may point away from anthropocentrism as well. I have in mind the story in which God calls upon Abraham to be willing to sacrifice his son in order to obey God. I have in mind the story of Uzzah who died because he reached out to keep the ark of God from falling. And I have in mind the passages in which God commands his people to kill the inhabitants of territory they conquer. These are very difficult passages to read. They are particularly difficult because we believe deeply in our hearts and minds that human life is sacred, and we hear God commanding the people to destroy human life. I do not want, for a minute, to minimize the value of human life or the horror of violence. But these passages do suggest that in the biblical economy there is something of more value than human well-being.

What do these passages have to do with environmental value? The first group of passages suggest that the world is of value to God in ways that do not depend upon human beings. It is more accurate to the Scriptures to say that the created order is valuable because it is of value to God than it is to say that it is valuable because it is of value to humankind. Thus a theory of value that views nonhuman creation as valuable only because nonhuman creatures are good for humans misses something crucial to the biblical outlook. The second group of passages, to which I have only alluded here, suggest that humankind, although valuable to God, may not be valuable in the absolute way that we sometimes assume. In other words, the theory of value we see in Scripture is more theocentric than anthropocentric. Each and every thing that is of value, is of value because of its relationship to God, not because of its relationship to humankind.

Key Elements of a Theocentric Approach to Environmental Value

Although the Bible does not speak in the philosophical language of inherent worth and intrinsic value, I suggest that a theocentric approach to the value of our natural environment should be built on four basic tenets. First, in the theocentric perspective, it is God who is the source of all value. God is the greatest good and the primary valuer of all that is good. I have never thought of myself as a Platonist. But from the theocentric perspective it is very natural to say that God is the Form of the Good, in virtue of which all good things are good, and that one cannot fully understand any good thing unless one understands God. But God is not only the source all goodness, God is also a being who "beholds" all that is good. God is the all-seeing valuer.

A theocentric approach will be committed to the thesis that all value, all good things, have value because of their relationship to God.

Second, A theocentric approach will be committed to the thesis that all value, all good things, have value because of their relationship to God.If one takes the concept of "inherent worth" to imply that the possessor of such worth has value independently of its relationship to anything else, then the theocentric perspective holds that only God has inherent worth. There is no other reality whose value is not relative to that of God.

Third, this approach will see all of God's creation as having a value derived directly from its relationship to its creator. There is no part of creation that has the unique place of being that for which everything else in creation exists. All of creation is part of a single, created order that has been created for God's pleasure, and the elements within that order can bring pleasure to God apart from the involvement of humankind.

Finally, an anthropocentric approach to environmental value can make room for the truth that humans reflect God in a way that other creatures do not. Within the view that I am suggesting, humans have much in common with other creatures. They are part of the physical world of creation. Their existence can bring pleasure to their creator. Not only are they valued by God, they have the capacity to value other beings as well. As we have seen, this is a feature that many other animals share as well. But as the image of God, humankind is not only of valuable, but is held accountable by God to acknowledge and respect the value of others. As such, humankind does have a special relationship to God, which gives humans greater value that other aspects of creation. But this value is not at the center of earthly value in the sense that humans are valuable in themselves, and everything else is not. Humans are no more valuable in themselves than the rest of creation. All of creation is valuable because it is God's. The place I have given humankind in my account of theocentric ethics is the one aspect of all that I have said that will be most objectionable to nontheistic environmental philosophers. Yet it acknowledges the fact that humans are accountable in ways that other living things are not.

I will conclude with a few additional observations on the strength of the theocentric perspective. The theocentric perspective makes room for various levels of moral value. It can recognize that, in numerous ways, living things reflect the nature and design of their creator. Included among these are the fact that living things are teleological centers of life, the fact that some animals are the subjects-of-a-life, and the fact that humans are morally accountable in ways that other animals are not. All of these aspects of creation reflect the nature of God in some degree. So although humankind has a special value, the theocentric perspective places us clearly within the created order. Unlike the dualistic strand of Western thought, which has put humankind with God on one side of a great divide, and nature on the other side (which has often been seen as anti-God and evil), the theocentric approach places humankind clearly on the side of creation, with all of creation reflecting God's goodness. Finally, the theocentric perspective has plenty of room to acknowledge that whole ecosystems, and the entire ecosphere, are valuable as wholes, the wonderful work of the divine artist to bring pleasure to God and to be enjoyed as habitats for God's creatures. In short, theocentrism enables us to bring many insights from our exploration of environmental value into a coherent whole.

Final Project

FINAL "MY ENVIRONMENTAL PERSPECTIVE" PAPER

Throughout this book, we have been exploring the concept of nature and humankind's relationship to it. We have seen many ways of approaching the following questions:

- What is nature, or the environment?
- How is humankind related to it?
- Why is nature, or aspects of it, valuable?
- What implications does your view of nature have for how you live your life?

How you answer these questions defines your "environmental perspective," which is part of your overall worldview. Your environmental ethic should be based upon this perspective and you worldview in general. As your final project for the course, you are to complete one of the following:

1. Write a 4 to 6 page essay in which you state and explain your answer to the above questions. As you explain your view, you should show how it is similar to and different from other religious and philosophical views discussed in the course.
2. As an alternative, you may express your perspective in a different, creative way, such as through a song, a poem, a piece of artwork, a film, a PowerPoint presentation. However you express it, the project should in some way address one or more of the questions above and include a brief explanation of how your work expresses your environmental view.